T0396573

Biofuels and Biorefineries

Volume 1

For further volumes:
http://www.springer.com/series/11687

Aims and Scope of the series

Book Series in Biofuels and Biorefineries aims at being a powerful and integrative source of information on biomass, bioenergy, biofuels, bioproducts and biorefinery. It represents leading global research advances and opinions on converting biomass to biofuels and chemicals; presents critical evidence to further explain the scientific and engineering problems in biomass production and conversion; and presents the technological advances and approaches for creating a new bio-economy and building a clean and sustainable society to industrialists and policy-makers.

Book Series in Biofuels and Biorefineries provides the readers with clear and concisely-written chapters on significant topics in biomass production, biofuels, bioproducts, chemicals, catalysts, energy policy and processing technologies. The text covers areas of plant science, green chemistry, economy, biotechnology, microbiology, chemical engineering, mechanical engineering and energy studies.

Series description

Annual global biomass production is about 220 billion dry tons or 4,500 EJ, equivalent to 8.5 times the world's energy consumption in 2008 (532 EJ). On the other hand, the world's proven oil reserves at the end of 2011 amounted to 1652.6 billion barrels, which can only meet 54.2 years of global production. Therefore, alternative resources are needed to both supplement and replace fossil oils as the raw material for transportation fuels, chemicals and materials in petroleum-based industries. Renewable biomass is a likely candidate, because it is prevalent throughout the world and can readily be converted to other products. Compared with coal, the advantages of using biomass are: (i) it is carbon-neutral and sustainable when properly managed; (ii) it is hydrolysable and can be converted by biological conversion (e.g., biogas, ethanol); (iii) it can be used to produce bio-oil with high yield (up to 75%) by fast pyrolysis because it contains highly volatile compounds or oxygen; (iv) biofuel is clean because it contains little sulfur and its residues are recyclable; (v) it is evenly distributed geographically and can be grown close to where it is used, and (vi) it can create jobs in growing energy crops and building conversion plants. Many researchers, governments, research institutions and industries are developing projects to convert biomass (including forest woody and herbaceous biomass) into chemicals, biofuels and materials and the race is on to create new "biorefinery" processes. The development of biorefineries will create remarkable opportunities for the forestry sector, biotechnology, materials and the chemical processing industry, and it will stimulate advances in agriculture. It will help to create a sustainable society and industry based on renewable and carbon-neutral resources.

Zhen Fang • Richard L. Smith, Jr. • Xinhua Qi
Editors

Production of Biofuels and Chemicals with Ionic Liquids

Editors
Zhen Fang
Biomass Group
Xishuangbanna Tropical
Botanical Garden
Chinese Academy of Sciences
Kunming, China

Richard L. Smith, Jr.
Graduate School of Environmental Studies
Research Center of Supercritical Fluid
Technology
Tohoku University
Aoba-ku, Sendai, Japan

Xinhua Qi
College of Environmental Science
and Engineering
Nankai University
Tianjin, China

ISSN 2214-1537 ISSN 2214-1545 (electronic)
ISBN 978-94-007-7710-1 ISBN 978-94-007-7711-8 (eBook)
DOI 10.1007/978-94-007-7711-8
Springer Dordrecht Heidelberg New York London

Printed on acid-free paper

Springer is part of Springer Science+Business Media (www.springer.com)

Preface

Interest in biofuels and value-added chemicals that can be produced from biomass is increasing daily as societies look to sustainable sources of energy. Ionic liquids are used for the pretreatment and chemical transformation of biomass due to their unique ability for dissolving lignocellulosic materials. Although there are many books on the topic of either biomass conversion or ionic liquids, the unique feature of this book is that it links biomass conversion with ionic liquids and chemicals such that processing, chemistry, biofuel production, enzyme compatibility and environmental treatment are covered for conceptual design of a biorefinery. This book is the first book of the series entitled *Biofuels and Biorefineries*.

This book consists of 12 chapters contributed by leading world-experts on biomass conversion with ionic liquids. Each chapter was subjected to peer-review and carefully revised by the authors and editors so that the quality of the material could be improved. The chapters are arranged in five parts:

Part I: Synthesis and Fundamentals of Ionic Liquids for Biomass Conversion (Chaps. 1, 2, and 3).
Part II: Dissolution and Derivation of Cellulose and Fractionation of Lignocellulosic Materials with Ionic Liquids (Chaps. 4, 5, and 6).
Part III: Production of Biofuels and Chemicals in Ionic Liquids (Chaps. 7, 8, and 9).
Part IV: Compatibility of Ionic Liquids with Enzymes in Biomass Treatment (Chaps. 10 and 11).
Part V: Ionic Liquids for Absorption and Biodegradation of Organic Pollutants in Multiphase Systems (Chap. 12).

Chapter 1 introduces the fundamentals of ionic liquids related to biomass treatment. Chapter 2 gives an outline of design and synthesis of ionic liquids for cellulose dissolution and plant biomass treatment. Chapter 3 overviews the recent advances made with choline-chloride (ChCl) not only for the activation of biomass but also for its conversion to value-added chemicals. Chapter 4 summarizes approaches to design of ionic liquids that have good capability for dissolving cellulose and discusses factors for realizing efficient room-temperature dissolution of cellulose dissolution and subsequent enzymatic hydrolysis. Chapter 5 provides a comprehensive overview about the use of ionic liquids for the chemical derivatization of cellulose. Chapter 6 reviews the current state of knowledge and process development in the area of ionic liquid fractionation of wood, and reports findings on factors that control the solubility of wood in ionic liquids. Chapter 7 provides an overview on the biodiesel production in ionic liquids, ionic liquids-catalyzed biodiesel production, ionic liquids-modified enzymes for biodiesel production, purification of bio-alcohols with ionic liquids, and prospects. Chapter 8 focuses on catalytic transformations of biomass into fuels and chemicals in ionic liquids. Chapter 9 describes the efficient methods for producing the platform chemical, 5-hydroxymethylfurfural with ionic liquids. Chapter 10 covers the biocompatibility issues of ionic liquids, for example, biocatalysts in ionic liquids media, the effect of ionic liquids properties on the activity and stability of enzymes, approaches to enhance the activity and stability of enzymes in the ionic liquids containing medium, and rational design of ionic liquids for use with enzymatic reactions. Chapter 11 focuses on the application of enzyme technology in ionic liquids. Chapter 12 introduces the potential of ionic liquids for hydrophobic organic pollutants absorption and biodegradation in multiphase systems.

This book reviews many aspects of the ionic liquids techniques necessary for efficient development of biomass resources. The text should be of interest to students, researchers, academicians and industrialists in the area of ionic liquids and biomass conversion.

Beijing, People's Republic of China — Zhen Fang
Sendai, Japan — Richard L. Smith, Jr.
Tianjin, People's Republic of China — Xinhua Qi

Acknowledgements

First and foremost, we would like to thank all the contributing authors for their many efforts to insure the reliability of the information given in the chapters. Contributing authors have really made this project realizable.

Apart from the efforts of authors, we would also like to acknowledge the individuals listed below for carefully reading the book chapters and giving constructive comments that significantly improved the quality of many aspects of the text:

Dr. Leigh Aldous, the University of New South Wales, Australia;
Dr. Agnieszka Brandt, Imperial College London, UK;
Prof. Johann Görgens, Stellenbosch University, South Africa;
Prof. Yanlong Gu, Huazhong University of Science and Technology, China;
Prof. Mohd Ali Hashim, University of Malaya, Malaysia;
Prof. Noriho Kamiya, Kyushu University, Japan;
Dr. Takao Kishimoto, Toyama Prefectural University, Japan;
Dr. Jong Min Lee, Nanyang Technological University, Singapore;
Dr. Sang Hyun Lee, Konkuk University, South Korea;
Prof. Jean-Marc Lévêque, Université de Savoie, France,
Dr. Ruigang Liu, Chinese Academy of Sciences, China;
Prof. Rafael Luque, Universidad de Córdoba, Spain;
Dr. Patrick Navard, Ecole des Mines de Paris/CNRS, France;
Dr. Vladimír Raus, Academy of Sciences of the Czech Republic;
Prof. Robin D. Rogers, The University of Alabama, USA.;
Prof. Roger A. Sheldon, Delft University of Technology, The Netherlands;
Prof. Run-Cang Sun, Beijing Forestry University;
Dr. Yugen Zhang, Institute of Bioengineering and Nanotechnology, Singapore;

We are also grateful to Ms. Becky Zhao (senior editor) and Ms. Abbey Huang (editorial assistant) for their encouragement and assistant with the guidelines during preparation of the book.

Finally, we would like to express our deepest gratitude towards our family for their kind cooperation and encouragement, which helped us in completion of this project.

Zhen FANG, August 1, 2013 in Kunming
Richard L. Smith, Jr., August 1, 2013 in Sendai
Xinhua Qi, August 1, 2013 in Tianjin

Contents

Contributors

Mitsuru Abe Department of Biotechnology, Tokyo University of Agriculture and Technology, Koganei, Tokyo, Japan

Abdeltif Amrane Ecole Nationale Supérieure de Chimie de Rennes, CNRS, UMR 6226, Rennes Cedex 7, France

Université européenne de Bretagne, Rennes Cedex, France

Dimitris S. Argyropoulos Departments of Forest Biomaterials and Chemistry, North Carolina State University, Raleigh, NC, USA

Center of Excellence for Advanced Materials Research (CEAMR), King Abdulaziz University, Jeddah, Saudi Arabia

Pierre-François Biard Ecole Nationale Supérieure de Chimie de Rennes, CNRS, UMR 6226, Rennes Cedex 7, France

Université européenne de Bretagne, Rennes Cedex, France

Alfredo Santiago Rodriguez Castillo Ecole Nationale Supérieure de Chimie de Rennes, CNRS, UMR 6226, Rennes Cedex 7, France

Université européenne de Bretagne, Rennes Cedex, France

Annabelle Couvert Ecole Nationale Supérieure de Chimie de Rennes, CNRS, UMR 6226, Rennes Cedex 7, France

Université européenne de Bretagne, Rennes Cedex, France

Blair J. Cox Department of Chemical Engineering, The University of Texas at Austin, Austin, TX, USA

UT Dallas Venture Development Center, Cyclewood Solutions Inc., Richardson, TX, USA

Teresa de Diego Department of Biochemistry and Molecular Biology B and Immunology, University of Murcia, Murcia, Spain

Karine De Oliveira Vigier Institut de Chimie des Milieux et Matériaux de Poitiers (IC2MP), CNRS-University of Poitiers, ENSIP, Poitiers, France

John G. Ekerdt Department of Chemical Engineering, The University of Texas at Austin, Austin, TX, USA

Zhen Fang Xishuangbanna Tropical Botanical Garden, Chinese Academy of Sciences, Kunming, China

Martin Gericke Institute of Organic Chemistry and Macromolecular Chemistry, Centre of Excellence for Polysaccharide Research, Friedrich Schiller University of Jena, Jena, Germany

Solène Guihéneuf Université européenne de Bretagne, Rennes Cedex, France

Université de Rennes 1, Sciences Chimiques de Rennes, UMR CNRS 6226, Groupe Ingénierie Chimique & Molécules pour le Vivant (ICMV), Rennes Cedex, France

Thomas Heinze Institute of Organic Chemistry and Macromolecular Chemistry, Centre of Excellence for Polysaccharide Research, Friedrich Schiller University of Jena, Jena, Germany

José Luis Iborra Department of Biochemistry and Molecular Biology B and Immunology, University of Murcia, Murcia, Spain

Toshiyuki Itoh Department of Chemistry and Biotechnology, Graduate School of Engineering, Tottori University, Tottori, Japan

François Jérôme Institut de Chimie des Milieux et Matériaux de Poitiers (IC2MP), CNRS-University of Poitiers, ENSIP, Poitiers, France

Alistair W.T. King Department of Chemistry, University of Helsinki, Helsinki, Finland

Yoon-Mo Koo Department of Marine Science and Biological Engineering, Inha University, Incheon, Republic of Korea

Timo Leskinen Departments of Forest Biomaterials and Chemistry, North Carolina State University, Raleigh, NC, USA

Xingmei Lu Beijing Key Laboratory of ILs Clean Process, Key Laboratory of Green Process and Engineering, State Key Laboratory of Multiphase Complex Systems, Institute of Process Engineering, Chinese Academy of Sciences, Beijing, People's Republic of China

Ngoc Lan Mai Department of Marine Science and Biological Engineering, Inha University, Incheon, Republic of Korea

Arturo Manjón Department of Biochemistry and Molecular Biology B and Immunology, University of Murcia, Murcia, Spain

Hiroyuki Ohno Department of Biotechnology, Tokyo University of Agriculture and Technology, Koganei, Tokyo, Japan

Ludovic Paquin Université européenne de Bretagne, Rennes Cedex, France

Université de Rennes 1, Sciences Chimiques de Rennes, UMR CNRS 6226, Groupe Ingénierie Chimique & Molécules pour le Vivant (ICMV), Rennes Cedex, France

Xinhua Qi College of Environmental Science and Engineering, Nankai University, Tianjin, China

Richard L. Smith Jr. Research Center of Supercritical Fluid Technology, Tohoku University, Sendai, Japan

Xinxin Wang Beijing Key Laboratory of ILs Clean Process, Key Laboratory of Green Process and Engineering, State Key Laboratory of Multiphase Complex Systems, Institute of Process Engineering, Chinese Academy of Sciences, Beijing, People's Republic of China

Haibo Xie Division of Biotechnology, Dalian National Laboratory for Clean Energy, Dalian Institute of Chemical Physics, Chinese Academy of Sciences (CAS), Dalian, People's Republic of China

Junli Xu Beijing Key Laboratory of ILs Clean Process, Key Laboratory of Green Process and Engineering, State Key Laboratory of Multiphase Complex Systems, Institute of Process Engineering, Chinese Academy of Sciences, Beijing, People's Republic of China

College of Chemistry and Chemical Engineering, University of Chinese Academy of Sciences, Beijing, People's Republic of China

Suojiang Zhang Beijing Key Laboratory of ILs Clean Process, Key Laboratory of Green Process and Engineering, State Key Laboratory of Multiphase Complex Systems, Institute of Process Engineering, Chinese Academy of Sciences, Beijing, People's Republic of China

Zongbao Kent Zhao Division of Biotechnology, Dalian National Laboratory for Clean Energy, Dalian Institute of Chemical Physics, Chinese Academy of Sciences (CAS), Dalian, People's Republic of China

Qing Zhou Beijing Key Laboratory of ILs Clean Process, Key Laboratory of Green Process and Engineering, State Key Laboratory of Multiphase Complex Systems, Institute of Process Engineering, Chinese Academy of Sciences, Beijing, People's Republic of China

Part I
Synthesis and Fundamentals of Ionic Liquids for Biomass Conversion

Chapter 1
Fundamentals of Ionic Liquids

Junli Xu, Qing Zhou, Xinxin Wang, Xingmei Lu, and Suojiang Zhang

Abstract Ionic liquids (ILs) are composed of cations and anions that exist as liquids at relatively low temperatures (<100 °C). They have many attractive properties, such as chemical and thermal stability, low flammability, and immeasurably low vapor pressures. This review provides a summary of the fundamental structural features of ionic liquids, the physical properties, and their applications as solvents for biomass.

Keywords ILs • Properties • Cellulose • Biomass

1.1 Introduction

The energy crisis has caused great pressure on the economic development and environmental sustainability worldwide, resulting in renewable energy, such as, solar, wind, and biomass, receiving significant attention [1]. Especially, as a resource of fuel and chemicals, biomass is developed greatly due to its large potential and universality as an energy resource. Biomass pretreatment is a key procedure for efficient processing. Biomass pretreatment was first conducted with acid or alkali, as well as some organic solvents. Gradually, considering the environmental and economic influence, ionic liquids (ILs) were introduced for biomass

J. Xu
Beijing Key Laboratory of ILs Clean Process, Key Laboratory of Green Process and Engineering, State Key Laboratory of Multiphase Complex Systems, Institute of Process Engineering, Chinese Academy of Sciences, Beijing 100190, People's Republic of China

College of Chemistry and Chemical Engineering, University of Chinese Academy of Sciences, Beijing 100049, People's Republic of China

Q. Zhou • X. Wang • X. Lu (✉) • S. Zhang (✉)
Beijing Key Laboratory of ILs Clean Process, Key Laboratory of Green Process and Engineering, State Key Laboratory of Multiphase Complex Systems, Institute of Process Engineering, Chinese Academy of Sciences, Beijing 100190, People's Republic of China
e-mail: xmlu@home.ipe.ac.cn; sjzhang@home.ipe.ac.cn

Z. Fang et al. (eds.), *Production of Biofuels and Chemicals with Ionic Liquids*, Biofuels and Biorefineries 1, DOI 10.1007/978-94-007-7711-8_1,

pretreatment and biomass conversion with ILs was followed since ILs have many unique and excellent properties.

This part aims to briefly introduce the definition of ILs, the structures and classification of ILs, meanwhile, the properties of ILs will be discussed in detail, including melting point, viscosity, density, and thermal stability. Then the history, advantage and current status of ILs applied to cellulose/biomass, the quantities and kinds of ILs are used in dissolution and separation of cellulose from biomass are also summarized.

1.2 Overview

Ionic liquids (ILs), are organic compounds containing salts with many attractive properties like extremely low vaporization pressure and melting point, excellent thermal stability and wide liquid ranges [2–5]. ILs have widely been used in many areas, for example, chemical synthesis, catalysis, biocatalysts and electrochemical devices [6–10]. ILs can be chosen to have different anions and cations so that one can form IL with the desired properties. Especially, some kinds of ILs with special functional groups have been designed for application in many industrial processes, such as imidazolium-based ILs, phosphonium-based ILs, amino-based ILs, acid-based ILs, and biodegradable ILs [11].

Lignocellulosic biomass is an abundant plant material and widely available, so that it has attracted much attention for conversion to fuels and chemicals [12, 13]. The main components of biomass are cellulose, hemicelluloses, lignin and other extractives. However, the complex structure of biomass makes its chemical degradation and biological conversion difficult to realize [14]. Pretreatment to disrupt the structures is necessary and a key procedure for biomass utilization.

Cellulose, as an important component of biomass, is composed of thousands of β- (1–4) – linked glucose units [15], which form many intermolecular or intramolecular hydrogen bonds [16]. Cellulose is widely treated with several organic solvents, such as N, N-dimethylformamide/nitrous tetroxide (DMF/N_2O_4) [17], N, N-dimethylacetamide lithium chloride (DMAc/LICl) [18], N-methylmorpholine (NMMO) [19] and dimethyl sulfoxide (DMSO)/tetrabutylammonium fluoride (TBAF) [20]. These traditional solvents suffer from volatility, toxicity, and solvent recovery issues [21, 22], so novel solvents, such as ILs, have received attention for cellulose dissolution in recent years. Cellulose dissolution with present ILs dates back to 2002 [23], before which, a solvent system for dissolving cellulose was discovered by Graenacher [24].

1.2.1 The Structures and Classification of ILs

There are a large number of ILs that can be produced theoretically, while the synthesized and reported ones are very limited [25]. By the end of 2009, more than 1,800 available ionic liquids which were composed of 714 different cations and 189 different anions have been reported in the book named ionic liquid: physicochemical properties have reported [26]. According to our current ionic liquids

Table 1.1 Common used cations of ILs

Types	Cations	Structures
Imidazolium based cations	1-alkyl-3-methylimidazolium	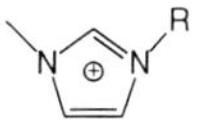
	1-alkyl-2, 3-dimethylimidazolium	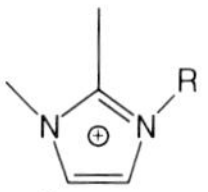
Pyridinium based cations	1-alkylpyridinium	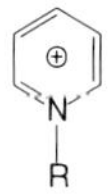
	1-alkyl-1-methylpyridinium	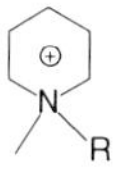
Pyrrolidinium based cation	1-alkyl-1-methylpyrrolidinium	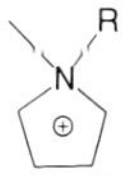
Phosphonium, ammonium, sulfonium based cations	Tetraalkylphosphonium	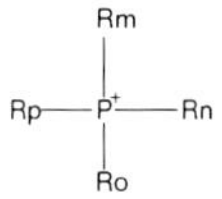
	Tetraakylammonium	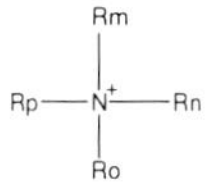
	Trialkylsulfonium	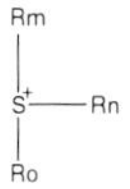
Metal based cations	M = Co^{2+}, Ni^{2+}	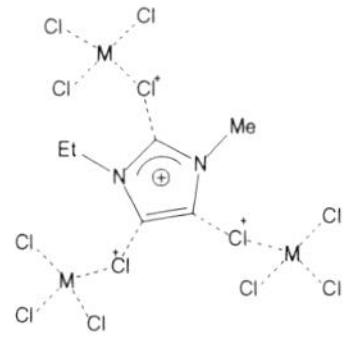
Functionalized cations	1-phenylethanoyl-3-methylimidazolium	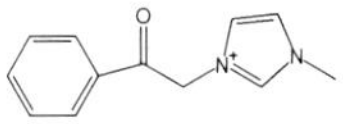
	N-propane sulfuricpyridiniumdihydrogen	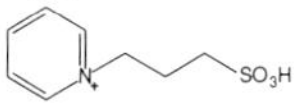
	1-methyl-4-(2-azidoethyl)-1, 2, 4-triazolium	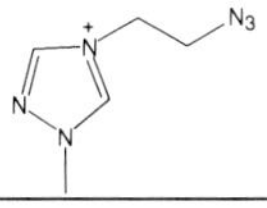

database, there are more than 2,300 kinds of ILs, including varieties of 229 anions, 907 cations. In these ILs, the most common used cations are imidazolium, pyridinium, piperidinium, tetraalkylphosphonium, tetraalkylammonium, trialkylsulfonium and metal based, functional cations, the cations are listed in Table 1.1.

However, the generally used ions are either inorganic or organic, for example, hexafluorophosphate, bis (trifluoromethylsulfonyl) imide, tetrafluoroborate, trifluoromethanesulfonate, dicyanamide, halide, formate, acetate, and alkyl-phosphate and so on (Table 1.2).

Table 1.2 Common used anions of ILs

Entry	Anions	Structures
1	Hexafluorophosphate	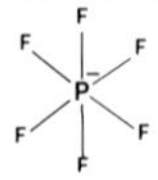
2	Tetrafluoroborate	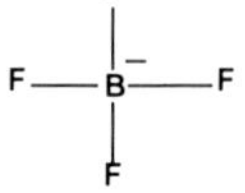
3	Trifluoromethanesulfonate	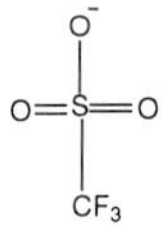
4[a]	Methyl sulfate	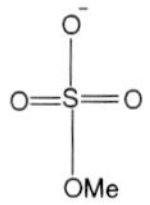
5	Acetate	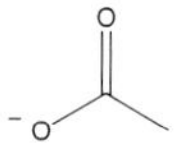
6	$[N(CN_2)]^-$ chloride/bromide/iodide	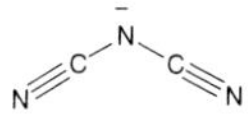
7	Chloride/bromide/iodide	Cl^-, Br^-, I^-
8[a]	Dimethyl phosphate	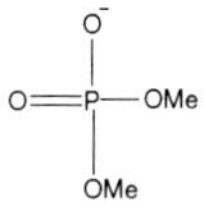
9	Hydrogen maleate	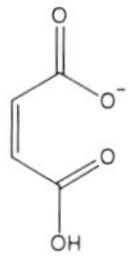
10	Tyrosine	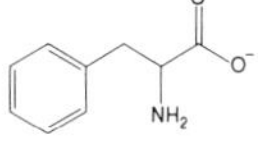
11	Valine	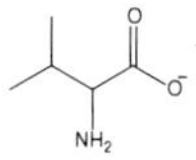
12[b]	Tetrakis [3, 5-bis (trifluoromethyl) phenyl] borate	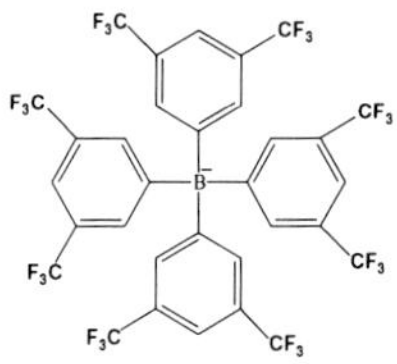
13[c]	bis(trifluoromethanesulfonyl)amide	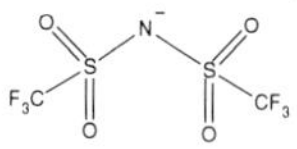
14	Trifluoroacetate	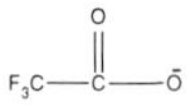
15[d]	$AlCl_4$-	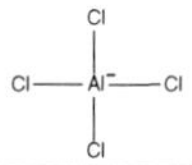

[a] the methyl also could be changed into ethyl et al.

[b] the other borate and borane anions are fluoroacetoxyborate, bis(oxalato)borate, alkyl carborane et al.

[c] the similar anions are bis(perfluoroethylsulfonyl)amide, 2,2,2-trifluoro-N-(trifluoromethanesulfonyl) acetamide, tris(trifluoromethanesulfonyl)methanide.

[d] the similar metal salts based anion like $FeCl_4$-.

1.3 Synthesis of ILs

The synthetic routs are greatly related to the structures and composition of ILs, such as metathesis, protic synthesis, halogen free synthesis and other special methods, in this review, the synthesis procedures are summarized into one step method, two step method, enhanced methods and others especial methods, such as the synthetic method of chiral ILs.

1. One step method
   The ILs synthesized in one step method are mainly produced by the nucleophilic solvents reacting with the alkyl halide or esters, and the tertiary amine neutralized with acid. For example, the alkylimidazole halide, quaternary ammonium halide, alkyl sulfate, alkyl phosphate and neutralization reaction [27–31]. The alkylimidazole based ILs are synthesized according to Eq. 1.1.

$$\mathrm{Rim} + \mathrm{R'X} \rightarrow \left[\mathrm{RR'im}\right]\mathrm{X} \tag{1.1}$$

2. Two step method
   In the two step method, the alkylimidazole halide was first synthesized, then the halide was changed to targeted anions by complex reaction,metathesis reaction, ions exchange and electrolytic method. Many common used ILs are produced with this method, for examples, the $[\mathrm{Bmim}][\mathrm{FeCl_3}]$, $[\mathrm{Bmim}][\mathrm{PF_6}]$, $[\mathrm{Bmim}]_2[\mathrm{SO_4}]$ and $[\mathrm{bpy}][\mathrm{NO_3}]$ [32–36]. The synthetic routs are shown in the Eq. 1.2.

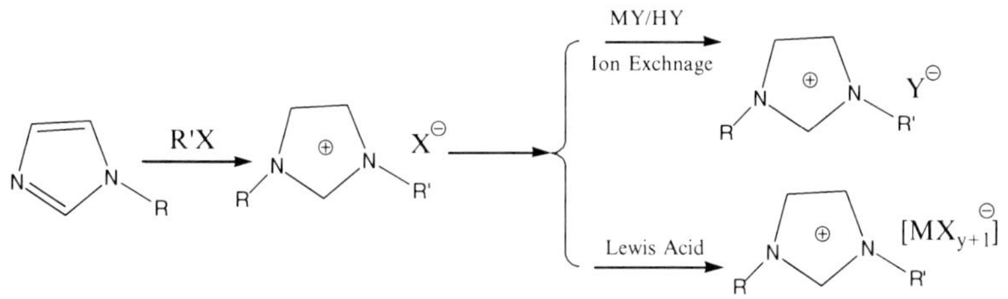

(1.2)

R'X

MY/HY

Ion Exchnage

Lewis Acid

$X^{\ominus}$ $Y^{\ominus}$ $[MX_{y+1}]^{\ominus}$

3. Enhanced methods
   In the synthesis process, microwave or ultrasonic are used to enhance the reaction to increase the reaction and conversion rate. For example, the [Bmim] $[\mathrm{BF_4}]$ is synthesized in the route as shown in Eq. 1.3 [37].

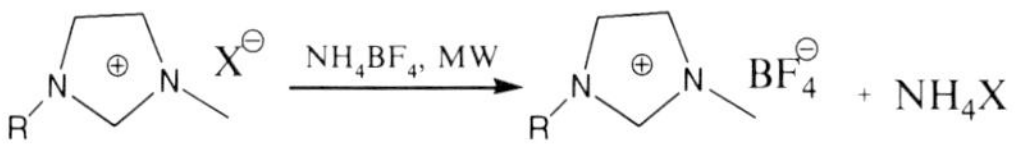

$$X^{\ominus} \xrightarrow{\mathrm{NH_4BF_4,\ MW}} BF_4^{\ominus} + \mathrm{NH_4X} \tag{1.3}$$

4. The synthetic method of chiral ILs [38, 39]
   Chiral ILs have more unique properties than that of common ILs, which combining the advantages of chiral material and ILs, and the chiral materials or asymmetric synthesis are both used in the procedure. For example, the chiral

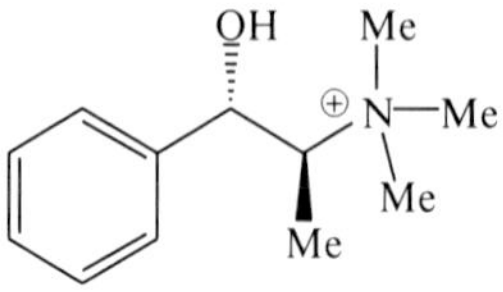

Fig. 1.1 (1S, 2R)-(+)-N, N-dimethylephdrinium

Table 1.3 Companies that produce ILs

Entry	Classification	Companies	Website
1	Range of ILs	Merck	http://www.merck.de/de/index.html
		Sigma-Aldrich	http://www.sigmaaldrich.com/united-states.html
		CAS	http://www.casact.org/
2	Imidazolium ILs	BASF	http://www.basf.com/group/corporate/us/en/
		io-li-tec	http://www.iolitec-usa.com/
3	Phosphonium ILs	Cytec	http://www.cytec.com/index.htm
4	Typically functional ILs	Frontier Scientific	http://www.frontiersci.com/
		Linzhou Keneng Materials Technology Co. Ltd.	http://lzkn.atobo.com.cn/
5	Ammonium ILs	Bioniqs	http://www.ipgroupplc.com/

compound (1S, 2R)-(+)-N, N-dimethylephdrinium cation(Fig. 1.1) can be used to synthesized ephedrinium based chiral ILs [39], and the nicotine based chiral ILs are synthesized according to the Eq. 1.4 [40, 41].

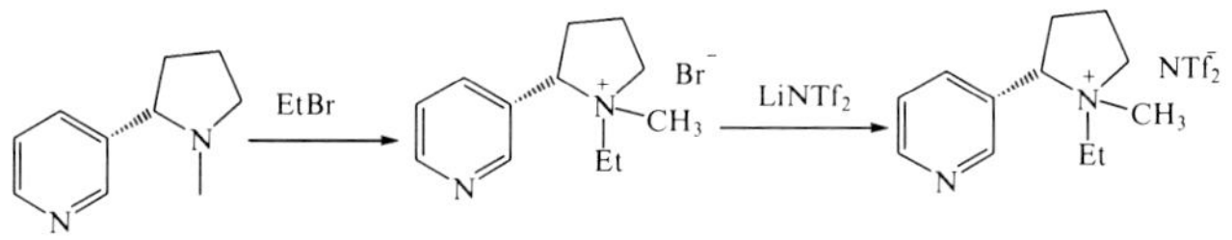

EtBr LiNTf$_2$ (1.4)

Now ILs are applied in a variety of fields, so many companies that produce and sale ILs are occurred all over the world. The common companies that produce ILs are listed in Table 1.3.

1.4 Characterization and Purification

After the ILs synthesis, the characterizations are followed to confirm the structures and purities. Nuclear magnetic resonance (NMR) spectrometry, Fourier transform infrared spectroscopy (FT-IR), the X-ray diffraction (XRD), elementary analysis (EA) and mass spectroscopy (MS) are common used techniques in ILs analysis. The

NMR included ^{1}H NMR and ^{13}C NMR, together with the FT-IR are widely to confirm the desired structures and functional groups of ILs. Meanwhile, the EA and MS are usually used to detect the ILs purities.

The common impurities are water, metathesis byproducts, sorbents and chemical drying agents, especially the water, which almost exit in all kinds of ILs, the water in ILs has various effects on the ILs applications, for the water sensitive reaction, the water removal is very necessary, while, sometimes a little amount of water in ILs may enhance the reaction, so the water content in ILs can be controlled by evaporated or vacuum drying according to the ILs applications. Some halide salts, alkali metals and heavy metal precipitation as byproducts in ILs, some of them are reduced by passing the ILs through silica gel [42], the precipitations are common filtered by millipore filters. In addition, some other methods, such as sorbents, distillation, zone melting and clean synthetic routes are also developed to obtain more pure ILs [39]. Sometimes the color of ILs would be dark when the reaction temperature is high, activated carbon is introduced to use in ILs discoloration. Meanwhile, the ILs those are solid at room temperature would be recrystallized to get better quality.

The separation and purification methods of ILs can be decided by different water solubility of ILs. The cation and anion composed of ILs both influence the water solubility of ILs, of course, the increased alkyl chain (n) length of the cation decreases the water solubility of ILs, and $n = 10$ is a boundary of liquid and solid phase of ILs. With a short alkyl chain cation, the ILs containing halide, acetate, sulfate, or phosphate are generally liquids insoluble with water, while the ILs with BF_4 or PF_6 are mostly water-immiscible [39].

1.5 Physicochemical Properties of ILs

ILs used in cellulose and biomass pretreatment has attracted much interest due to their excellent properties, such as the low melting point, high thermal stability and solvation capacity, especially the hydrogen-bonds among the cations and anions [43]. Moreover, the promising diversity of ILs suggests that appropriate ILs will be non-volatile polar solvents for carbohydrate dissolution and biomass pretreatment.

1.5.1 Melting Point

The melting point of a compound represents the lower limit of the liquid range and together with thermal stability defines the temperature window in which it can be used as a solvent [44].

The melting point is defined as the temperature of equilibrium between solid and liquid state in thermodynamic. Since the change of the Gibbs free energy equals zero at the equilibrium, the melting point (T_m) is defined as follows [45]:

$$T_m = \frac{\Delta H}{\Delta S} \quad (1.5)$$

Recently, many researchers study the effect of structural features on melting point because the melting point is an important factor for employment of ILs as a reaction media. Factors influencing in ILs melting point are charge, size and the distribution of charge on the constituent ions. For the same class of ILs, small changes in the shape of the uncharged, covalent regions of the ions need to be considered [46]. For example, the melting point of imidazolium based ILs is influenced by four factors: electron delocalization, H-bonding ability, the symmetry of the ions, and van der Waals interactions. Both alkyl substitutions on the cation and ion asymmetry have been shown to interfere with the packing efficiency of ions in the crystalline lattice [47]. Researchers are always try to find that how the chemical structure affects the melting point by many different measuring methods. The following trends can be concluded from open literature:

1. The sizes and shapes of cations of ILs are important factors influencing the melting points of ILs. In general, as the size of the cation increases, the melting point of the salts decreases.
2. With increasing the size of the anion, the melting point of the salts decreases, which reflects the weaker columbic interactions in the crystal lattice. For instance, from Cl^- to $[BF_4]^-$ to $[PF_6]^-$ to $[AlCl_4]^-$, the melting points of the sodium salts decrease from 801 to 185 °C with increasing thermochemical radius of the anion [47].
3. With the same anion, symmetry of the alkyl substitution also affects the melting point of ILs. Generally, highly asymmetric alkyl substitution has been identified as important for obtaining high melting point. For example, the melting point of [Mmim]Cl is 124.5 °C while the melting point of [Bmim]Cl is 65 °C.
4. The alkyl chain length of cations also affect melting point of ILs. For example, as for 1-alkyl-3-methylimidazolium tetrafluoroborate, with an increase in the alkyl chain length (up to $n = 8$), the melting point decrease, where n is the number of carbon atoms. But the melting point of ILs increases gradually with increasing chain length when $n > 8$. The same condition occurs in 1-alkyl-3-methylimidazoliumbis (trifyl) imide [46].
5. For ILs in which the only difference is the degree of branching within the alkyl chain at the imidazolium ring, higher degree of branching within the alkyl chain, higher melting point of ILs.

Owing to their unique properties, ILs are widely used as a kind of versatile solvents in biomass separation/conversion. Most ILs used for biomass pretreatment have low melting points, so that they are liquid at room temperature, the melting points of the ILs listed in Fig. 1.2 are all lower than 350 K [26]. That's because

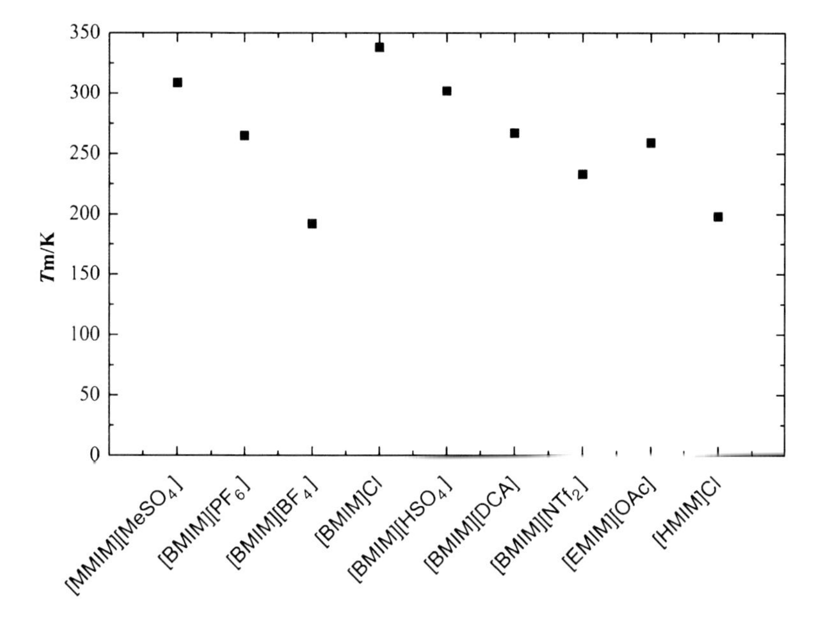

Fig. 1.2 The melting points of ILs used for biomass pretreatment

lower melting points of ILs make them easy to handle and are known as a technical advantage on the recyclability of the solvents. In recent years, some new types of ILs that have lower melting points and a sufficient polarity to further process carbohydrates have been claimed to replace chloride-based ILs [48]. These ILs are formate, acetate or phosphate salts with imidazolium based cations. They have been proven to be potential solvents to dissolve cellulose under mild condition.

1.5.2 Viscosity

Viscosity is one of the most important physical properties when considering ionic liquid applications. The viscosities of many ILs are much higher than most organic solvents at room temperature. Generally, the viscosity of ILs is 10–1,000 mPa s. A low viscosity is generally desired to use IL as a solvent, to minimize pumping costs and increase mass transfer rates while higher viscosities may be favorable for other applications such as lubrication or use in membranes [49]. Viscosity can be fitted with the Vogel-Tammann-Fulcher equation although it usually follows a non-Arrhenius behavior. Viscosities of ILs remain constant when the shear rate increases so that they have Newtonian and non-Newtonian behaviors [44].

The viscosity of ILs is usually affected by the kind of the anion, cation and substituents on the cation and anion of the imidazolium-based ILs. Generally, for ILs with the same anion, the alkyl substituents on the imidazolium cation is larger,

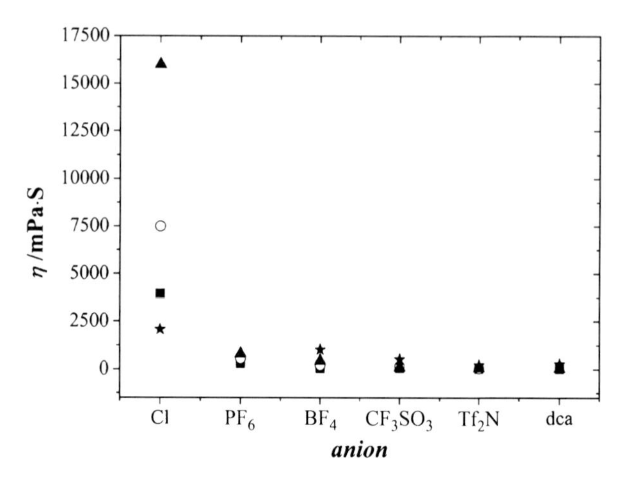

Fig. 1.3 Effect of anion on the viscosities of ILs. ■: $[BMIM]^+$ based ILs at 303.15 K; ○: $[HMIM]^+$ based ILs at 298.15 K; ▲: $[P_{6,6,6,14}]^+$ based ILs at 303.15 K; ★: $[OMIM]^+$ based ILs at 298.15 K

the viscosity of ILs is higher. For example, for the 1-alkyl-3-methylimidazolium hexafluorophosphate and bis((trifluoromethyl)sulfonyl)imide series ($[Rmim][PF_6]$ and $[Rmim][Tf_2N]$), viscosity increases with increasing the number of carbon atoms in the linear alkyl group [50]. Furthermore, branching of the alkyl chain in 1-alkyl-3-methylimidazolium salts usually result in lower viscosity. Finally, a reduction in van der Waals interactions can also attribute to the low viscosity of ILs bearing polyfluorinated anions. Hydrogen bonding between counter anions and symmetry can also affect viscosity. In short, the viscosity of ILs based on the most common anions decreases in the order $Cl^- > [PF_6]^- > [BF_4]^- > [TfO]^- > [Tf_2N]^- > [dca]^-$ (shown in Fig. 1.3) [26, 51–53].

The impurities in the ILs greatly affect their viscosities [4]. In one study [54], a series of ILs were prepared and purified by many kinds of techniques. Then their impurities were analyzed and physical properties were evaluated. The results showed that chloride concentrations of up to 6 wt% were found for some of the preparative methods whereas chloride concentrations of between 1.5 and 6 wt% increased the observed viscosity by between 30 and 600 %. Studies also found that the non-halo aluminate alkylimidazolium ILs absorbed water rapidly from the air. As little as 2 wt% (20 mol%) water could reduce the viscosity of $[BMIM][BF_4]$ by more than 50 %. Therefore, purities and handling should be carefully considered when viscosities of ILs are measured.

Owing to the widely application of ILs, the experimental measurement and theoretical modeling of viscosities of ILs and mixtures are essential in the development and design of processes [55]. There are several models used for the prediction of ILs viscosities.

Abbott [56] proposed a theoretical model for prediction of viscosities by modifying the "whole theory". In that model, 11 ILs mainly based on imidazolium at three temperatures (298, 303 and 364 K) were investigated. The model had low reliability despite its theoretical interpretation and therefore it has limited application for practical processes.

Han et al. [57] proposed a QSPR method for prediction of the viscosity of imidazolium based ILs. In that work, a database of 1,731 experimental data values at various temperatures and pressures were used for 255 ILs, that included 79 cations and 71 anions. As for the viscosity of imidazolium-based ILs, the cation-anion electrostatic interactions have important effects.

The most useful viscosity estimation models for complex molecules are those based on group contributions. The methods usually use some variation of temperature dependence proposed by de Guzman [58].

The Orrick–Erbarmethod [59] proposed employs a group contribution technique to estimate the A and B parameters in the following equation [49]:

$$\ln \frac{\eta}{\rho M} = A + \frac{B}{T} \tag{1.6}$$

Where the η and ρ are the viscosity in mPa · s units and density is in g · cm^{-3} units, respectively.

Viscosities calculated by the following method are in good agreement with experimental literature data. The model could predict the viscosity of new ILs in wide ranges of temperature and could be extended to a larger range of ILs as data for these become available. It is also shown that an Orrick–Erbar-type approach was successfully applied to estimate of the viscosity of ILs by a group contribution method.

In 2002, it was reported that ILs can dissolve biomass materials [23]. Viscosity plays a role in cellulose solvation, because it considered that ILs with low viscosity are more efficient and easier to handle in dissolving cellulose [60]. When an IL has a low viscosity, cellulose can be dissolved at room temperature. For example, microcrystalline cellulose was dissolved at a lower temperature in 1-ethyl-3-methyl imidazolium methylphosphonate [EMIM][CH_3PO_4] in compared with 1-ethyl-3-methylimidazolium dimethylphosphate [EMIM][$(CH_3)_2PO_4$] [61]. However, viscosities of ionic liquids are not the only important parameter in biomaterial dissolution. In 1-benzyl-3-methylimidazolium chloride [$PhCH_2MIM$]Cl, researchers have found that it was a rather powerful solvent no matter its dicyanamide anion, the cation-anion pair resulted in reasonably low viscosity [62]. Nevertheless, it was found that ILs containing alkyloxy or alkyloxyalkyl groups have low viscosities and that they are beneficial for dissolving cellulose. Especially, a powerful solvent for cellulose has been found to be 1-(3, 6, 9-trioxadecyl)-3-ethylimidazolium acetate [Me$(OEt)_3$-Et-Im][OAc] (in Table 1.6) [63].

1.5.3 Density

Many density correlations of ILs have been reported because it is an important fundamental property [49]. IL database, such as The UFT/Merck Ionic Liquids Biological Effects Database, IUPAC Ionic Liquids Database and Tohoku Molten

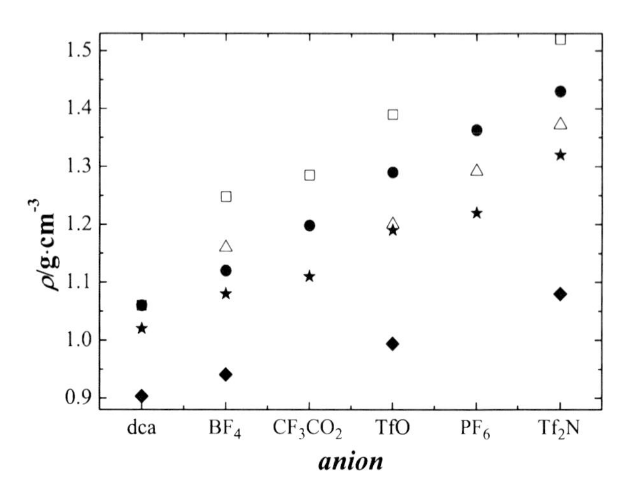

Fig. 1.4 Effect of anion on the densities of ILs. □: $[EMIM]^+$ based ILs at 298.15 K; ●: $[BMIM]^+$ based ILs at 298.15 K; △: $[HMIM]^+$ based ILs at 298.15 K; ★: $[OMIM]^+$ based ILs at 298.15 K; ◆: $[P_{6,6,6,14}]^+$ based ILs at 293.15 K

Salt Database, provide up-to-date information on densities of ILs [26]. Generally, imidazolium-based ILs are widely used as solvents for ILs applications owing to their excellent physical properties, such as low viscosities and high thermal and aqueous stability. Their densities are more available in the open literature than other properties ILs.

The usual densities of ILs vary between 1.12 g · cm^{-3} ($[(n\text{–}C_8H_{17})(C_4H_9)_3N]$ $[(CF_3SO_2)_2N]$) and 2.4 g · cm^{-3} (a 34–66 mol% $[(CH_3)_3S]Br/AlBr_3$ ionic liquid) [64, 65]. However, large cations have densities lower than water, such as aliquat (in Table 1.6), since their long alkyl chains have higher flexibility. From IPE IL database [26], the densities of most ionic liquids tend to have low sensitivity to variations in temperature. Furthermore, the impact of impurities on densities is much less dramatic than for viscosities.

One can make some conclusions on the cation effect and also to study the effect of alkyl chain length on the density and derived properties. Gardas et al. [66] found that as the alkyl chain length in the pyrrolidinium cation increases, the density of the corresponding IL decreases, similar to that observed for imidazolium-based ILs [67]. Generally, the order of increasing density for ILs with a common cation is $N(CN)_2^- < BF_4^- < CF_3CO_2^- < CF_3SO_3^- < PF_6^- < Tf_2N^-$ (shown in Fig. 1.4) [26, 53, 69]. The higher densities of Tf_2N^--containing ILs arise from the much higher mass of the anion [68].

Since it is difficult to measure densities of all ILs at different conditions, it is necessary to find a method to estimate densities of ILs. So far, a large amount of models about prediction of densities of ILs have been proposed in the open literature. The common models are group contribution methods (GCMs), quantitative structure property relationships (QSPRs) and artificial neural networks (ANNs) [70]. For instance, Lazzús [71] used the method of ANNs to estimation the density of imidazolium-based ionic liquids at different temperature and pressure. The method has a better estimation results, but it is not convenient to be used in prediction.

Table 1.4 The decomposition temperatures of ILs used in dissolving cellulose

Name of ILs	Abbreviation	Decomposition temperature/K
1-Butyl-3-methylimidazolium tetrafluoroborate	$[BMIM][BF_4]$	680.15 [26]
1-Butyl-3-methylimidazolium bis-(trifluoromethane-sulfonyl)imide	$[BMIM][NTf_2]$	675.15 [26]
1-Butyl-2,3-dimethylimidazolium Tetrafluoroborate	$[BM_2IM][BF_4]$	671.15 [75]
Trihexyltetradecyl phosphonium dicyanamide	$[P_{6,6,6,14}][dca]$	668.15 [26]
1-Butyl-3-methylimidazolium hexafluorophosphate	$[BMIM][PF_6]$	663.15 [26]
1,3-Dimethylimidazolium Methylsulfate	$[MMIM][MeSO_4]$	649.15 [79]
1-Butyl-3-methylimidazolium hydrogen sulfate	$[BMIM][HSO_4]$	603.15 [80]
1-Allyl-3-methylimidazolium chloride	[AMIM]Cl	558.95 [81]
1-Butyl-3-methylimidazolium carboxybenzene	$[BMIM][(C_6H_5)COO]$	552.75 [26]
1-Butyl-3-methylimidazolium chloride	[BMIM]Cl	527.15 [26]
1-Hexyl-3-methylimidazolium chloride	[HMIM]Cl	526.15 [26]
1-Ethyl-3-methylimidazolium acetate	[EMIM][OAc]	493.15 [82]
1-Butyl-3-methylimidazolium formate	[BMIM][dca]	473.15 [83]

The volumetric properties for the ILs can be estimated by the values of density. For example, $[C_nMIM][BF_4]$ (n = 2, 3, 4, 5, 6) are common ILs, many researchers have previously reported many properties in wide ranges of temperature. We can plot of values of in ρ against T, a straight line was obtained for given IL, and its empirical linear equation is [72]

$$\mathrm{Ln}\rho = b - \alpha T \tag{1.7}$$

Where b is an empirical constant, the negative value of slope, $\alpha = -(\partial \ln\rho/\partial T)_p$, is thermal expansion coefficient of the IL $[C_nMIM][BF_4]$.

We can also obtain the molecular volume (V_m) of ILs. The value of V_m was calculated using the following equation [72]

$$V_m = M/(N\rho) \tag{1.8}$$

Where M is molar mass of ILs, N is Avogadro's constant.

1.5.4 Thermal Stability

Compared with organic solvents, most ILs have relatively high thermal stability. The decomposition temperatures reported in the open literature are generally >200 °C, and they are liquid state in a wide range of temperatures (from 70 to 300–400 °C). The decomposition temperatures of ILs that dissolving cellulose are listed in Table 1.4.

Many literature works have investigated the thermal stability of ILs on imidazolium and anion structures. The onset of thermal decomposition is similar for the different cations but appears to decrease as the anion hydrophilicity increases. Ngo et al. [73] found that the thermal stability of the imidazolium-based ILs increases with increasing linear alkyl substitution. Owing to the facile elimination of the stabilized alkyl cations, the presence of nitrogen substituted secondary alkyl groups decreases the thermal stability of ILs. They also found that the stability dependence on the anion is $[PF_6]^- > [Tf_2N]^- \sim [BF_4]^- >$ halides. Fox et al. [74] found that the alkyl chain length does not have a large effect on the thermal stability of the ILs.

However, the thermal stability of ILs has been revised [4]. The range of thermal stability of ILs published in the open literature is overstated. The decomposition temperature of ILs calculated from fast thermo gravimetric analysis (TGA) scans in a protective atmosphere and does not imply a long-term thermal stability below those temperatures [44]. Fox and his group have done some nice study on the thermal stability of ILs [74–76]. Compared with the data from both isothermal and constant ramp rate programs for the decomposition of 1-butyl-2, 3-dimethylimidazolium tetrafluoroborate ([BMMIM][BF_4]) under N_2, they found that isothermal TGA experiments may be the more appropriate method for evaluating the thermal stabilities of ILs [75]. Based on TGA pyrolysis data of 1, 2, 3-trialkylimidazolium room temperature ILs, They also found that although the calculated onset temperatures were above 350 °C, significant decomposition does occur 100 °C or more below these temperatures.

Singh et al. [77] analyzed the thermal stability of imidazolium based ILs [BMIM][PF_6] in a confined geometry. They found that [BMIM][PF_6] in confined geometry starts at an earlier temperature than that for the unconfined ILs. The loss of alkyl chain end groups of [BMIM] cation of ILs assign to the early decomposition by using a phenomenological 'hinged spring model'. The idea of 'hinged spring' model is that the imidazolium ring is supposed to be 'hinged' to the SiO_2 matrix pore walls by surface oxygen interacting with the C-H group of the imidazolium ring.

Researchers have reported a new method to study the changes that occur during thermal aging of ILs. The method is potentiometric titration, which is precise, low-cost and quick analytically. To a small extent, they found that imidazolium salts start to decompose at much lower temperatures than those obtained from thermo gravimetric analysis by using this method. They also concluded that the stability of ILs is also influenced by water, except by their composition, such as anion type and alkyl substituent at the imidazolium ring. For instance, 2 wt% of water in ILs could bring about increased degradation of [BMIM]Cl at 140 °C. Furthermore, [BMIM][BF_4], [EMIM][CH_3SO_3], [BMIM][CH_3SO_3] and [BMIM][Tf_2N] are completely stable at 140 °C for 10 days [78]. Finally, from these data, long-term stability of ILs is a complicated problem with obvious and serious implications for their use as solvents media of chemical reactions.

1.6 Solubility of Cellulose/Biomass in ILs

Many kinds of materials can be dissolved in ILs, such as the metal salts, gases, carbohydrates, sugar alcohols, cellulose and even the biomass [36, 84, 85]. In this chapter, cellulose dissolution with various ILs would be discussed.

Cations and anions play an important role in the cellulose dissolution process [23]. The soluble ability of cellulose in ILs can be modified by changing the cations or anions. Anions that form hydrogen bonds with hydroxyl groups are effective for cellulose dissolution, small size and alkalinity of anions promote to increase cellulose soluble ability, for example, halide, acetate, formate and dialkyl phosphate [22, 61, 86]. Different imidazolium, pyridinium and pyrrolidinium-based cations are commonly used in cellulose dissolution together with the anions above mentioned. Cellulose soluble ability decreases in the ILs with length of the alky chain. Meanwhile, alkyl chains or anions with hydroxyl groups tend to be adverse to the cellulose dissolution in ILs due to the increase of hydrogen-bond acidity of ILs [22].

The Kamlet-Taft polarity parameters, for example, β is the hydrogen-bond basicity parameter, α is the measure of the hydrogen-bond acidity, which express the ability to donate and accept hydrogen bonds, respectively, and π^* is the parameter of the interactions through dipolarity and polarizability. The parameters α and β are similar to the acid and base characteristics according to the definition, while, the α and β are not completely consistent with the acidity and alkalinity of ILs in all conditions, the α and β emphasize the acceptance ability of hydrogen-bond. As for ILs, β has been the most useful parameter in predicting the solubility of cellulose in different ILs with various anions [87, 88].

With ILs, the higher β and dipolarity caused the better ability to dissolve cellulose [22]. With the cation $[Bmim]^+$, some anions showed different trends for dissolving cellulose due to the increasing hydrogen-bond acceptance ability, for example, the β value of some anions are in an order of $OAc^- > HCOO^- > (C_6H_5)COO^- > H_2NCH_2COO^- > dca^-$, and the cellulose solubility was about 16 wt% in [Bmim][OAc] ($\beta = 1.161$), which was higher than those in [Bmim][HCOO] ($\beta = 1.01$), [Bmim][$(C_6H_5)COO$] ($\beta = 0.98$) [8, 89–92]. For a given cation, the effect of anions on cellulose dissolution changes greatly, this result may contribute to the formula weight of ILs, which means that the cellulose dissolution is close to the mass percent in the given ILs.

The Kamlet-Taft parameters of selected ILs and examples of used in dissolving cellulose and biomass pretreatment are as followed in Table 1.5, and the properties of ILs examples and applications in biomass are shown in Table 1.6. Common ILs used for cellulose/biomass pretreatment are shown with details in Table 1.7.

Table 1.5 Kamlet-Taft parameters of selected ILs [61, 93–96]

ILs	α	β	π*
[BMIM][$MeSO_3$]	0.44	0.77	1.02
[BMIM][$MeSO_4$]	0.55	0.67	1.05
[BMIM][HSO_4]	–	0.67	1.09
[BMIM][$MeCO_2$]	0.47	1.20	0.97
[BMIM][$N(CN)_2$]	0.54	0.59	1.05
[BMIM][Me_2PO_4]	–	1.12	–
[BMIM]Cl	0.44	0.84	1.14
[AMIM]Cl	0.46	0.83	1.17
[Bmim][OAc]	0.57	1.16	0.89
[Bmim][HCOO]	–	1.01	–
[Bmim][(C6H5)COO]	–	0.98	–
[EMIM][OAc]	–	1.074	–
[EMIM][(MeO)HPO_2]	0.52	1.00	1.06
[EMIM][(MeO)$MePO_2$]	0.50	1.07	1.04
[EMIM][$(MeO)_2PO_2$]	0.51	1.00	1.06
[AMIM][HCOO]	0.48	0.99	1.08
[AEIM][HCOO]	0.47	0.99	1.06
[TMIM][HCOO]	0.46	0.99	1.06
[BMIM][OTf]	0.63	0.48	0.97
[BMIM][PF_6]	–	0.44	–

π* stands for the parameter of the interactions through dipolarity and polarisability

1.7 Conclusion and Prospects

ILs have many intriguing properties, such as low vapor pressure, high chemical and thermal stability, wide electrochemical window, non-flammability, wide liquid range and recognition ability of biomaterials. They are applied in a variety of fields, including extraction, organic synthesis, catalysis/biocatalysts, materials science, electrochemistry and separation technology. Furthermore, because ILs have ionic nature, they may interact with charged groups in the enzyme, either in the active site or at its periphery, causing changes in the enzyme's structure.

ILs are promising solvents, for reaction and separation, offers tremendous possibilities for the development of sustainable industry, advanced materials and chemicals. Up to date, cellulose dissolution with ILs has been well developed. And the problems of ILs like high cost, which is obstacle to the industrial scale of ILs application, and the future researches trends and orient of ILs in cellulose/biomass applications, such as the binary systems including complex ILs or solvents, additives and catalysts, will be described more clearly.

The recycle of ILs in an efficient way is also should be well developed. The process of dissolving cellulose or biomass applications in ILs should be optimized to reduce the loss of cellulose or biomass. Binary and ternary system of ILs perhaps will be more efficient for the dissolution of cellulose/biomass, meanwhile, the large scale of synthesis and functional design of ILs with high stabilities and low viscosities will be developed. Considering of the environmental effect and special purposes, bio-degradable ILs and chiral ILs, acid or base enhanced ILs with lower

Table 1.6 Physical properties of ILs that dissolve cellulose

No.	Name	Abbreviation	Viscosity/ (mPa · s)	Density/ (g · cm^{-3})	Biomass
1	1,3-Dimethylimidazolium methylsulfate	[MMIM][$MeSO_4$]	72.91[a] [97]	1.32725[a] [97]	Softwood, kraft lignin, maple wood flour [87]
2	1-Allyl-3-methylimidazolium formate	[AMIM][HCOO]	66[a] [98]		Cellulose [98]
3	1-Allyl-3-methylimidazolium chloride	[AMIM]Cl	821[a] [99]	1.1660[a] [99]	Southern pine powder, Norway spruce sawdust, spruce, fir, beech, chestnut, maple wood flour, cellulose, maple [62, 112, 113]
4	1-Butyl-3-methylimidazolium hexafluorophosphate	[BMIM][PF_6]	450[a] [100]	1.369[a] [101]	Does not dissolve [23, 114]
5	1-Butyl-3-methylimidazolium tetrafluoroborate	[BMIM][BF_4]	233[a] [102]	1.213[a] [103]	Does not dissolve [23, 114, 116]
6	1-Butyl-2,3-dimethylimidazolium tetrafluoroborate	[BM_2IM][BF_4]	780[a] [104]	1.2[a] [104]	Softwood kraft lignin [115]
7	1-Butyl-3-methylimidazolium chloride	[BMIM]Cl	3950[b] [26]	1.08[a] [26]	Cellulose, lignin, wood, xylan [23, 62, 112, 114, 116–119]
8	1-Butyl-3-methylimidazolium hydrogen sulfate	[BMIM][HSO_4]	1572[c] [80]	1.32[d] [80]	Pine sapwood, debarked mixed willow stems, miscanthus [94]
9	1-Butyl-3-methylimidazolium formate	[BMIM][HCOO]	38[a][105]		Sugars, starch and cellulose [63]
10	1-Butyl-3-methylimidazolium thiocyanate	[BMIM][SCN]	51.7[a] [106]	1.06979[a] [106]	Pulp cellulose [63]
11	1-Butyl-3-methylimidazolium dicyanamide	[BMIM][dca]	37[a] [104]	1.06[a] [104]	Avicel cellulose [22, 63]
12	1-Butyl-3-methylimidazolium bis-(trifluoromethanesulfonyl)imide	[BMIM][NTf_2]	54.5[a] [107]	1.433[a] [103]	Avicel [63]
13	1-Butyl-3-methylimidazolium carboxybenzene	[BMIM][(C_6H_5)COO]			Avicel [22]
14	1-Ethyl-3-methylimidazolium dialkybenzenesulfonate	[EMIM][ABS]			Sugarcane plant waste [120, 121]
15	1-Ethyl-3-methylimidazolium acetate	[EMIM][OAc]	91[a] [104]	1.03[a] [104]	Spruce, beech, chestnut, maple wood flour, red oak, southern yellow pine, kraft lignin, cellulose [23, 114, 118, 120]

(continued)

Table 1.6 (continued)

No.	Name	Abbreviation	Viscosity/ (mPa · s)	Density/ ($g \cdot cm^{-3}$)	Biomass
16	1-Ethyl-3-methylimidazolium diethyl phosphate	[EMIM][DEP]		1.1400^{a} [108]	Wheat straw, cellulose [61, 122]
17	1-Ethyl-3-methylimidazolium xylene sulfonate	[EMIM][XS]			Kraft wood lignin [120]
18	1-Hexyl-3-methylimidazolium Chloride	[HMIM]Cl	716^{a} [109]	1.03^{a} [109]	Cellulose dissolving pulps [114]
19	1-Hexyl-3-methylimidazolium trifluoromethanesulfonate	[HMIM] [CF_3SO_3]	160^{a} [104]	1.2^{a} [104]	Softwood kraft lignin [115]
20	N,N-Dimethylethanolammonium Acetate	[MM(EtOH)NH] [OAc]		0.8247^{a} [110]	Cellulose filter paper cotton [88]
21	Tetrabutylamimonium formate	[TBA][HCOO]			Cellulose [63]
22	1,8-Diazabicyclo[5.4.0]undec-7-enium hydrochloride	[HDBU]Cl			Cellulose [123]
23	1,8-Diazabicyclo[5.4.0]undec-7-enium formate	[HDBU][HCOO]			Cellulose [123]
24	1,8-Diazabicyclo[5.4.0]undec-7-enium acetate	[HDBU][OAc]			Cellulose [123]
25	1,8-Diazabicyclo[5.4.0]undec-7-enium thiocyanate	[HDBU][SCN]			Cellulose [123]
26	1,8-Diazabicyclo[5.4.0]undec-7-enium hydrogensufate	[HDBU][HSO_4]			Cellulose [123]
27	1,8-Diazabicyclo[5.4.0]undec-7-enium trifluoroacetate	[HDBU][TFA]			Cellulose [123]
28	1,8-Diazabicyclo[5.4.0]undec-7-enium methanesulfonate	[HDBU] [$MeSO_3$]			Cellulose [123]
29	1,8-Diazabicyclo[5.4.0]undec-7-enium lactate	[HDBU][Lac]			Cellulose [123]
30	1,8-Diazabicyclo[5.4.0]undec-7-enium tosylate	[HDBU][Tos]			Cellulose [134]

31	1,8-Diazabicyclo[5.4.0]undec-7-enium saccharinate	[HDBU][Sacch]			Cellulose [123]
32	1, 8-diazabicyclo [5.4.0] undec-7-enium bis(trifluoromethanesulfonyl)imide	[HDBU][NTf_2]			Cellulose [123]
33	8-Methyl-1,8-diazabicyclo[5.4.0]undec-7-enium hydrogensulfate	[MDBU][HSO_4]			Cellulose [123]
34	8-Methyl-1,8-diazabicyclo[5.4.0]undec-7-enium chloride	[MDBU]Cl			Cellulose [22]
35	8-Octyl-1,8-diazabicyclo[5.4.0]undec-7-enium chloride	[ODBU]Cl			Cellulose [22]
36	1-Butyl-4-methylpyridinium hexafluorophosphate	[BMPy][PF_6]			Insoluble [115]
37	1-Butyl-4-methylpyridinium chloride	[BMPy]Cl			Avicel, spruce sulfite, pulp, Cotton linters, cellulose [22]
38	AMMOENG 110 formate	[Amm110][HCOO]			Avicel [63]
39	AMMOENG 110 acetate	[Amm110][OAc]			Avicel [63]
40	Trihexyltetradecyl phosphonium dicyanamide	[P66614][dca]	378.0^{a} [111]	0.898^{a} [111]	Avicel [63]
41	AMMOENG 110 dicyanamide	[Amm110][dca]			Avicel [63]
42	Tetrabutylammonium formate	[Bu_4N][HCOO]			Avicel [63]
43	N,N,Ntriethyl-3,6,9-trioxadecylammonium acetate	[$Me(OEt)_3$-Et_3N][OAc]			Avicel [63, 117]
44	Benzyldimethyl(tetradecyl)ammonium chloride	BDTACl			Spruce sulfite, avicel, pulp, cotton linters [124, 125]
45	1-(2-Hydroxyethyl)-3-methylimidazolium chloride	[EOHMIM]Cl			Not mentioned [126]
46	1-Allyl-2,3-dimethylimidazolium chloride	[AdMIM]Br			Avicel, spruce sulfite, pulp, cotton linters[63, 127]
47	1-(3,6,9-Trioxadecyl)-3-ethylimidazolium	[$Me(OEt)_3$-Et-Im][OAc]			Avicel [63]

[a]at 298.15K; [b]at 303.15K; [c]at 296.65K; [d]at 293.15K.

Table 1.7 Common ILs used for cellulose/biomass pretreatment

ILs	Material	Solubility	Conditions	Refs.
[BMIM]Cl	Cellulose	10 wt%	100 °C	[23][a]
	Cellulose	25 wt%	Microwave heating 3–5 s	[23][a]
	Southern yellow pine (<0.125 mm)	52.6 %	110 °C, 16 h	[128][b]
	Wood chips	Partially soluble	130 °C, 15 h	[62]
	Silk	9.51 % (w/w)	100 °C	[129]
	Bird feather	23 %	100 °C, 48 h	[130]
	Chitin	10 wt%	110 °C, 5 h	[131]
[AMIM]Cl	Cellulose	14.5 wt%	80 °C	[132]
	Ball-milled Southern pine powder	8 wt%	80 °C, 8 h	[62]
	Norway spruce sawdust	8 wt%	110 °C, 8 h	[62]
	Pine	26 wt%	24 h	[133]
	Southern pine TMP	5 wt%	130 °C, 8 h	[62]
[EMIM][OAc]	Cellulose/lignin	4 wt%	75 °C, 1 h	[137]
	Southern yellow pine (<0.125 mm)	98.5 %	110 °C, 16 h	[128][b]
	Red oak (0.125–0.250 mm)	99.5 %	110 °C, 16 h	[128][b]
	Aspen wood	5 wt%	120 C	[135]
	Energy cane bagasse	5 wt%	120 °C	[136]
[BMIM][PF_6]	–	Insoluble	Microwave	[23]
[Ch][AA]	Sugar Cane Bagasse	5 wt%	Magnetic stirrer	[134]

[a]0.5–1.0 g of fibrous cellulose, 10 g [Bmim]Cl
[b]Wood load (0.50 g) in 10 g IL

viscosity, higher thermal stability would be designed and synthesized by calculating the structure-function relationship or predicting the properties with group contribution method or semi-rational formula.

Acknowledgements This research was supported financially by the Projects of International Cooperation and Exchanges NSFC (No. 21210006), Natural Science Foundation of Beijing of China (No.2131005, No.2132055) and National High Technology Research and Development Program of China (863 Program) (No. 2012AA063001).

References

1. Xie H, King A, Kilpelainen I, et al. Thorough chemical modification of wood-based lingocellulosic materials in ionic liquids. Biomacromolecules. 2007;8:3740–8.
2. Dupont J, de Souza RF, Suarez PAZ. Ionic liquids (molten salt) phase organ metallic catalysis. Chem Rev. 2002;102:3667–91.

3. Welton T. Room-temperature ionic liquids: solvents for synthesis and catalysis. Chem Rev. 1999;99:2071–83.
4. Kosmulski M, Gustafsson J, Rosenholm JB. Thermal stability of low temperature ionic liquids revisited. Thermochim Acta. 2004;412:47–53.
5. Hagiwara H, Sugawara Y, Isobe K, et al. Immobilization of Pd $(OAc)_2$ in ionic liquid on silica: application to sustainable Mizoroki–Heck reaction. Org Lett. 2004;6:2325–8.
6. Bosmann A, Datsevich L, Jess A, et al. Deep desulfurization of diesel fuel by extraction with ionic liquids. Chem Commun. 2001;7:2494–5.
7. Galinski M, Lewandowski A, Stepniak I. Ionic liquids as electrolytes. Electrochim Acta. 2006;51:5567–80.
8. Adams CJ, Earle MJ, Roberts G. Friedel-Crafts reactions in room temperature ionic liquids. Chem Commun. 1998;1998:2097–8.
9. Fischer T, Sethi A, Welton T, et al. Diels-Alder reactions in room-temperature ionic liquids. Tetrahedron Lett. 1999;40:793–6.
10. Snedden P, Cooper AI, Scott K, et al. Cross-linked polymer-ionic liquid composite materials. Macromolecules. 2003;36:4549–56.
11. Gurkan BE, Fuente JCL, Mindrup EM, et al. Equimolar CO_2 absorption by anion functionalized ionic liquids. J Am Chem Soc. 2010;132:2116–17.
12. Pickett J. Sustainable biofuels: prospects and challenge. London: The Royal Society; 2008.
13. Perlack RD, Wright LL, Turhollow AF, et al. Biomass as a feedstock for a bioenergy and bioproducts industry: the technical feasibility of a billion-ton annual supply. Oak Ridge: U.S. Department of Energy and U.S. Department of Agriculture; 2005.
14. Yang B, Wyman CE. Effect of xylan and lignin removal by batch and flow through pretreatment on the enzymatic digestibility of corn stover cellulose. Biotechnol Bioeng. 2004;86:88–95.
15. Updegraff DM. Semimicro determination of cellulose in biological materials. Anal Biochem. 1969;32:420–4.
16. Qian X, Ding SY, Nimlos MR, et al. Atomic and electronic structures of molecular crystalline cellulose I β: a first-principles investigation. Macromolecules. 2005;38:10580–9.
17. Hammer RB, Turbak AF. Abstracts of papers of the American Chemical Society 1977;173: 8–8.
18. McCormick CL, Dawsey TR. Preparation of cellulose derivatives via ring opening reaction with cyclic reagents in lithium chloride/N, N-dimethylacetamide. Macromolecules. 1990;23:3606–10.
19. Ciacco GT, Liebert TF, Frollini E, et al. Application of the solvent dimethylsulfoxide/tetrabutyl-ammonium fluoride trihydrate as reaction medium for the homogeneous acylation of sisal cellulose. Cellulose. 2003;10:125–32.
20. Fink HP, Weigel P, Purz HJ, et al. Structure formation of regenerated cellulose materials from NMMO-solutions. Prog Polym Sci. 2001;26:1473–524.
21. Heinze T, Liebert T. Unconventional methods in cellulose functionalization. Prog Polym Sci. 2001;26:1689–762.
22. Xu AR, Wang JJ, Wang HY. Effects of an ionic structure and lithium salts addition on the dissolution of cellulose in 1-butyl-3-methylimidazolium-based ionic liquid solvent systems. Green Chem. 2010;12:268–75.
23. Swatloski RP, Spear SK, Holbrey JD, et al. Dissolution of cellulose with ionic liquids. J Am Chem Soc. 2002;124:4974–5.
24. Graenacher C. Cellulose solution. US Patent, 1943176; 1934.
25. Rogers RD. Materials science – reflections on ionic liquids. Nature. 2007;447:917–18.
26. Zhang SJ, Lu XM, Zhou Q, et al. Ionic liquids: physicochemical properties. Oxford: Elsevier; 2009.
27. Nishida T, Tashiro Y, Yamamoto M. Physical and electrochemical properties of 1-alkyl-3-methylimidazolium tretrafluoroborate for electrolyte. J Fluor Chem. 2003;120:135–41.

28. MacFarlane DR, Sun J, Golding J, et al. High conductivity molten salts based on the imide ion. Electrochim Acta. 2000;45:1271–8.
29. Holbery JD, Reichert WM. Efficient, halide free synthesis of new, low cost ionic liquids: 1, 3-dialkylimidazolium salts containing methyl- and ethyl-sulfate anions. Green Chem. 2002;4:407–13.
30. Zhou YH, Roberston AJ, Hillhouse JH. Phosphonium and imidazolium slats and methods of their preparation. WO Patent 2004/016631; 2004.
31. Hirao M, Sugimoto H, Ohno H. Preparation of novel room-temperature molten salts by neutralization of amines. J Electro Chem Soc. 2000;147:4168–72.
32. Zhang QG, Yang JZ, Lu XM, et al. Studies on an ionic liquid based on $FeCl_3$ and its properties. Fluid Phase Equilibr. 2004;226:207–11.
33. Cull SG, Holbrey JD, Vargas-Mora V, et al. Room temperature ionic liquids as replacements for organic solvents in multiphase bioprocess operations. Biotechnol Bioeng. 2000;69:227–33.
34. Paul A, Mandal PK, Samanta A. How transparent are the imidazolium ionic liquids? A case study with 1-methyl-3-buthylimidazolium hexafluorophosphate, [Bmim][PF_6]. Chem Phys Lett. 2005;402:375–9.
35. Wasserscheid P, Sesing M, Korth W. Hydrogensulfate and tetrakis (hydrogensufato) borate ionic liquids: synthesis and catalytic application in highly Bronsted-acidic systems for Friedel-Crafts alkylation. Green Chem. 2002;4:134–8.
36. Roger M. Electrochemical 497 process for producing ionic liquids. US Patent 2003/0094380A1; 2003.
37. Namboodiri VV, Varma RS. An improved preparation of 1, 3-dialkylimidazolium tetrafluoroborate ionic liquids using microwaves. Tetrahedron Lett. 2002;43:5381–3.
38. Ding J, Armstrong DW. Chiral ionic liquids: synthesis and applications. Chirality. 2005;17:281–92.
39. Clare B, Sirwardana A, MacFalane DR. Synthesis, purification and characterization of ionic liquids. In: Kirchner B, editor. Ionic liquids. New York: Springer; 2009.
40. Baudequin C, Baudoux J, Levillain J, et al. Ionic liquids and chirality: opportunities and changelings. Tetrahedron Asymmetry. 2003;14:3081–3.
41. Kitazume T. Preparation of optically active ionic liquid of nicotiniumbis(trifluoromethylsulfonyl) amides for solvents. US Patent 2001/0031875; 2001.
42. Scammells PJ, Scott JL, Singer RD. Ionic liquids: the neglected issues. Aust J Chem. 2005;58:155–69.
43. Dong K, Zhang SJ. Hydrogen bonds: a structure insight into ionic liquids. Chem Eur J. 2012;18:2748–61.
44. Chiappe C, Pieraccini D. Ionic liquids: solvent properties and organic reactivity. J Phys Org Chem. 2005;18:275–97.
45. Farahani N, Gharagheizi F, Mirkhani SA, et al. Ionic liquids: prediction of melting point by molecular-based model. Thermochim Acta. 2012;549:17–34.
46. Aguirre CL, Cisternas LA, Valderrama JO. Melting-point estimation of ionic liquids by a group contribution method. Int J Thermophys. 2012;33:34–46.
47. Holbrey JD, Rogers RD. Melting points and phase diagrams. In: Wasserscheid P, editor. Ionic liquids in synthesis. New York: Wiley; 2003.
48. Cao Y, Wu J, Zhang J, et al. Room temperature ionic liquids (RTionic liquids): a new and versatile platform for cellulose processing and derivatization. Chem Eng J. 2009;147:13–21.
49. Ramesh LG, Coutinho JAP. A group contribution method for viscosity estimation of ionic liquids. Fluid Phase Equilibr. 2008;266:195–201.
50. Dzyuba S, Bartsch RA. Influence of structural variations in 1-alkyl (aralkyl)-3-methylimidazolium hexafluorophosphates and bis(trifluoromethylsulfonyl)imides on physical properties of the ionic liquids. Chem Phys Chem. 2002;3:161–6.

51. Lovelock KRJ, Cowling FN, Taylor AW. Effect of viscosity on steady-state voltammetry and scanning electrochemical microscopy in room temperature ionic liquids. J Phys Chem B. 2010;114:4442–50.
52. Ge ML, Zhao RS, Yi YF, et al. Densities and viscosities of 1-butyl-3-methylimidazolium trifluoromethanesulfonate + H_2O binary mixtures at T = (303.15 to 343.15) K. J Chem Eng Data. 2008;53:2408–11.
53. Kulkarni PS, Branco LC, Crespo JG, et al. Comparison of physicochemical properties of new ionic liquids based on imidazolium, quaternary ammonium, and guanidinium cations. Chem Eur J. 2007;13:8478–88.
54. Seddon R, Stark A, Torres MJ. Influence of chloride, water, and organic solvents on the physical properties of ionic liquids. Pure Appl Chem. 2000;72:2275–87.
55. Arce A, Rodil E, Soto A. Volumetric and viscosity study for the mixtures of 2-ethoxy-2-methylpropane, ethanol, and 1-ethyl-3-methylimidazolium ethylsulfate ionic liquid. J Chem Eng Data. 2006;51:1453–7.
56. Abbott AP. Application of hole theory to the viscosity of ionic and molecular liquids. Chem Phys Chem. 2004;5:1242–6.
57. Han C, Yu G, Wen L, et al. Data and QSPR study for viscosity of imidazolium-based ionic liquids. Fluid Phase Equilibr. 2011;300:95–104.
58. de Guzman J. Relation between fluidity and heat of fusion. Anales Soc Espan Fis Quim. 1913;11:353–62.
59. Reid RC, Prausnitz JM, Sherwood TK. The properties of gases and liquids. New York: McGraw-Hill; 1987.
60. Wang H, Gurau G, Rogers RD. Ionic liquid processing of cellulose. Chem Soc Rev. 2012;41:1519–37.
61. Fukayaa Y, Hayashia K, Wadab M, et al. Cellulose dissolution with polar ionic liquids under mild conditions: required factors for anions. Green Chem. 2008;10:44–6.
62. Kilpeläinen I, Xie H, King A, et al. Dissolution of wood in ionic liquids. J Agric Food Chem. 2007;55:9142–8.
63. Zhao H, Baker GA, Song Z, et al. Designing enzyme-compatible ionic liquids that can dissolve carbohydrates. Green Chem. 2008;10:696–705.
64. Ma M, Johnson KE. Proceedings of the ninth international symposium on Molten salts. Hussey CL, Newman DS, Mamantov G, Ito Y, et al. editors. Pennington: The Electrochemical Society; 1994. p. 94–13: 179–186.
65. Sun J, Forsyth M, Mac-Farlane DR. Room temperature molten salts based on the quaternary ammonium ion. J Phys Chem B. 1998;102:8858–64.
66. Gardas RL, Costa HF, Freire MG, et al. Densities and derived thermodynamic properties of imidazolium-, pyridinium-, pyrrolidinium-, and piperidinium-based ionic liquids. J Chem Eng Data. 2008;53:805–11.
67. Gardas RL, Freire MG, Carvalho PJ, et al. P F T measurements of imidazolium based ionic liquids. J Chem Eng Data. 2007;52:1881–8.
68. Ye CF, Shreeve JM. Rapid and accurate estimation of densities of room-temperature ionic liquids and salts. J Phys Chem A. 2007;111:1456–61.
69. Del Sesto RE, Corley C, Robertson A, et al. Tetraalkylphosphonium-based ionic liquids. J Organomet Chem. 2005;690:2536–42.
70. Paduszyński K, Domanska UA. New group contribution method for prediction of density of pure ionic liquids over a wide range of temperature and pressure. Ind Eng Chem Res. 2012; 51:591–604.
71. Lazzús J. A group contribution method to predict ρ-T-P of ionic liquids. Chem Eng Commun. 2010;197:974–1015.
72. Xu WG, Li L, Ma XX, et al. Density, surface tension, and refractive index of ionic liquids homologue of 1-alkyl-3-methylimidazolium tetrafluoroborate $[C_nmim][BF_4]$ (n = 2,3,4,5,6). J Chem Eng Data. 2012;57:2177–84.

73. Ngo HL, LeCompte K, Hargens L, et al. Thermal properties of imidazolium ionic liquids. Thermochim Acta. 2000;357:97–102.
74. Fox DM, Awad WH, Gilman JW, et al. Flammability, thermal stability, and phase change characteristics of several trialkylimidazolium salts. Green Chem. 2003;5:724–7.
75. Fox DM, Gilman W, De Long HC, et al. TGA decomposition kinetics of 1-butyl-2, 3-dimethylimidazolium tetrafluoroborate and the thermal effects of contaminants. J Chem Thermodyn. 2005;37:900–5.
76. Fox DM, Gilman JW, Morgan AB, et al. Flammability and thermal analysis characterization of imidazolium-based ionic liquids. Ind Eng Chem Res. 2008;47:6327–32.
77. Singh MP, Singh RK, Chandra S. Thermal stability of ionic liquid in confined geometry. J Phys D Appl Phys. 2010;43:1–4.
78. Meine N, Benedito F, Rinaldi R. Thermal stability of ionic liquids assessed by potentiometric titration. Green Chem. 2010;12:1711–14.
79. Domańska U, Pobudkowska A, Eckert F. Liquid–liquid equilibria in the binary systems (1, 3-dimethylimidazolium, or 1-butyl-3-methylimidazolium methylsulfate + hydrocarbons). Green Chem. 2006;8:268–76.
80. Grishina EP, Ramenskaya LM, Gruzdev MS. Water effect on physicochemical properties of 1-butyl-3-methylimidazolium based ionic liquids with inorganic anions. J Mol Liq. 2013;177:267–72.
81. Li WZ, Ju MT, Wang YN. Separation and recovery of cellulose from Zoysia japonica by 1-allyl-3-methylimidazolium chloride. Carbohydr Polym. 2013;92:228–35.
82. Wendler F, Todi LN, Meister F. Thermostability of imidazolium ionic liquids as direct solvents for cellulose. Thermochim Acta. 2012;528:76–84.
83. Liang R, Yang MR, Xuan XP. Thermal stability and thermal decomposition kinetics of 1-butyl-3-methylimidazolium dicyanamide. Chin J Chem Eng. 2010;18:736–41.
84. Hu YF, Liu ZC, Xu CM, et al. The molecular characteristics dominating the solubility of gases in ionic liquids. Chem Soc Rev. 2011;40:3802–23.
85. Lucinda JAC, Ewa BŁ, Rafał BŁ. A new outlook on solubility of carbohydrates and sugar alcohols in ionic liquids. RSC Adv. 2012;2:1846–55.
86. King AWT, Asikkala J, Mutikainen I, et al. Distillable acid–base conjugate ionic liquids for cellulose dissolution and processing. Angew Chem Int Ed. 2011;50:6301–5.
87. Abrani MA, Brandt A, Crowhurst L, et al. Understanding the polarity of ionic liquids. Phys Chem Chem Phys. 2011;13:16831–40.
88. Zhao H, Jones CIL, Baker GA, et al. Regenerating cellulose from ionic liquids for an accelerated enzymatic hydrolysis. J Biotechnol. 2009;139:47–54.
89. Edgar KJ, Buchanan CM, Debenham JS, et al. Advances in cellulose ester performance and application. Prog Polym Sci. 2001;26:1605–88.
90. Cai T, Zhang HH, Guo QH, et al. Structure and properties of cellulose fibers from ionic liquids. J Appl Polym Sci. 2010;115:1047–53.
91. Rahatekar SS, Rasheed A, Jain R, et al. Solution spinning of cellulose carbon nanotube composites using room temperature ionic liquids. Polymer. 2009;50:4577–83.
92. Fort DA, Remsing RC, Swatloski RP, et al. Can ionic liquids dissolve wood? Processing and analysis of lingocellulosic materials with 1-n-butyl-3-methylimidazolium chloride. Green Chem. 2007;9:63–9.
93. Brandt A, Hallett JP, Leak DJ, et al. The effect of the ionic liquid anion in the pretreatment of pine wood chips. Green Chem. 2010;12:672–9.
94. Brandt A, Ray MJ, To TQ, et al. Ionic liquid pretreatment of lingocellulosic biomass with ionic liquid-water mixtures. Green Chem. 2011;13:2489–99.
95. Padmanabhan S, Kim M, Blanch HW, et al. Solubility and rate of dissolution for miscanthus in hydrophilic ionic liquids. Fluid Phase Equilibr. 2011;309:89–96.
96. Doherty TV, Mora-Pale M, Foley SE, et al. Ionic liquid solvent properties as predictors of lignocellulose pretreatment efficacy. Green Chem. 2010;12:1967–75.
97. Pereiro AB, Verdía P, Tojo E, et al. Physical properties of 1-butyl-3-methylimidazolium methyl sulfate as a function of temperature. J Chem Eng Data. 2007;52:377–80.

98. Fukaya Y, Sugimoto A, Ohno H. Superior solubility of polysaccharides in low viscosity, polar, and halogen-free, 1,3-dialkylimidazlium formats. Biomacromolecules. 2006;7:3295–7.
99. Wu D, Wu B, Zhang YM, et al. Density, viscosity, refractive index and conductivity of 1-allyl-3-methylimidazolium chloride + water mixture. J Chem Eng Data. 2010;55:621–4.
100. Huddleston JG, Visser AE, Reichert WM, et al. Characterization and comparison of hydrophilic and hydrophobic room temperature ionic liquids incorporating the imidazolium cation. Green Chem. 2001;3:156–64.
101. Lee SH, Lee SB. The hilde brand solubility parameters, cohesive energy densities and internal energies of 1-alkyl-3-methylimidazolium-based room temperature ionic liquids. Chem Commun. 2005;2005:3469–71.
102. Olivier-Bourbigou H, Magna L. Ionic liquids: perspectives for organic and catalytic reactions. J Mol Catal A Chem. 2002;182–183:419–37.
103. Ohlin CA, Dyson PJ, Laurenczy G. Carbon monoxide solubility in ionic liquids: determination, prediction and relevance to hydroformylation. Chem Commun. 2004;2004:1070–1.
104. Berthod A, Ruiz-Angel M, Carda-Broch S. Ionic liquids in separation techniques. J Chromatogr A. 2008;1184:6–18.
105. Chen Y, Zhang YM, Ke FY, et al. Solubility of neutral and charged polymers in ionic liquids studied by laser light scattering. Polymer. 2011;52:481–8.
106. Domańska U, Laskowska M. Temperature and composition dependence of the density and viscosity of binary mixtures of 1-butyl-3-methylimidazolium thiocyanate + 1-alcohols. J Chem Eng Data. 2009;54:2113–19.
107. Wu B, Reddy RG, Rogers RD. Novel ionic liquid thermal storage for solar thermal electric power system. Proceedings of solar forum, solar energy: the power to choose. Washington, DC; 2001.
108. Wang JY, Zhao FY, Liu RJ, et al. Thermophysical properties of 1-methyl-3-methylimidazolium-dimethylphosphate and 1-ethyl-3-methylimidazolium diethylphosphate. J Chem Thermodyn. 2011;43:47–50.
109. Ann E, Visser W, Matthew R, et al. Characterization of hydrophilic and hydrophobic ionic liquids: alternatives to volatile organic compounds for liquid-liquid separations. Ionic liquids industrial applications to green chemistry. ACS Symp Ser. 2002;818:289–308.
110. Yokozeki A, Shiflett MB. Vapor–liquid equilibria of ammonia + ionic liquid mixtures. Appl Energy. 2007;84:1258–73.
111. Diogo JCF, Caetano FJP, Fareleira JMNA. Viscosity measurements of the ionic liquid trihexyl(tetradecyl)phosphoniumdicyanamide $[P_{6,6,6,14}]$[dca] using the vibrating wire technique. J Chem Eng Data. 2012;57:1015–25.
112. Aantharam PD, Constance AS, Sasidhar V. Mitigation of cellulose recalcitrance to enzymatic hydrolysis by ionic liquid pretreatment. Appl Biochem Biotech. 2007;136–140:407–21.
113. Zavrel M, Bross D, Funke M, et al. High-throughput screening for ionic liquids dissolving (ligno-) cellulose. Bioresour Technol. 2009;100:2580–7.
114. Lee SH, Doherty TV, Linhardt Robert J, et al. Ionic liquid-mediated selective extraction of lignin from wood leading to enhanced enzymatic cellulose hydrolysis. Biotechnol Bioeng. 2009;102:1368–76.
115. Pu YP, Jiang N, Ragauskas AJ. Ionic liquid as a green solvent for lignin. J Wood Chem Technol. 2007;27:23–33.
116. Anantharam PD, Sasidhar V, Constance AS. Enhancement of cellulose saccharification kinetics using an ionic liquid pretreatment step. Biotechnol Bioeng. 2006;95:904–10.
117. D'Ippolito G, Dipasquale L, Vella FM, et al. Hydrogen metabolism in the extreme thermophile Thermotoga neapolitana. Int J Hydrog Energy. 2010;35:2290–5.
118. Li Q, He YC, Xian M, et al. Improving enzymatic hydrolysis of wheat straw using ionic liquid 1-ethyl-3-methylimidazolium diethyl phosphate pretreatment. Bioresour Technol. 2009;100:3570–5.
119. Shill K, Padmanabhan S, Xin Q, et al. Ionic liquid pretreatment of cellulosic biomass: enzymatic hydrolysis and ionic liquid recycle. Biotechnol Bioeng. 2011;108:510–20.

120. Tan SSY, Macfarlane DR, Upfal J, et al. Extraction of lignin from lignocellulose at atmospheric pressure using alkylbenzenesulfonate ionic liquid. Green Chem. 2009;11:339–45.
121. Tan SY. Studies in bagasse fractionation using ionic liquids. PhD thesis, Monash University, Australia; 2009.
122. Jurgen V, Erdmenger T, Claudia H, et al. Extended dissolution studies of cellulose in imidazolium based ionic liquids. Green Chem. 2009;11:417–24.
123. D'Andola G, Szarvas L, Massonne K et al. Ionic liquids for solubilizing polymers. WO 2008/043837; 2008.
124. Omar AES, Andreas K, Ludmila CF, et al. Applications of ionic liquids in carbohydrate chemistry: a window of opportunities. Biomacromolecules. 2007;8:2629–47.
125. Thomas H, Katrin S, Susann B. Ionic liquids as reaction medium in cellulose functionalization. Macromol Biosci. 2005;5:520–5.
126. Luo HM, Li YQ, Zhou CR. Study on the dissolubility of the cellulose in the functionalized ionic liquid. Polym Mater Sci Eng. 2005;21:233–5.
127. Barthel S, Heinze T. Acylation and carbanilation of cellulose in ionic liquids. Green Chem. 2006;8:301–6.
128. Sun N, Rahman M, Qin Y, et al. Complete dissolution and partial delignification of wood in the ionic liquid 1-ethyl-3-methylimidazolium acetate. Green Chem. 2009;11:646–55.
129. Phillips DM, Drummy LF, Conrady DG, et al. Dissolution and regeneration of Bombyxmori silk fibroin using ionic liquids. J Am Chem Soc. 2004;126:14350–1.
130. Sun P, Liu ZT, Liu ZW. Particles from bird feather: a novel application of an ionic liquid and waste resource. J Hazard Mater. 2009;170:786–90.
131. Xie HB, Zhang SB, Li SH. Chitin and chitosan dissolved in ionic liquids as reversible sorbents of CO_2. Green Chem. 2006;8:630–3.
132. Zhang H, Wu J, Zhang J, et al. 1-allyl-3-methylimidazolium chloride room temperature ionic liquid: a new and powerful nonderivatizing solvent for cellulose. Macromolecules. 2005;38:8272–7.
133. Wang XJ, Li HQ, Cao Y, et al. Cellulose extraction from wood chip in an ionic liquid 1-allyl-3-methylimidazolium chloride (AmimCl). Bioresour Technol. 2011;102:7959–65.
134. Hou XD, Li N, Zong MH. Facile and simple pretreatment of sugar cane bagasse without size reduction using renewable ionic liquids – water mixtures. ACS Sustain Chem Eng. 2013;1:519–26.
135. Goshadrou A, Karimi K, Lefsrud M. Characterization of ionic liquid pretreated as pen wood using semi-quantitative methods for ethanol production. Carbohydr Polym. 2013;96:440–9.
136. Qiu ZH, Aita GM. Pretreatment of energy cane bagasse with recycled ionic liquid for enzymatic hydrolysis. Bioresour Technol. 2013;129:532–7.
137. FitzPatrick M, Champagne P, Cunningham MF, et al. Application of optical microscopy as a screening technique for cellulose and lignin solvent systems. Can J Chem Eng. 2012;90:1142–52.

Chapter 2
Solubilization of Biomass Components with Ionic Liquids Toward Biomass Energy Conversions

Mitsuru Abe and Hiroyuki Ohno

Abstract Ionic liquids (ILs) are collecting keen interest as novel solvents for plant biomass, especially for cellulose. ILs have several unique properties and they dissolve cellulose under milder condition than existing procedures. Here, we give an outline of the development of biomass dissolving ILs together with their physico-chemical properties. Dissolution and/or extraction of not only cellulose but also lignin with ILs are overviewed. The extracted biomass is expected to be converted into other energies. For this purpose, energy-saving biomass treatment is inevitable, and ILs are one of the most potential media for this. This chapter will deliver further ideas on the design of ILs for cellulose dissolution or plant biomass treatment in the near future.

Keywords Ionic liquid • Polarity • Hydrogen bond • Cellulose • Lignin • Dissolution • Design of ions • Save energy

2.1 Introduction

Ionic liquids (ILs) are organic salts with melting point below 100 °C, and especially those with the melting point at and below room temperature are called "room temperature Ionic liquids" [1]. There are a few important properties required for solvents such as non-volatility, non-flammability, and thermal stability in a wide temperature range. Although there are many solvents that have some of these properties, there are few solvents that have all of the above-mentioned properties. ILs have unique properties different from molecular solvents. Many ILs have non-volatility, non-flammability, and stability in a wide temperature range. Furthermore, there is a potential chance to design new ILs through unlimited

M. Abe • H. Ohno (✉)
Department of Biotechnology, Tokyo University of Agriculture and Technology, 2-24-16 Naka-cho, Koganei, Tokyo 184-8588, Japan
e-mail: ohnoh@cc.tuat.ac.jp

Z. Fang et al. (eds.), *Production of Biofuels and Chemicals with Ionic Liquids*, Biofuels and Biorefineries 1, DOI 10.1007/978-94-007-7711-8_2,

possibility of combination of ion pairs. ILs are accordingly known as "designer solvents". It is easy to change their physico-chemical properties by the selection of suitable ions.

One of successful examples on the design of ILs is cellulose dissolution. Plant biomass is one example of a renewable and abundant natural material. These materials can be considered to be the embodied energy of sunlight and so is one possible method to produce energy on earth. Considering the limit of fossil fuels, there are increasing trials to convert plant biomass into user-friendly energy. There are many industrial plants for bioethanol production from corn starch or sugar cane in US and other countries. There are established methods to convert starch into sugar in our human life. However, since these processes compete with food industry, there are ethical concerns about the use of edible plant biomass as raw materials for fuel production [2]. Cellulosic biomass therefore is attracting attention as energy sources because they are inedible materials for human beings.

Cellulosic biomass essentially consists of cellulose, hemicellulose, and lignin. To obtain energy from cellulosic biomass with minimum given energy, following three steps are required, namely (1) extraction of cellulose from biomass, (2) hydrolysis of the cellulose into glucose or other oligosaccharides, and (3) oxidation or fermentation. However, cellulosic biomass is scarcely used for bioenergy production because of its very poor solubility in common molecular solvents. The chemical and physical stability of cellulose are known to be derived from many intra- and inter-molecular hydrogen bonds [3, 4]. Since ordinary molecular solvents have not enough power to dissolve cellulose, it is required to heat the mixture or stir it for a long time which is inefficient for energy conversion. The energy cost for dissolution and extraction processes for cellulose should be very low.

Many scientists recognize that ILs have great potential as solvents for cellulose and are paying particular attention to ILs as novel solvents for cellulose under mild conditions. Design of ILs to dissolve cellulose with low energy cost is therefore indispensable for energy conversion. Without this step, it is difficult to use cellulosic biomass as valuable materials as well as fossil fuel substitutes. The discussion in this chapter concentrates on the dissolution of biomass in ILs.

2.2 Chloride Type Salts for Cellulose Dissolution

Concerted attempts to dissolve cellulose do not have such a long history. The first reported study on cellulose dissolution using an ionic material was reported in 1934 by Graenacher and co-workers [5]. They used a mixture of amine and a pyridinium salt to dissolve cellulose. At this stage, pyridinium salt was not used as an IL but as an added salt. Thus, the "first" study of cellulose dissolution with ILs that was reported was by Swatloski et al. in 2002 [6]. They reported that 1-butyl-3-methylimidazolium chloride ([C4mim]Cl) dissolved pulp cellulose. This IL dissolved 3 % cellulose at 70 °C, and 10 % cellulose at 100 °C. They also clarified that the cellulose dissolving degree was improved with a combination of IL soaking and

Table 2.1 Solubility of dissolving pulp cellulose in ionic liquids [6]

Ionic liquid	Method	Solubility (wt%)
[C4mim]Cl	Heat (100 °C)	10
	Heat (70 °C)	3
	Heat (80 °C) + sonication	5
	Microwave irradiation	25
[C4mim]Br	Microwave irradiation	5–7
[C4mim]SCN	Microwave irradiation	5–7
[C4mim][BF_4]	Microwave irradiation	Insoluble
[C4mim][PF_6]	Microwave irradiation	Insoluble
[C6mim]Cl	Heat (100 °C)	5
[C8mim]Cl	Heat (100 °C)	Slightly soluble

other physicochemical treatments such as sonication or microwave irradiation. On the other hand, tetrafluoroborate-type salts and hexafluorophosphate-type salts did not dissolve cellulose unlike [C4mim]Cl. The data shown in Table 2.1 provides that the properties of ILs that deeply affect the solubility of cellulose in the corresponding ILs.

[C4mim]Cl has a high melting point (T_m) of 73 °C and high viscosity, thus it is hard to use as a solvent at ambient condition [7, 8]. Heinze and co-workers reported that some chloride salts which have pyridinium or ammonium cations also dissolve cellulose (Scheme 2.1, Table 2.2) [8]. 3-Methyl-*n*-butylpyridinium chloride ([C4mpy]Cl) dissolved cellulose much better than [C4mim]Cl, and benzyldimethyl(tetradecyl)ammonium chloride (BDTAC) has a lower T_m (52 °C). Sometimes, the degree of polymerization of cellulose (DP) decreases after dissolution in ILs (see Table 2.2). In the case of energy conversion, changes in the DP are a less important factor. Some studies also require dissolving cellulose without lowering DP from the viewpoint of cellulose application. In both cases, the dissolution of cellulose under mild conditions is suitable considering efficient processing. Then there are some studies on reducing the T_m of these chloride salts. Physical chemistry tells us that small anions such as chloride anion interact strongly with cations due to higher charge density resulting high T_m of the salts. It is therefore important to design cations to lower the T_m. Mizumo et al. developed liquid state chloride salts using imidazolium cations having allyl group(s) [9]. Allyl group is effective to show conformational change or rotation of the group, that induces to lower the T_m of the imidazolium salt. They clarified that the adopting allyl groups into imidazoliulm cation is a valid way to lower the T_m of the chloride salts. After this report, Zhang and co-workers reported that one of room temperature ILs, 1-allyl-3-methylimidazolium chloride ([Amim]Cl), has a good ability to dissolve cellulose [10]. [Amim]Cl dissolved no cellulose at room temperature, but it dissolved cellulose at 60 °C under stirring. With increasing temperature, cellulose could be dissolved easily in [Amim]Cl.

With an increase in the variety of chloride type cellulose-dissolving ILs (CDILs), cellulose dissolving mechanisms as well as dominant properties of the ILs for cellulose dissolution have been gradually clarified. Those studies suggest that the chloride anion works dominantly to dissolve cellulose by breaking the hydrogen bonding networks of cellulose fibrils. Remsing and co-workers have

3-Methyl-N-butylpyridinium chloride Benzyldimethyl(tetradecyl)ammonium chloride

Scheme 2.1 Structure of chloride type ILs to dissolve cellulose under heating [8]

Table 2.2 Solubility of cellulose samples in ILs

Cellulose		Solubility of cellulose					
		[C4mim]Cl		[C4mpy]Cl		BDTAC	
Type	DP	%	DP[a]	%	DP[a]	%	DP[a]
Avicel	286	18	307	39	172	5	327
Spruce sulfite pulp	593	13	544	37	412	2	527
Cotton linters	1,198	10	812	12	368	1	966

Reprinted with permission from Heinze et al. [8], Copyright (2005) John Wiley and Sons
[a]After regeneration, DP: Degree of polymerization of cellulose, T_m of [C4mim]Cl, [C4mpy]Cl, and BDTAC was 73, 95, and 52 °C, respectively

clarified that [C4mim]Cl makes hydrogen bonding between the carbohydrate hydroxyl protons and the chloride ions in a 1:1 stoichiometric ratio using ^{13}C and $^{35/37}$Cl NMR relaxation measurements [11]. The relaxation time of the imidazolium cation and chloride anion in [C4mim]Cl was analyzed as a function of concentration (wt%) of cellobiose as a model compound of cellulose. The relaxation time of the cation was almost constant regardless of cellobiose concentration. This means that there are no specific interaction between cations and cellobiose. On the other hand, there was a clear relationship between the $^{35/37}$Cl relaxation time and cellobiose concentration. This suggests that the chloride anion interacts strongly with the dissolved carbohydrate. They analyzed the interaction between the chloride anions and non-derivatized carbohydrates. This study clarified that the chloride ions interact in a 1:1 ratio with the carbohydrate hydroxyl protons.

Some simulation studies have also been reported on carbohydrate dissolution in dialkylimidazolium chloride-type ILs. Youngs and co-workers reported about the molecular dynamics simulations of glucose solvation by 1,3-dimethylimidazolium chloride [C1mim]Cl [12]. They found that the primary solvation shell around the glucose consists predominantly of chloride anions hydrogen bonding with the hydroxyl groups of glucose ring. This is the predominant interaction between glucose rings and chloride-type ILs. There is a small contribution of cations on the carbohydrate-IL interaction. Cations were however also found near the glucose, and a hydrogen at the 2-position of the imidazolium ring interacted with an oxygen atom of the secondary hydroxyl group of the glucose. A weak contribution of van der Waals force was also seen between the glucose and the cations. Even at high glucose concentrations (16.7 wt%), the anion-cation interactions and overall liquid structure of [C1mim]Cl were found not to be significantly changed. This means that the glucose is readily solubilized by the IL even under high concentration. Gross and co-workers reported on the thermodynamics of cellulose solvation in [C4mim]Cl [13]. All-atom molecular dynamics (MD) simulations were conducted to analyze the thermodynamic driving force of the cellulose dissolving process and to clarify the role of both anions and cations in the process. They suggested that the dissociated cellulose has higher potential energy in water than that in [C4mim]Cl. They suggested that the cellulose insolubility in water is mostly derived from the entropy reduction of the solvent. In addition, they also suggested that both the anion and cation of the IL interact with the glucan residues. In the case of Cl^- anions, they form hydrogen bonds with the hydroxyl groups of cellulose from either equatorial or axial directions. On the other hand, for the cations, the contact with cellulose along the axial directions was closer than that along the equatorial directions. They concluded that interacting with cellulose along axial directions and disrupting the cellulose fibrils is an important step of cellulose dissolution.

2.3 Carboxylate Type Salts with Low Melting Points and Low Viscosity

As described in the previous section, chloride type ILs have a strong ability to dissolve cellulose, and it is predominantly attributed to the anion to form hydrogen bonding with the hydroxyl groups of cellulose. However, most chloride salts have both a high melting point and high viscosity. These properties are not suitable for the improvement of cellulose solubilization. Various attempts have been made to reduce the melting point of chloride-based ILs, as discussed above. Despite the prepared chloride salts being in their liquid state at room temperature, heating is necessary to dissolve cellulose. Since the necessity of the continuous heating requires an excessive amount of energy consumption, this leads the increase in the total cost of the cellulose treatment process. It is therefore strongly desired to develop novel ILs to dissolve cellulose with low energy cost. Design of anion structure is required because there is a limitation to overcome the problem by only optimization of cations of chloride-based salts.

To design novel CDILs, we should have an analytical method to evaluate the hydrogen bonding basicity of ILs, because chloride type ILs dissolve cellulose through making favorable hydrogen bonds with hydroxyl groups of cellulose. The analysis of physicochemical properties of ILs is essential for the design of CDILs. There are many ways to investigate or predict the proton accepting ability, in other words, hydrogen bond basicity. For example, Hansen solubility parameters [14], COSMO-RS [15], and the Kamlet-Taft parameters [16, 17] are known as useful empirical or semi-empirical polarity scales. Especially, it is known that Kamlet-Taft parameters are very useful, which requires three solvatochromic dyes

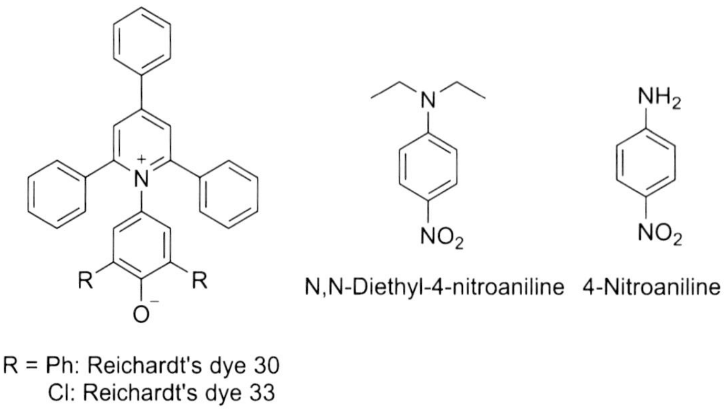

N,N-Diethyl-4-nitroaniline 4-Nitroaniline

R = Ph: Reichardt's dye 30
Cl: Reichardt's dye 33

Scheme 2.2 Structure of prove dyes for Kamlet-Taft parameter measurements [16]

(Scheme 2.2). From the shift of the absorption maximum wavelength of the individual dye molecules shown in Scheme 2.2, three Kamlet-Taft parameters such as α, β, and π values are calculated. These three parameters, α, β, and π values represent hydrogen bond acidity, hydrogen bond basicity, and polarizability, respectively [16, 17]. Since ILs are conductive materials, it is not easy to determine the polarity with conventional electrochemical methods. Considering this, Kamlet-Taft parameters are quite useful to evaluate the polarity of ILs.

Brandt and co-workers compiled the correlation between cellulose dissolving ability and the Kamlet-Taft β value of several ILs (Fig. 2.1) [18]. Although the plotted data were measured at different conditions (*e.g.* different temperature, dissolution time, degree of polymerization (DP) of cellulose, moisture content, purity of ILs, etc.), there is a certain correlation between cellulose solubility and the β value of the ILs. ILs with β value of less than 0.6 have no power to dissolve cellulose under any condition. The ILs having a β value of more than 0.6 start to dissolve cellulose and solubility increases with an increase of their β value. Here, it should be noted that the β value is not only the factor to govern the cellulose solubility. There are still many ILs that cannot solubilize cellulose in spite of their larger β value [19]. Other factors such as α value and ion structure should also be considered for the design of cellulose solvents. Although the β value does not entirely determine the cellulose dissolving ability, it is a useful design parameter for CDILs.

According to the data compiled by Ohno and co-workers, a series of carboxylate salts (Scheme 2.3) were confirmed to have strong hydrogen bond basicity (Table 2.3) [20]. Since there are a wide variety of carboxylic acid derivatives, carboxylate anions have been selected as good anions to construct CDILs [21].

From the structures listed in a patent by Swatloski and co-workers, BASF reported that imidazolium ILs bearing acetate anions are effective for the dissolution of cellulose [22]. Since 1-ethyl-3-methyl-imidazolium acetate ([C2mim]OAc) is less toxic, and less viscous, this IL is a favorable solvent for cellulose. Fukaya and

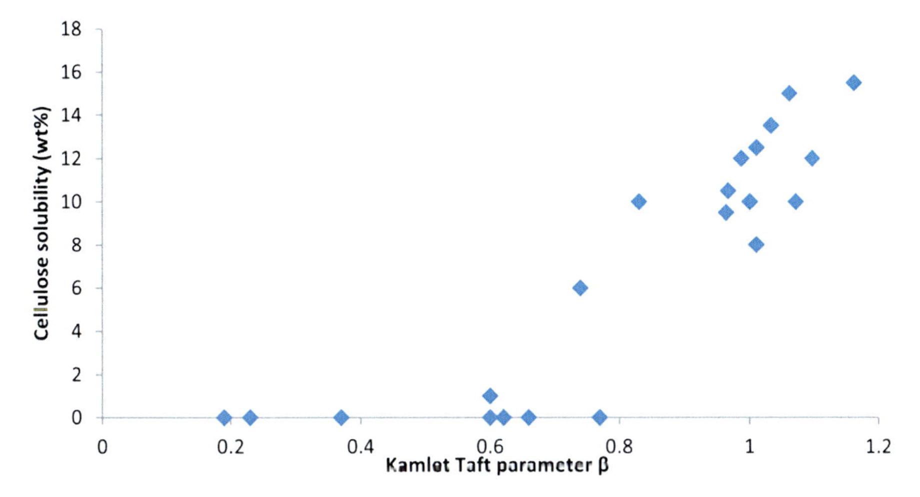

Fig. 2.1 Cellulose solubility in [C2mim] and [C4mim] type ILs as a function of the Kamlet-Taft β value of the ILs. The *plotted data* were measured under different conditions (Reproduced from Brandt et al. [18] with permission from The Royal Society of Chemistry)

$[C_4mim][RCOO]$ $[Amim][HCOO]$

R = H : [HCOO], C_nH_{2n+1} (n = 1~3) : $[C_nCOO]$, $C(CH_3)_3$: $[t\text{-}C_4COO]$

Scheme 2.3 Structure of carboxylate type salts (Reprinted with permission from Ohno and Fukaya [20], Copyright (2009) The Chemical Society of Japan)

Table 2.3 Kamlet-Taft parameters of a series of [C4mim] carboxylate-type salts

	Kamlet-Taft parameters		
Anion	α	β	π^*
[HCOO]	0.56	1.01	1.03
$[C_1COO]$	0.55	1.09	0.99
$[C_2COO]$	0.57	1.10	0.96
$[C_3COO]$	0.56	1.10	0.94
$[t\text{-}C_4COO]$	0.54	1.19	0.91
Cl	0.47	0.87	1.10

Reprinted with permission from Ohno and Fukaya [20], Copyright (2009) The Chemical Society of Japan

co-workers also reported that a series of carboxylate-type ILs for cellulose dissolution [23]. They suggested that 1-allyl-3-methylimidazolium formate ([Amim] formate, IL<u>3</u> in Scheme 2.4) is a good solvent to dissolve cellulose. This ionic liquid shows no melting temperature but low glass transition temperature (−76 °C) and low viscosity (66 cP at 25 °C) (Table 2.4). The hydrogen bond basicity of IL<u>3</u>

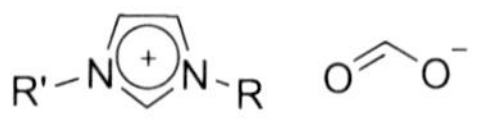

1: R = C_2H_5 R' = CH_3 3: R = CH_2CHCH_2 R' = CH_3
2: R = C_3H_7 R' = CH_3 4: R = CH_2CHCH_2 R' = C_2H_5

Scheme 2.4 Structure of formate salts with imidazolium cations which have different length of alkyl chains (Reprinted with permission from Fukaya et al. [23], Copyright (2006) American Chemical Society)

Table 2.4 Physicochemical properties and Kamlet-Taft parameters of the ILs

					Kamlet-Taft parameters		
IL	T_g (°C)[a]	T_m (°C)[a]	T_d (°C)[b]	η (cP) (at 25 °C)	α	β	π^*
1	[c]	52	212	[d]	[d]	[d]	[d]
2	−73	c	213	117	0.46	0.99	1.06
3	−76	c	205	66	0.48	0.99	1.08
4	−76	c	205	67	0.47	0.99	1.06
[Amim]Cl	−51	c	256	2,090	0.46	0.83	1.17
[C4mim]Cl	[c]	66	262	[d]	0.44[e]	0.84[e]	1.14[e]

Reprinted with permission from Fukaya et al. [23], Copyright (2006) American Chemical Society
[a]Temperature at signal peak
[b]Temperature for 10 wt% loss under N_2 gas
[c]Not observed
[d]Not measured
[e]Measured under a supercooled state

was higher than that of chloride salts. The IL3 was confirmed to have a good ability to dissolve cellulose under mild condition. It solubilized 10 wt% cellulose at 60 °C though [Amim]Cl required 100 °C to dissolve the same concentration of cellulose (Fig. 2.2).

After the appearance of these carboxylate type CDILs, many studies were reported about the cellulose dissolving mechanism by carboxylate salts. Remsing and co-workers analyzed the solvation mechanisms of acetate and chloride type salts, such as [C4mim]Cl, [Amim]Cl, and 1-ethyl-3-methylimidazolium acetate ([C2mim][OAc]) using $^{35/37}$Cl and ^{13}C NMR relaxation [24]. The $^{35/37}$Cl and ^{13}C relaxation rates of anions showed a strong dependency on the carbohydrate concentration in the ILs having acetate or chloride anions. Especially, in the case of [C2mim][OAc], with the increase of carbohydrate concentration, the reorientation rate of the anion decreased faster than that of cations. They suggested that the interactions between the cations and carbohydrates are nonspecific, and concluded that the solvation mechanism was almost the same regardless of the structure of the anions.

Zhang and co-workers also analyzed the interaction between [C2mim][OAc] and cellobiose, a repeating unit of cellulose (Scheme 2.5), using ^{1}H-NMR spectroscopy [25]. The acetate anion made hydrogen bonds with hydroxyl groups of cellobiose, and the imidazolium cation also interacted with the oxygen atom of

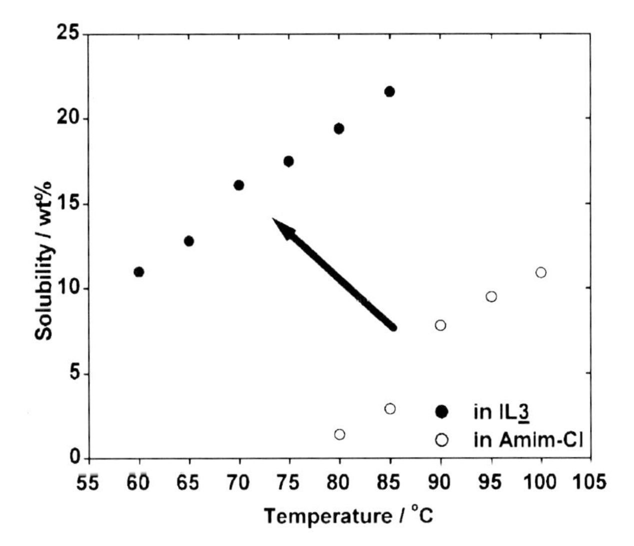

Fig. 2.2 Cellulose solubility as a function of temperature in IL3 and [Amim]Cl (Reprinted with permission from Fukaya et al. [23], Copyright (2006) American Chemical Society)

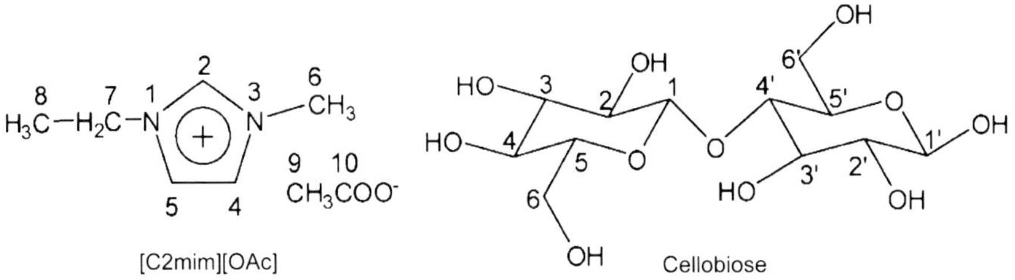

[C2mim][OAc] Cellobiose

Scheme 2.5 Structure and atom number of [C2mim][OAc] and cellobiose (Reproduced from Zhang et al. [25] with permission from the PCCP Owner Societies)

hydroxyl group of cellobiose, especially via the most acidic proton in the C-2 position (Fig. 2.3).

Liu and co-workers carried out molecular dynamics simulations to clarify the interaction of cellulose and ILs [26]. They suggested that the interaction energy between a series of (1–4) linked β-D-glucose oligomers and [C2mim][OAc] was stronger than that with water or methanol. The estimated energy for hydrogen bonding between the hydroxide group on glucose unit and water or ethanol was estimated to be around 5 kcal mol^{-1}, whereas that in [C2mim][OAc] was estimated to be 14 kcal mol^{-1}. Furthermore, some of these cations interacted with these polysaccharides through hydrophobic interactions. Xu and co-workers reported that the cellulose solubility of [C4mim][OAc] was certainly improved by addition of lithium salts [27]. They have suggested that lithium cation interacts with an oxygen atom of C3-hydroxyl group of cellulose, and it causes cleavage of the O(6)H-O(3) inter-molecular hydrogen bonding. This result means that cations also make a certain contribution to dissolve cellulose depending on their structure.

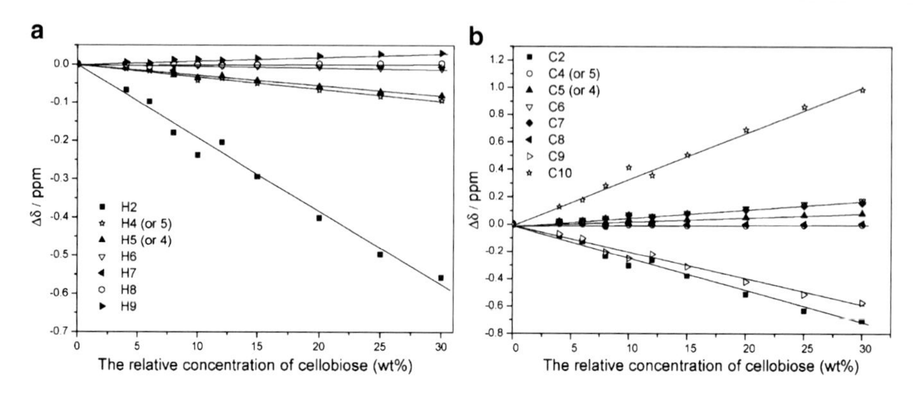

Fig. 2.3 Chemical shift of proton (**a**) and carbon (**b**) in [C2mim][OAc] as the function of concentration of cellobiose in DMSO-d_6 ($\Delta\delta = \delta - \delta_{neat}$) (Reproduced from Zhang et al. [25] with permission from the PCCP Owner Societies)

2.4 Phosphonate Type Salts for Cellulose Dissolution Without Heating

As described above, carboxylate salts have a good ability to dissolve cellulose and the dissolution mechanism has been analyzed. However these carboxylate type ILs also have a drawback in terms of thermal stability. These ILs still require heat to dissolve certain amounts of cellulose. To overcome these problems, methylphosphonate salts were proposed as stable CDILs. Fukaya and co-workers have synthesized 1-ethyl-3-methylimidazolium methylphosphonate ([C2mim][$(CH_3)(H)PO_2$]; IL5 in Scheme 2.6), and found that this IL5 had a good stability and dissolved cellulose without heating [28].

This IL5 has very low glass transition temperature (−86 °C), low viscosity (107 cP at 25 °C), and high Kamlet-Taft β value (1.00). The physicochemical properties of IL5 allow dissolution of 6 wt% cellulose within 1 h at 30 °C (Fig. 2.4), and allow it to dissolve 4 wt% cellulose without heating (at 25 °C) within 5 h.

Considering the above mentioned results, the structure of CDILs and cellulose solubility data are summarized in Table 2.5.

2.5 Functional Ionic Liquids for Cellulose Dissolution

Recently, a novel type of CDIL has been developed that has high added-value as well as cellulose dissolving ability. Ito and co-workers reported that some amino acid type ILs dissolved cellulose [19]. Especially, *N*,*N*-diethyl-*N*-(2-methoxyethyl)-*N*-methylammonium alanine dissolved cellulose well at 100 °C. These amino acid-based ILs are halogen-free and polar ILs [34, 35]. Since amino acids are biomolecules, cheap products, and environmentally-friendly materials.

Scheme 2.6 Structure of phosphonate type salts [28]

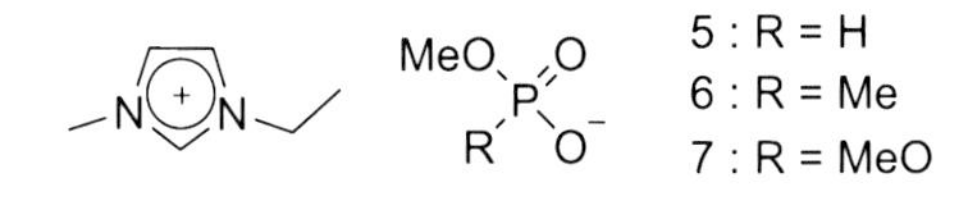

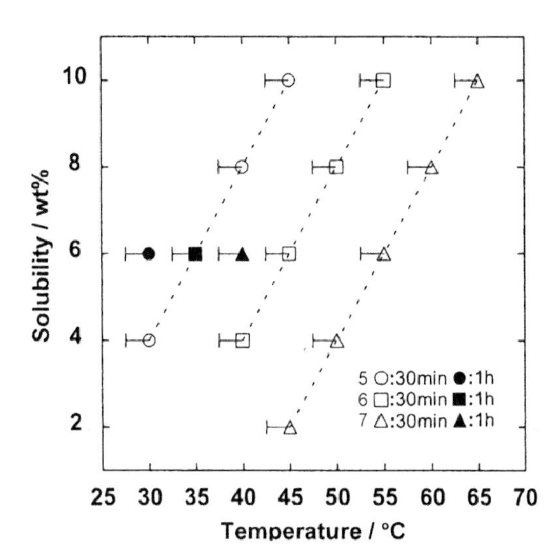

Fig. 2.4 Cellulose solubility (wt%) in phosphonate type salts (ILs**5**, **6**, and **7**) as a function of temperature [28]

ILs composed of amino acids are expected to generate more interest as potential solvents in the near future [36–38]. More recently, they reported that the alanine containing salt dissolved 23 wt% cellulose at room temperature with the aid of DMSO [39]. Mixtures of ILs and molecular liquids are gaining attention of IL researchers as new liquids containing the advantages of both IL and molecular liquids. The properties of these mixtures are of course a function of the mixing ratio. In other words, it is not difficult to control the fluid properties by adjusting the mixing ratio when adequate IL and molecular liquids are chosen.

For most processes, highly pure ILs are needed to maintain the efficiency of the cellulose dissolution due to keep their unique properties. ILs have a characteristic properties about vapor pressure, namely ILs are non- or very low-volatile liquids. In other words, it is quite difficult to purify ILs by distillation. Polar and distillable ILs are expected to improve some processes. One of solutions is the use of ILs prepared by the neutralization [40]. King and co-workers reported the distillable acid–base conjugate ILs which has cellulose dissolving ability [41]. They found that the neutralized salt of 1,1,3,3-tetramethylguanidine (TMG) with propionic acid ([TMGH][CO_2Et]) has been shown to be technically distillable, and it dissolved 5 wt% cellulose within 10 min at 100 °C. This dissolution capability and distillable property are dependent upon the relative basicity of the competing base, and the equilibrium is temperature dependent.

Phosphonate type salts require low energy cost to dissolve cellulose, and they would be potential solvents for cellulose technology. However, most CDILs have a

Table 2.5 IL structure and cellulose solubility

Ionic liquids	Raw material	Solubility (wt%)	Condition	Refs.
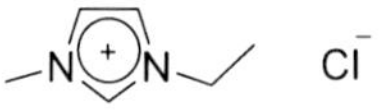	Cellulose	15.8	85 °C	[29]
	Avicell	10–14	100 °C, 1 h	[7]
[C2mim]Cl	Avicell	5	90 °C	[30]
	Cellulose, DP = 268	12	90 °C, 12 h	[31]
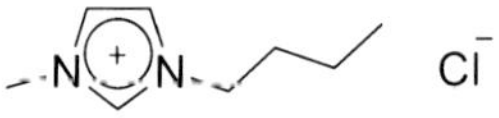	Pulp cellulose, DP = 1,000	10	100 °C	[6]
[C4mim]Cl		3	70 °C	[6]
	Avicel, DP = 286	18	83 °C, 12 h	[8]
	Avicel	20	100 °C, 1 h	[7]
	Cellulose	1.8	80–90 °C, 20 min	[32]
	Cellulose	13.6	85 °C	[29]
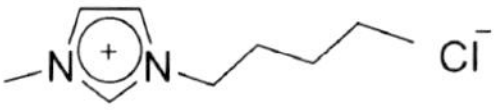	Avicel	1.5	100 °C	[33]
	Avicel	1	100 °C, 1 h	[7]
[C5mim]Cl				
	Pulp cellulose	5	100 °C	[6]
	Avicel	7	100 °C	[33]
[C6mim]Cl	Avicel	5	50 °C	[30]
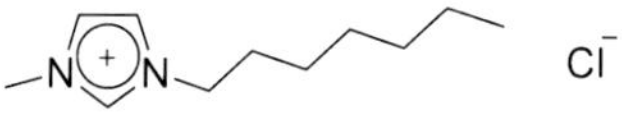	Pulp cellulose	Slightly soluble	100 °C	[6]
[C7mim]Cl	Avicel	5	100 °C	[33]
		5	100 °C, 1 h	[7]
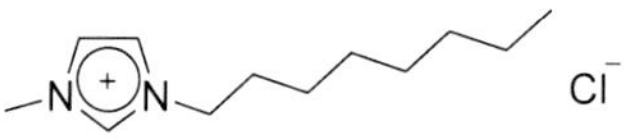	Avicel	4.5	100 °C	[33]
		4	100 °C, 1 h	[7]
[C8mim]Cl		Slightly soluble	100 °C	[30]
	Cellulose, DP = 250	10	100 °C	[23]
[Amim]Cl	Avicel	5	90 °C	[30]
	Cellulose pulp	14.5	80 °C	[10]
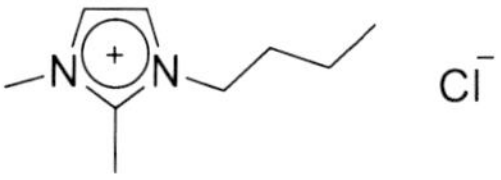	Cellulose, DP = 286	9	90 °C	[31]
[C4dmim]Cl	Cellulose, DP = 593	6	90 °C	[31]
	Cellulose, DP = 1,198	4	90 °C	[31]
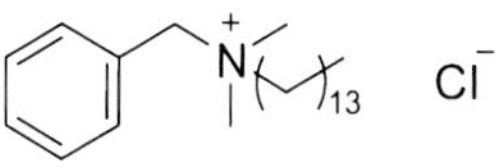	Cellulose, DP = 286	5	62 °C	[8]
[BzDTA]Cl	Cellulose, DP = 593	2	62 °C	[8]
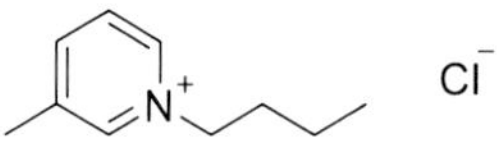	Cellulose, DP = 286	39	105 °C, 12 h	[8]
	Avicel	5	105 °C	[30]
[3MBPy]Cl				

(continued)

Table 2.5 (continued)

Ionic liquids	Raw material	Solubility (wt%)	Condition	Refs.
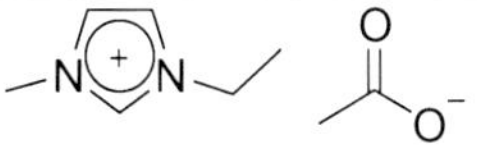 [C2mim][OAc]	Cellulose	13.5	85 °C	[29]
	Avicel	8	100 °C, 1 h	[7]
	Avicel	5	90 °C	[30]
 [C4mim][OAc]	Celulose, DP = 569	13.2		[29]
	Avicel	12	100 °C, 1 h	[7]
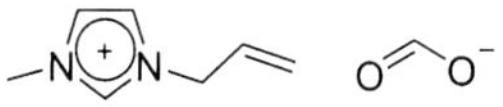 [Amim][HCOO]	Cellulose, DP = 250	21.5	85 °C	[23]
	Cellulose, DP = 250	10	60 °C	[23]
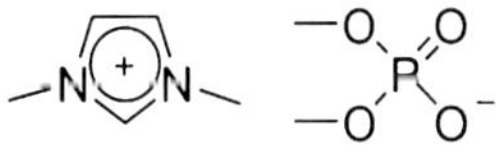 [C1mim][$(MeO)_2PO_2$]	Avicel	10	100 °C, 1 h	[7]
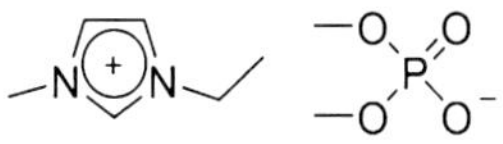 [C2mim][$(MeO)_2PO_2$]	MCC	10	65 °C	[28]
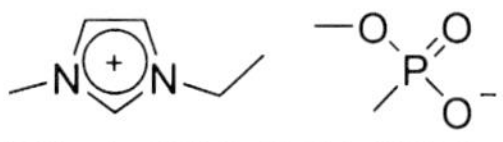 [C2mim][$(MeO)(Me)PO_2$]	Avicel	10	100 °C, 1 h	[7]
 [C2mim][$(MeO)(H)PO_2$]	MCC	6	30 °C, 1 h	[28]

critical drawback for biomass treatment process, especially for an energy conversion system. Addition of a small amount of water to the ILs certainly decreases the cellulose dissolving ability. These ILs cannot dissolve cellulose in the presence of a certain amount of water. Mazza and co-workers reported that the influence of water on the precipitation of cellulose in ILs [42]. Addition of a small amount of water was reported to greatly decrease the cellulose dissolving ability of CDILs. Gericke and co-workers analyzed cellulose precipitation from CDILs by addition of several anti-solvents including water [43]. According to the paper, once dissolved cellulose was easily precipitated from CDILs by the addition of 20 wt% water. This precipitation was found in all ILs used in the study, namely [C4mim]Cl, [Amim]Cl, and [C2mim][OAc]. Hauru and co-worker also reported the cellulose precipitation from CDILs [44]. The cellulose solutions became turbid by the addition of 2–3 equivalents of water, which is equivalent to 20–25 wt% water content. ILs easily absorb water from air [45], and especially CDILs have a high water absorption rate because they are very polar. Generally polar materials are hydrophilic. Troshenkova and co-workers reported on the water absorbability of a CDIL, [C2mim][OAc] [46]. This IL adsorbed up to 27 wt% of water from air at 25 °C. [C2mim][OAc] was hydrated by the water exothermically (11 kJ mol^{-1}), such values being

Table 2.6 Correlation between water content of TBPH and cellulose solubility (wt%) at 25 °C

Water content (wt%)	Cellulose (wt%)[a]	Dissolution time (min)
60	0.5	>2 weeks
50	15	5
	20	–[b]
40	1	1
	15	3
	20	5
	25	–[c]
30	15	5
	20	–[b]
20	5	7
	10	–[b]
10	0.5	Not dissolved

Reproduced from Abe et al. [47] with permission from The Royal Society of Chemistry
[a]Final concentration
[b]Most of the cellulose were dissolved within 30 min, but complete dissolution was not confirmed even after 1 h
[c]Difficult to stir

comparable to the thermal effect of chemical reactions. This means that CDILs should be sufficiently dried before cellulose treatment, and this might require a considerable amount of energy.

Quite recently, a novel IL derivative was reported as a cellulose solvent which dissolves cellulose without heating even in the presence of water. Abe and co-workers reported that tetra-*n*-butylphosphonium hydroxide (TBPH) containing 30–50 wt% water dissolved cellulose (15–20 wt% at final concentration) without heating at 25 °C (Table 2.6) [47]. Since this solution contained water, we do not need to dry the cellulose materials before dissolution process. TBPH/water mixture is expected as a potential solvent for cellulose regardless of water content.

2.6 Ionic Liquids for Lignocellulose Dissolution

ILs are being investigated as a solvent for not only pure cellulose but also other cellulosic biomass. Cellulosic biomass, such as wood, is composed of several hardly soluble polymers and many other materials. Other polysaccharides are also an attractive target to be extracted from biomass. In 2007, Fort and co-workers reported wood biomass treatment by ILs, and they clarified that a mixture of [C4mim]Cl and DMSO partially dissolved wood biomass at 100 °C (Fig. 2.5) [48]. The dissolving degree was achieved to about 70 % (wt/wt at added biomass). They analyzed the extracted materials and clarified that they were a mixture of polysaccharides and lignin. Shortly after this, Kilpeläinen and co-workers reported on wood dissolution by [C4mim]Cl and [Amim]Cl [49]. They treated soft- and hard-wood such as Norway spruce sawdust and southern pine thermomechanical pulp at temperatures between 80 and 130 °C for 8 and 13 h, respectively, and

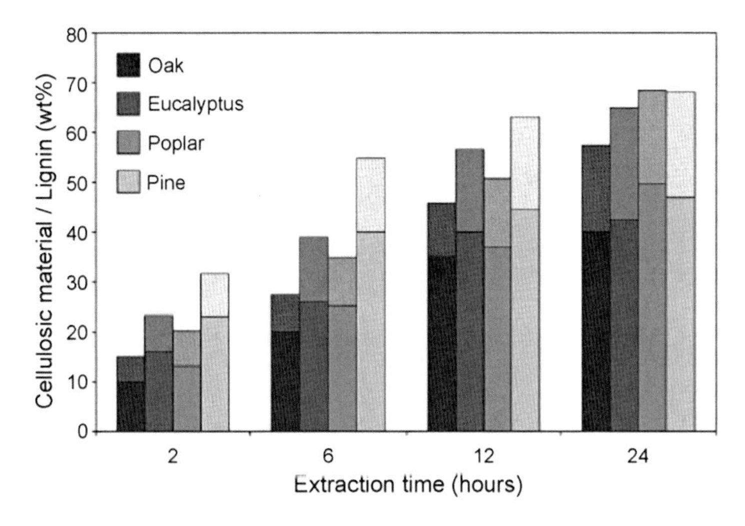

Fig. 2.5 Cellulosic material (*solid bars*) and lignin material (*dashed bars*) extraction profiles in [C4mim]Cl/DMSO-d_6 at 100 °C for the different wood (Reproduced from Fort et al. [48] with permission from The Royal Society of Chemistry)

observed that the biomass samples were partially dissolved. When the dissolution of the same lignocellulosic samples was soaked in 1-benzyl-3-methylimidazolium chloride ([Bnmim]Cl), transparent amber solutions were obtained. Wang and co-workers used a room temperature IL, [Amim]Cl to extract cellulose-rich material from several wood chips such as pine, poplar, Chinese parasol, and catalpa [50]. They showed that pine was one of the most suitable wood species for cellulose extraction with ILs, and its cellulose extraction degree reached to 62 %.

Miyafuji and co-worker observed the state of woodchips from softwood, *Cryptomeria japonica*, during the ILs treatment using light microscope [51]. Figure 2.6 shows micrographs of latewood, earlywood, and the latewood/earlywood boundary after treatment with [C2mim]Cl at 120 °C. The cell walls in latewood became disordered after 0.5 h treatment. In addition, some destruction or flaking was observed in the cell walls after 4 h treatment. By contrast, no significant change was observed in earlywood even after 4 h treatment. They suggested that latewood swells easier than earlywood because of the difference in the density.

Although ILs could dissolve only a part of wood biomass in an early stage, the complete dissolution of wood was achieved by Sun and co-workers in 2009 with carboxylate salts under heating [52]. After that the following separation methods were also investigated. Kilpeläinen and co-workers also reported the complete dissolution of lignocellulose materials [49]. That process helps to break some of interchain chemical bonds such as lignin-carbohydrate bond, and the lignocellulose material was used after mechanical pulping. Sun and co-workers clarified that [C2mim][OAc] completely dissolved softwood (southern yellow pine) and hardwood (red oak) after 46 and 25 h heating at 110 °C for pine and oak, respectively. In addition, they suggested that carbohydrate-free lignin and cellulose-rich materials

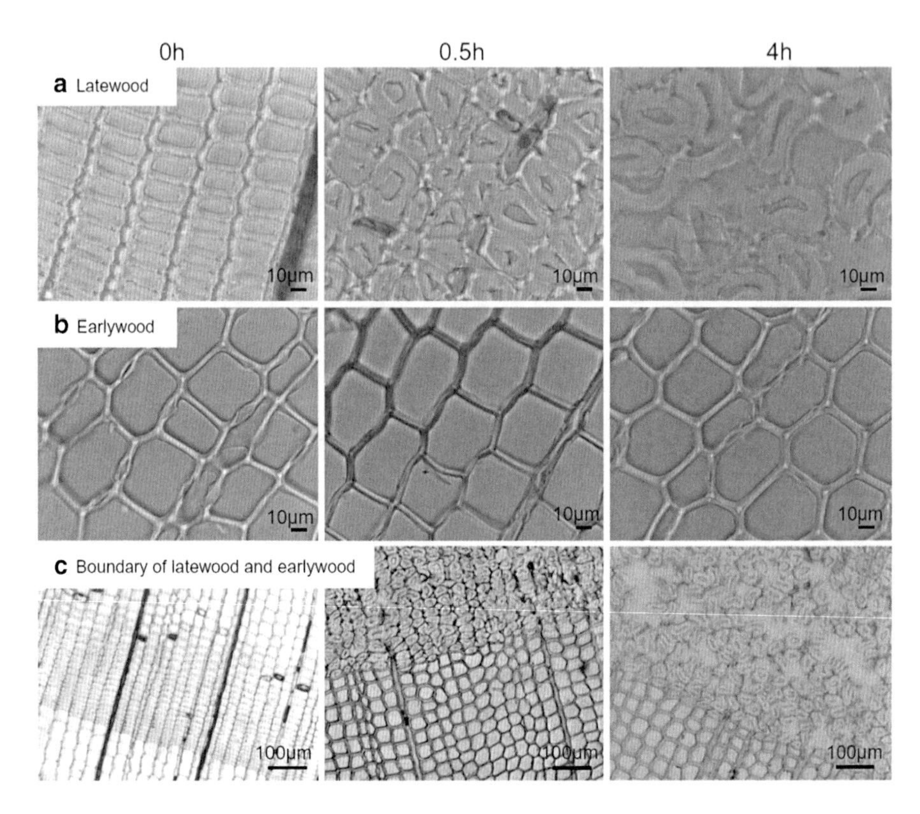

Fig. 2.6 Light microscopic images of wood ((**a**) latewood, (**b**) earlywood, (**c**) boundary of latewood and earlywood) treated with [C2mim]Cl at 120 °C [51] (With kind permission from Springer Science+Business Media)

were obtained by adequate precipitating process by the addition of acetone and water. On this basis, they developed the biomass treatment process as shown in Scheme 2.7.

Regarding lignin regeneration, Casas and co-workers also studied and reported some interesting results [53]. They collected regenerated lignin from *Pinus radiata* and *Eucalyptus globulus* woods dissolved in imidazolium-type ILs. Lignin was successfully regenerated by precipitation with methanol from wood solutions in [Amim]Cl, [C4mim]Cl, or [C2mim]Cl. Against this, lignin was not regenerated from acetate-type ILs. In addition, contents of different functional groups in the regenerated lignin were found to depend on the species of IL employed as well as wood species dissolved.

In the next section, direct lignin extraction from wood is mentioned. Sun and co-workers investigated the effect of particle size of the added biomass [52]. For [C4mim]Cl, the particle size was observed to have a significant influence on the extraction of lignin. The IL dissolved 52.6 % of the finely milled biomass (<0.125 mm), but only 26.0 % of coarser biomass (0.25–0.50 mm). It is easy to comprehend that smaller particles have larger gross surface area and lignin is easier to be solubilized. On the other hand, for [C2mim][OAc], the particle size of biomass did

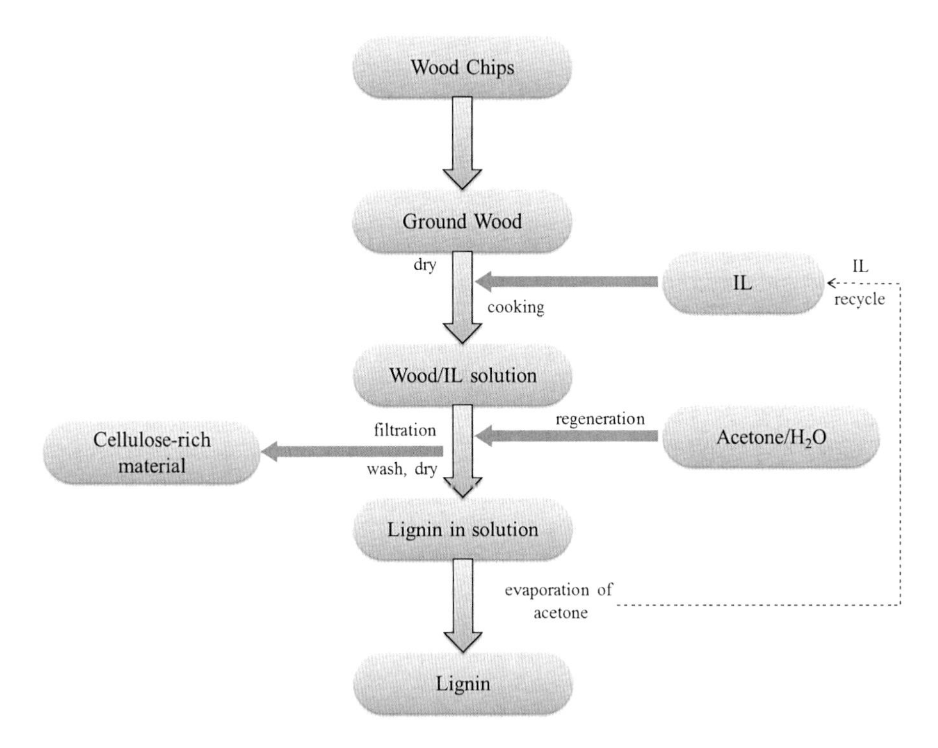

Scheme 2.7 Suggested wood biomass treatment process using IL (Reproduced from Sun et al. [52] with permission from The Royal Society of Chemistry)

not affect the results significantly. The [C2mim][OAc] dissolved more than 90 % of the added wood even from the particles as large as 0.5–1.0 mm. Sun et al. also evaluated the effects of some pretreatments, i.e., microwave or ultrasound irradiation (Table 2.7) [52]. These pretreatments accelerated the lignocellulose dissolution. With 60 × 3 s microwave pulses, the time for complete dissolution (t_{cd}) was reduced to shorter than half of that without pretreatment. As seen in Table 2.7, ultrasound pretreatment also accelerated the dissolution. In spite that these pretreatments are effective, it should not be ignored that these steps also consume energy.

In 2011, Sun et al. reported that complete dissolution of lignocellulose was carried out with shorter mixing time at temperature above the glass transition temperature of lignin [54]. Complete dissolution of 0.5 g bagasse in 10 g of [C2mim][OAc] requires more than 15 h heating at 110 °C, by contrast, it dissolves completely in the IL within 5–15 min heating at 175–195 °C. In addition, processing bagasse in the IL at 185 °C for 10 min gave higher yields of both recovered lignin and carbohydrate than the previous methods using lower temperatures and longer times (e.g., 110 °C, 16 h). There was an associated problem with the thermal stability of [C2mim][OAc], because about 15 % of the IL degraded after processing at the higher temperature.

Miyafuji and co-workers reported that cellulose dissolving ILs work as not only a solvent for plant biomass but also a reaction medium. They found that [C2mim]Cl

Table 2.7 Effect of pretreatment on the time required to achieve complete dissolution (t_{cd}) of 0.50 g of southern yellow pine sawdust (particle size 0.125–0.250 mm) in 10 g of [C2mim][OAc] at 110 °C [52]

Pretreatment method	Pretreatment condition	t_{cd} (h)
None	None	46
Microwave	30 × 3 s pulses	45
Microwave	60 × 3 s pulses	21
Microwave	100 × 3 s pulses	16
Ultrasound irradiation	1 h at 40 °C	23

dissolved wood and that the solubilized polymers such as cellulose were depolymerized to low molecular weight compounds just by mixing [55]. Japanese beech wood flours (0.09 g) were added to 3 g of [C2mim]Cl, and the mixture was heated to 90–120 °C under gentle stirring. After that, the molecular weight distribution of the solubilized compounds in [C2mim]Cl was studied by gel permeation chromatography. As a result, the molecular weight of the solubilized compounds was found to decrease as the treatment time was extended, and such depolymerization was more enhanced at higher temperature. They suggested that [C2mim]Cl penetrated into wood and liquefied polysaccharides such as cellulose at the initial stage of the reaction, and the crystal structure was gradually broken down.

The cellulosic biomass dissolving ability of several CDILs is summarized in Table 2.8.

2.7 Selective Extraction from Plant Biomass

In the previous section, the dissolution of plant biomass was mentioned. Those studies dissolve cellulose, hemicellulose, lignin, and some other materials altogether. On the other hand, some researchers are interested in isolating target or useful materials selectively from plant biomass. In the case of dissolution and collection process of cellulose, some separation processes of lignin and other polysaccharides are required. When one can design ILs suitable to extract only lignin selectively, the ILs should become a promising solvent in the pulping method of plant biomass.

Pu and co-workers reported about lignin dissolution using several ILs in 2007 [57], and Lee and co-workers reported on a similar process in 2009 [58]. Pu and co-workers investigated several imidazolium type ILs have a good ability to dissolve softwood kraft pulp lignin. In their study, 344 g/L lignin was dissolved in methylsulfate type ILs, and it was suggested that the selection of anions for ILs is important to dissolve lignin. In the case of [C4mim] salts, the order of lignin solubility was the function of anion species as follows: $MeOSO_3^- > Br^- > Cl^- > BF_4^-$. Lee and co-workers also investigated the lignin dissolving ability of imidazolium salts having several anions. According to their study, CDILs such as [C2mim][OAc] have good ability to dissolve lignin, and less polar ILs which have non-coordinating anions such as BF_4^- dissolve a small amount of Kraft lignin.

Table 2.8 Cellulosic biomass treatment ability of several CDILs

Ionic liquid	Biomass	Dissolution	Condition	Refs.
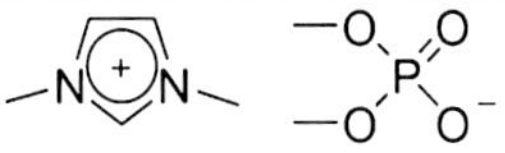 [C1mim][(MeO)$_2$PO$_2$]	Spruce chips	5 wt%, partial	90 °C	[30]
 [C2mim]Cl	Spruce chips	5 wt%, partial	90 °C	[30]
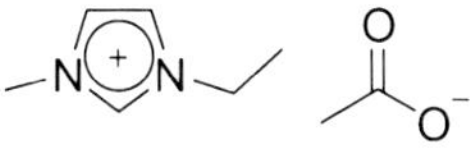	Spruce chips	5 wt%, Complete	90 °C	[30]
[C2mim][OAc]	Southern yellow pine, chips	92.6%[a]	110 °C, 16 h	[52]
	Southern yellow pine, chips	Complete	110 °C, 46 h	[52]
	Norway spruce saw dust	8 wt%	80 °C, 8 h	[49]
[Amim]Cl	Norway spruce TMP	7 wt%	130 °C, 8 h	[49]
	Pine	67%[a]	120 °C	[50]
	Poplar	30%[a]	120 °C	[50]
	Catalpa	23%[a]	120 °C	[50]
	Spruce chips	5 wt%, Complete	90 °C	[30]
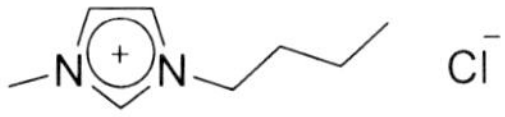	Pine	67%[a]	100 °C, 24 h, with DMSO-d_6	[48]
[C4mim]Cl	Poplar	68%[a]	100 °C, 24 h, with DMSO-d_6	[48]
	Oak	56%[a]	100 °C, 24 h, with DMSO-d_6	[48]
	Eucalyptus	64%[a]	100 °C, 24 h, with DMSO-d_6	[48]
	Wood chips	Partially soluble	130 °C, 15 h	[49]
	Norway spruce TMP	7 wt%	130 °C, 8 h	[49]
	Spruce chips	5 wt%, partial	90 °C	[30]
	Southern yellow pine, chips	26%[a]	110 °C, 16 h	[52]
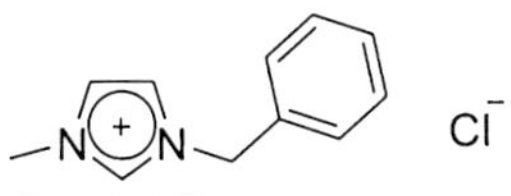 [Bnmim]Cl	Southern pine TMP	2 wt%	130 °C, 8 h	[49]
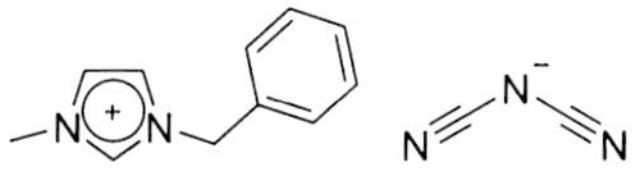 [Bnmim][DCA]	Aspen wood chips or powder	Dissolved	150 °C, 24 h	[49]

(continued)

Table 2.8 (continued)

Ionic liquid	Biomass	Dissolution	Condition	Refs.
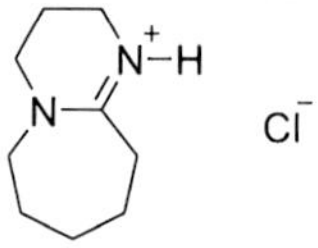 [HDBU]Cl	Aspen wood chips or powder	Partially dissolved	100 °C, 24 h	[56]
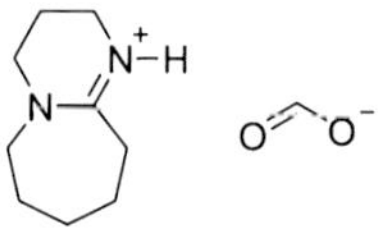 [HDBU][HCOO]	Aspen wood chips or powder	Partially dissolved	100 °C, 48 h	[56]
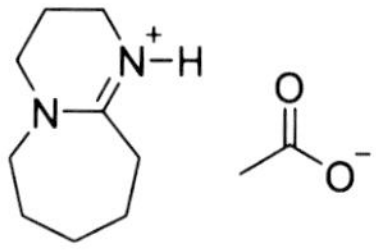 [HDBU][OAc]	Aspen wood chips or powder	Partially dissolved	100 °C, 48 h	[56]
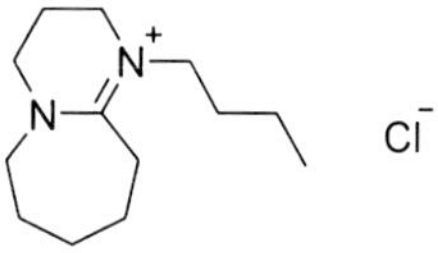 [C4DBU]Cl	Aspen wood chips or powder	Partially dissolved	100 °C, 48 h	[56]

Other values in wt% mean the amount of biomass against the ILs used as solvents
[a]The value was wt/wt of added biomass. The biomass was added to become 5 wt% against ILs

Based on the effectiveness of sodium xylenesulfonate as an agent for hydrotropic pulping of lignocellulose, Tan and co-workers studied the use of IL mixtures containing 1-ethyl-3-methylimidazolium alkylbenzenesulfonates ([C2mim][ABS]) to selectively dissolve lignin from sugarcane bagasse at atmospheric pressure and elevated temperature [59]. An extraction yield of up to 93 % was achieved. Compared with conventional lignin extraction methods, this system has several advantages such as no emission of toxic gases. But, a certain amount of carbohydrate losses (about 55 %) was caused during the biomass treatment.

Pinkert and co-workers reported that a class of food-additive derived ILs have a great ability to dissolve lignin without dissolving or degrading cellulose [60]. They suggested that 1-ethyl-3-methylimidazolium acesulfamate ([C2mim][Ace]) extracted 0.38 mass fraction of lignin of the added biomass in gentle extraction step (100 °C, 2 h), and the presence of a co-solvent (DMSO) increased the extraction degree to 0.56 mass fraction. Since this IL dissolves lignin, but not wood cellulose, it should become a promising solvent for pulping methods of lignocellulose materials.

Table 2.9 shows the lignin dissolving ability of several ILs.

Table 2.9 Lignin dissolving ability of several ILs

Ionic liquid	Lignin	Dissolution	Condition	Refs.
[C1mim][MeOSO$_3$]	Softwood kraft pulp lignin	344 g/L	50 °C	[57]
	Kraft lignin	>500 g/kg	90 °C, 24 h	[58]
[Cyanomim]Br	Alkali, low sulfonate content lignin	9.5 wt%	80–90 °C, 20 min	[32]
[C2mim][OAc]	Kraft lignin	300 g/kg	90 °C, 24 h	[58]
	Triticale straw	52.7 % of acid insoluble lignin	150 °C, 90 min	[61]
[Amim]Cl	Kraft lignin	>300 g/kg	90 °C, 24 h	[58]
[C3mim]Br	Alkali, low sulfonate content lignin	6.2 wt%	80–90 °C, 20 min	[32]
[C4mim]Cl	Softwood kraft pulp lignin	13.9 g/L	75 °C	[57]
	Alkali, low sulfonate content lignin	8.8 wt%	80–90 °C, 20 min	[32]
[C4mim]Br	Softwood kraft pulp lignin	17.5 g/L	75 °C	[57]
[C4mim][BF$_4$]	Kraft lignin	40 g/kg	90 °C, 24 h	[58]
[C4mim][PF$_6$]	Kraft lignin	1 g/kg	90 °C, 24 h	[58]
[C4mim][OTf]	Kraft lignin	>500 g/kg	90 °C, 24 h	[58]
[C4mim][MeOSO$_3$]	Softwood kraft pulp lignin	312 g/L	50 °C	[57]
[C4dmim][BF$_4$]	Softwood kraft pulp lignin	14.5 g/L	70–100 °C	[57]

(continued)

Table 2.9 (continued)

Ionic liquid	Lignin	Dissolution	Condition	Refs.
[C6mim][OTf]	Softwood kraft pulp lignin	275 g/L	50 °C	[57]
[Bnmim]Cl	Kraft lignin	>100 g/kg	90 °C, 24 h	[58]
[HDBU]Cl	Sulfur free lignin	Dissolved	150 °C, 24 h	[56]
[HDBU][HCOO]	Sulfur free lignin	Dissolved	100 °C, 24 h	[56]
[HDBU][OAc]	Sulfur free lignin	Dissolved	100 °C, 24 h	[56]
[HDBU][NTf_2]	Sulfur free lignin	Dissolved	100 °C, 24 h	[56]
[C8DBU][NTf_2]	Sulfur free lignin	Dissolved	100 °C, 48 h	[56]
[C2mim][ABS]	Sugarcane bagasse	97 % of lignin Extracted ~23 g/kg	190 °C, 90 min	[59]
	Pinus radiate wood flour	38 wt% of added lignin (>5.7 g/kg)	100 °C, 2 h	[60]
	Pinus radiate wood flour	51 wt% of added lignin (>7.7 g/kg)	100 °C, 16 h	[60]

(continued)

Table 2.9 (continued)

Ionic liquid	Lignin	Dissolution	Condition	Refs.
[C4mim][Ace]	Eucalyptus nitens	38 wt% of added lignin (>6.1 g/kg)	100 °C, 2 h	[60]
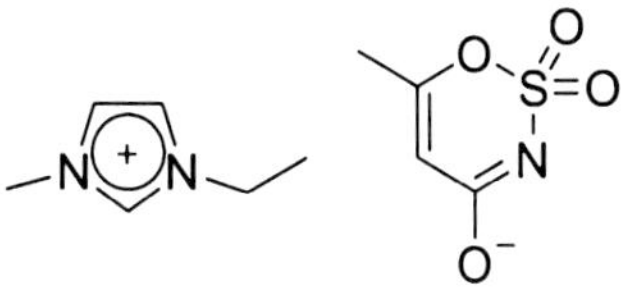 [C2mim][Ace]	Pinus radiate wood flour	43 wt% of added lignin (>6.5 g/kg)	100 °C, 2 h	[60]

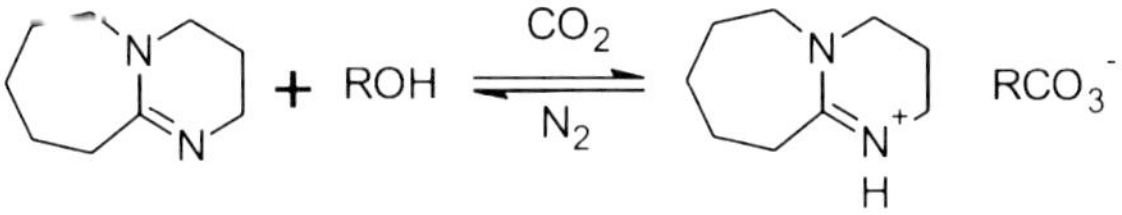

Scheme 2.8 Reaction scheme of the switchable ionic liquid (Reprinted from Anugwom et al. [62], Copyright (2012), with permission from Elsevier)

Selective extraction of not only lignin but also other components from biomass is important to construct an effective biomass conversion process. Anugwom and co-workers constructed the selective extraction process for hemicellulose from spruce, a typical plant biomass, using switchable ILs [62]. A switchable IL (Scheme 2.8) was investigated as dissolution/fractionation solvents for plant biomass. After the treatment for 5 days without stirring, the amount of hemicellulose in the undissolved fraction was reduced by 38 wt% as compared with that before treatment. They stated that the recovered hemicelluloses were very important in many industrial fields, because the spruce hemicellulose was mainly galactoglucomannans, which could be used as bioactive polymers, hydrocolloids, or papermaking chemicals [63].

ILs are useful to extract not only the main components of biomass such as polysaccharides and lignin but also some valuable materials. For example, a pharmaceutical ingredient, shikimic acid, was extracted from *Ginkgo biloba* with [C4mim]Cl [64]. Shikimic acid is the starting material for the synthesis of oseltamivir phosphate (Tamiflu®), which is used as an antiviral agent for the H5N1 strain of influenza [65]. As seen in Fig. 2.7, using [C4mim]Cl at 150 °C, the extraction yield of shikimic acid reached to 2.3 wt%, which was 2.5 times higher than that extracted with methanol at 80 °C. Usuki and co-workers also established the isolation process using an anion-exchange resin. They clarified that CDILs are useful and important to collect valuable materials from plant biomass.

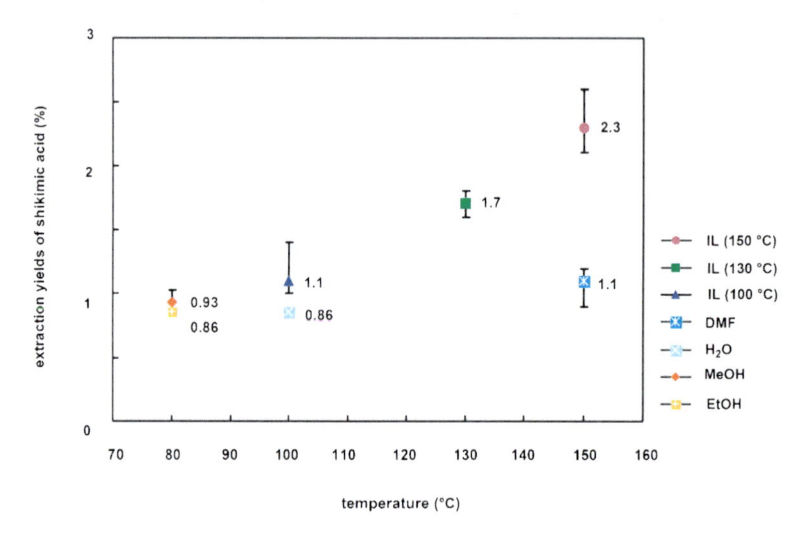

Fig. 2.7 Extraction degree of shikimic acid using [C4mim]Cl (100, 130, or 150 °C), DMF (150 °C), deionized H_2O (100 °C), methanol (80 °C), or ethanol (80 °C) (Reproduced from Usuki et al. [64] with permission from The Royal Society of Chemistry)

2.8 Energy Saving Dissolution of Plant Biomass

As described in the above two sections, ILs are promising solvents to treat plant biomass. The most important thing is the treatment capability of plant biomass under mild conditions. CDILs dissolve polysaccharides and a part of lignin at about 100 °C without any pressurization; this means that the energy-cost to treat biomass was reduced compared to other existing biomass treatment processes under heating above 150 °C and pressurization. It is very important for any industrial fields because energy-cost is directly linked to the price of the final product. The consumption of energy becomes of particular importance for the energy-producing industry, because the use of excess amount of energy to get comparable or less energy is meaningless. Although ILs can treat plant biomass under mild condition compared to some other methods such as kraft-pulping method, heating and long mixing time are still needed to dissolve plant biomass in ILs, too. These requirements should further be improved to reduce energy consumption.

Some researcher use ILs for just pretreatment of plant biomass and not for dissolution of cellulose. Li and co-workers reported that enzymatic hydrolysis was significantly improved by the use of CDIL, 1-ethyl-3-methylimidazolium diethyl phosphate ([C2mim][DEP]) [66]. They investigated the effect of temperature and time of the IL treatment on the hydrolysis efficiency. The pretreatment temperature was changed from 25 to 150 °C for 1 h stirring, and the hydrolysis efficiency of the pretreated wheat straw was significantly improved when the temperature was changed from 70 to 100 °C. On the other hand, the difference of the pretreatment time slightly affects the hydrolysis degree, and they reached the conclusion that only 30 min treatment was enough to accelerate the following hydrolysis. The yield of reducing sugars from wheat straw reached 54.8 % when the wheat straw was pretreated with [C2mim][DEP] at 130 °C for only 30 min. It remained only 20 % when the straw was enzymatically hydrolyzed in water for 12 h. In addition, the hydrolysis products did not show a negative effect on *S. cerevisiae* fermentation. Tan and co-workers reported the IL pretreatment of palm frond after extracting the palm oil for improving conversion of cellulose into reducing sugar through subsequent enzymatic hydrolysis [67]. During the pretreatment, lignin was partly decomposed and was dissolved in [C4mim]Cl and remained in the solution after regeneration process of cellulose. In addition, hemicellulose was autohydrolyzed during the pretreatment. Apart from crystallinity of cellulose, cellulose digestibility should also be influenced by other factors such as DP, surface area of cellulose, as well as state of cellulose protected by lignin and hemicellulose complexes. Uju and co-workers also studied the effect of pretreatment with ILs for plant biomass [68]. They used [C4mpy]Cl as a pretreatment IL for bagasse or *Eucalyptus*. The pretreatment of the biomass resulted in up to eightfold increase in the enzymatic saccharification compared with the untreated biomass. At short time pretreatment, [C4mpy]Cl showed higher potential to increase the initial degree of cellulose conversion than that in [C2mim][OAc]. They suggested that the significant acceleration of enzymatic saccharification was possibly caused by the reducing of DP of cellulose by the [C4mpy]Cl pretreatment. Bahcegul and co-workers studied the correlation between the particle size of plant biomass in detail and the pretreatment efficiency with ILs for subsequent enzymatic saccharification [69]. They used cotton stalks with four different particle size pretreated in [C2mim][OAc] or [C2mim]Cl. For [C2mim]Cl, the highest glucose yield (49 %) was obtained when the biomass had the smallest particle size, while cotton stalks with larger particle size gave lower glucose yield (33 %). On the contrary, for [C2mim][OAc], the lowest glucose yield (57 %) was obtained when the cotton stalks with the smallest particle size was examined, while cotton stalks with larger particle size gave higher glucose yield (71 %). Simply considering the overall surface area of the biomass particles, smaller particles gave higher glucose yield. Other unknown factor(s) should exist to affect the enzymatic saccharification. They suggested that the most suitable particle size of lignocellulosic biomass prior to pretreatment may change depending on the IL species.

For pretreatment of lignocellulosic biomass, it is not necessary to completely dissolve cellulose but heating is still a necessary step. On the other hand, some researchers are trying to dissolve plant biomass without heating. Abe et al. found that phosphinate-type ILs dissolved plant biomass and extracted polysaccharides from plant biomass without heating [70]. Since some phosphonate type ILs as seen in Scheme 2.9 have a good ability to dissolve cellulose at ambient temperature [28], we have prepared several phosphonate type ILs and evaluated their biomass treatment ability. As a result, polysaccharide extraction degree was found to be closely related to the viscosity. This means that the IL with low viscosity had good

Scheme 2.9 Structure of alkylphosphonate type salts [70]

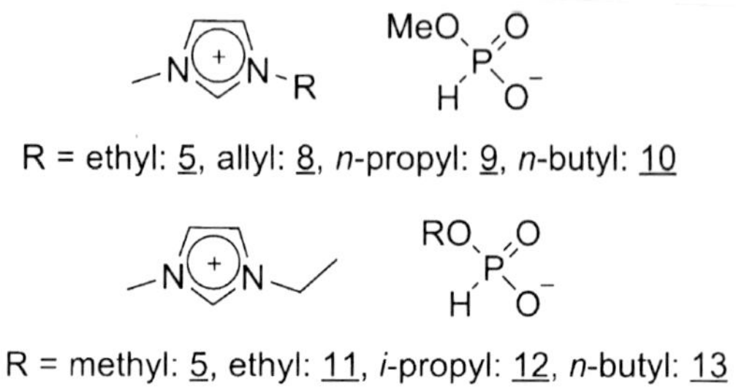

R = ethyl: 5, allyl: 8, *n*-propyl: 9, *n*-butyl: 10

R = methyl: 5, ethyl: 11, *i*-propyl: 12, *n*-butyl: 13

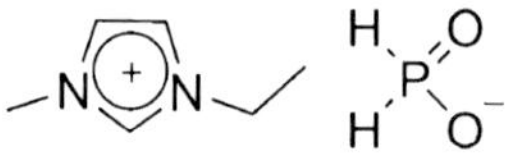

Scheme 2.10 Structure of 1-ethyl-3-methylimidazolium phosphinate (IL**14**) (Reproduced from Abe et al. [70] with permission from The Royal Society of Chemistry)

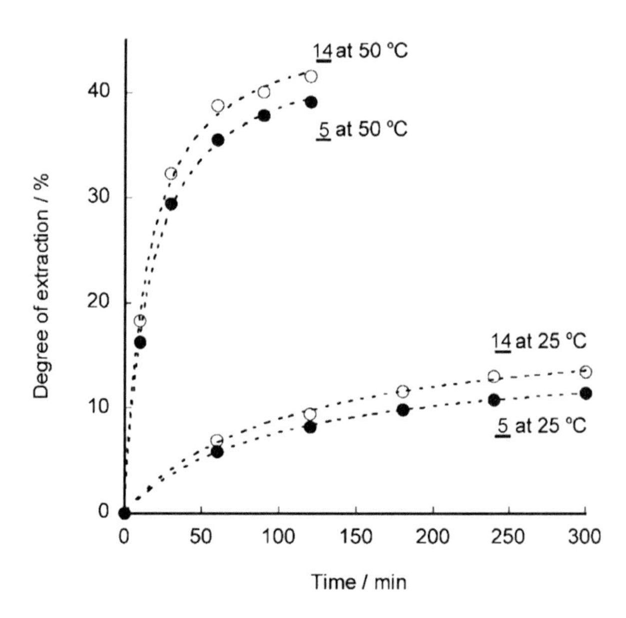

Fig. 2.8 Extraction degree of polysaccharides from bran using phophinate salt (IL**14**) or methylphosphonate salt (IL**5**) [70]

capacity to dissolve plant biomass within a short period of mixing time under mild condition when ILs have sufficiently high polarity. We accordingly designed a low viscosity and highly polar IL; 1-ethyl-3-methylimidazolium phosphinate (Scheme 2.10). With this IL, it became easy to extract polysaccharides rapidly from plant biomass under mild conditions (Fig. 2.8). Since this IL did not require any heating to extract polysaccharides from biomass, the energy-cost was reduced and this IL should be a promising solvent for plant biomass treatment. In addition, this IL is stable and recyclable. Thus, a closed system for biomass treatment as seen in Fig. 2.9 can be proposed.

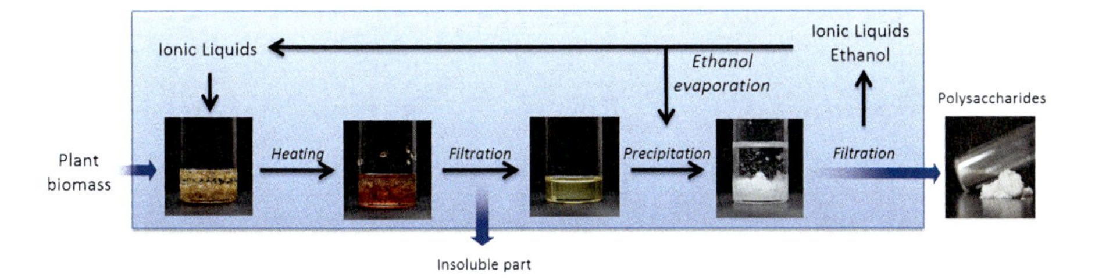

Fig. 2.9 Closed and energy-saving system (scheme) to extract polysaccharides from plant biomass [70]

Table 2.10 Correlation between water content of TBPH and extraction degree of polysaccharides after 1 h stirring at 25 °C

Water content of TBPH solution (wt%)	Polysaccharides extraction degree (%)
70	4.9
60	28
50	36
40	37
30	24

The extraction degree was calculated from the weight of the added poplar powder (5 wt% against the TBPH solution)

There are a few reports about the effect of water addition on the solubility of cellulose. Padmanabhan and co-workers reported about the influence of water on the lignocellulose solubility [71]. Prior to solubility measurements, 3–5 wt% water was added to cellulose dissolving ILs, namely chloride, acetate and phosphate-based ILs. After that, powder of *Miscanthus*, a lignocellulosic material, was added to the ILs, and stirred the mixture at over 100 °C. However, no cellulose was extracted. This result strongly suggested that water considerably suppressed the dissolution of lignocellulose in wet ILs. They concluded that ILs should be dried well in advance to extract cellulose from biomass. Since plant biomasses also contain a certain amount of water, the biomasses have to be dried before treatment with ILs. On the other hand, as mentioned above, TBPH has a great ability to dissolve cellulose without heating even in the presence of water [47]. So, we next tried to use this novel solvent to treat wood biomass. As expected, polysaccharides such as cellulose were extracted from wood powder without heating (Table 2.10). Poplar powder was used as a wood sample, and the powder was added to reach 5 wt% against TBPH solution. In the presence of 70 wt% water, TBPH could extract polysaccharides only 4.9 % of the weight of the added poplar. On the other hand, TBPH containing 40–50 wt% water successfully extracted cellulose and other polysaccharides for 36–37 % of the weight of the added popular. These results indicate that the extraction of cellulose from powder of wood such as poplar could be carried out even in the presence of considerable amounts of water.

For the development of sustainable human societies, we have to develop new energy conversion methods based on renewable energy sources instead of fossil fuels. ILs, which dissolve renewable cellulosic biomass with low energy cost, should serve as the foundation for future development of sustainable world, especially for the development of bioenergy production.

Acknowledgement Our research results mentioned here were obtained under the support of a Grant-in-Aid for Scientific Research from the Japan Society for the Promotion of Science (No. 21225007). It was also partly supported by Japan Science and Technology Agency (JST) through the CREST program.

References

1. Welton T. Room-temperature ionic liquids. Solvents for synthesis and catalysis. Chem Rev. 1999;99:2071–83.
2. Limayem A, Ricke SC. Lignocellulosic biomass for bioethanol production: current perspectives, potential issues and future prospects. Prog Energy Comb Sci. 2012;38:449–67.
3. Zugenmaier P. Conformation and packing of various crystalline cellulose fibers. Prog Polym Sci. 2001;26:1341–417.
4. Bochek AM. Effect of hydrogen bonding on cellulose solubility in aqueous and nonaqueous solvents. Russ J Appl Chem. 2003;76:1711–19.
5. Graenacher C. Cellulose solution. US Patent, No. 1943176.
6. Swatloski RP, Spear SK, Holbrey JD, Rogers RD. Dissolution of cellose with ionic liquids. J Am Chem Soc. 2002;124:4974–5.
7. Vitz J, Erdmenger T, Haensch C, Schubert US. Extended dissolution studies of cellulose in imidazolium based ionic liquids. Green Chem. 2009;11:417–24.
8. Heinze T, Schwikal K, Barthel S. Ionic liquids as reaction medium in cellulose functionalization. Macromol Biosci. 2005;5:520–5.
9. Mizumo T, Marwanta E, Matsumi N, Ohno H. Allylimidazolium halides as novel room temperature ionic liquids. Chem Lett. 2004;33:1360–1.
10. Zhang H, Wu J, Zhang J, He J. 1-Allyl-3-methylimidazolium chloride room temperature ionic liquid: a new and powerful nonderivatizing solvent for cellulose. Macromolecules. 2005;38:8272–7.
11. Remsing RC, Swatloski RP, Rogers RD, Moyna G. Mechanism of cellulose dissolution in the ionic liquid 1-n-butyl-3-methylimidazolium chloride: a C-13 and Cl-35/37 NMR relaxation study on model systems. Chem Commun. 2006;2006:1271–3.
12. Youngs TGA, Hardacre C, Holbrey JD. Glucose solvation by the ionic liquid 1,3-dimethylimidazolium chloride: a simulation study. J Phys Chem B. 2007;111:13765–74.
13. Gross AS, Bell AT, Chu J-W. Thermodynamics of cellulose solvation in water and the ionic liquid 1-butyl-3-methylimidazolim chloride. J Phys Chem B. 2011;115:13433–40.
14. Hansen CM. 50 years with solubility parameters – past and future. Prog Org Coat. 2004;51:77–84.
15. Klamt A. Conductor-like screening model for real solvents: a new approach to the quantitative calculation of solvation phenomena. J Phys Chem. 1995;99:2224–35.
16. Kamlet MJ, Taft RW. The solvatochromic comparison method. I. The beta.-scale of solvent hydrogen-bond acceptor (HBA) basicities. J Am Chem Soc. 1976;98:377–83.
17. Crowhurst L, Mawdsley PR, Perez-Arlandis JM, Salter PA, Welton T. Solvent-solute interactions in ionic liquids. Phys Chem Chem Phys. 2003;5:2790–4.
18. Brandt A, Gräsvik J, Hallett JP, Welton T. Deconstruction of lignocellulosic biomass with ionic liquids. Green Chem. 2013;15:550–83.
19. Ohira K, Abe Y, Kawatsura M, Suzuki K, Mizuno M, Amano Y, Itoh T. Design of cellulose dissolving ionic liquids inspired by nature. ChemSusChem. 2012;5:388.

20. Ohno H, Fukaya Y. Task specific ionic liquids for cellulose technology. Chem Lett. 2009;38:2–7.
21. Hermanutz F, Gähr F, Uerdingen E, Meister F, Kosan B. New developments in dissolving and processing of cellulose in ionic liquids. Macromol Symp. 2008;262:23–7.
22. Swatloski RP, Rogers RD, Holbrey JD. Dissolution and processing of cellulose using ionic liquids. 2003, WO 029329.
23. Fukaya Y, Sugimoto A, Ohno H. Superior solubility of polysaccharides in low viscosity, polar, and halogen-free 1,3-dialkylimidazolium formats. Biomacromolecules. 2006;7:3295–7.
24. Remsing RC, Hernandez G, Swatloski RP, Massefski WW, Rogers RD, Moyna G. Solvation of carbohydrates in N, N′-dialkylimidazolium ionic liquids: a multinuclear NMR spectroscopy study. J Phys Chem B. 2008;112:11071–8.
25. Zhang J, Zhang H, Wu J, Zhang J, He J, Xiang J. NMR spectroscopic studies of cellobiose solvation in EmimAc aimed to understand the dissolution mechanism of cellulose in ionic liquids. Phys Chem Chem Phys. 2010;12:1941–7.
26. Liu H, Sale KL, Holmes BM, Simmons BA, Singh S. Understanding the interactions of cellulose with ionic liquids: a molecular dynamics study. J Phys Chem B. 2010;114:4293–301.
27. Xu A, Wang J, Wang H. Effects of anionic structure and lithium salts addition on the dissolution of cellulose in 1-butyl-3-methylimidazolium-based ionic liquid solvent systems. Green Chem. 2010;12:268–75.
28. Fukaya Y, Hayashi K, Wada M, Ohno H. Cellulose dissolution with polar ionic liquids under mild conditions: required factors for anions. Green Chem. 2008;10:44–6.
29. Kosan B, Michels C, Meister F. Dissolution and forming of cellulose with ionic liquids. Cellulose. 2008;15:59–66.
30. Zavrel M, Bross D, Funke M, Buchs J, Spiess AC. High-throughput screening for ionic liquids dissolving (ligno-)cellulose. Bioresour Technol. 2009;100:2580–7.
31. Barthel S, Heinze T. Acylation and carbanilation of cellulose in ionic liquids. Green Chem. 2006;8:301–6.
32. Lateef H, Grimes S, Kewcharoenwong P, Feinberg B. Separation and recovery of cellulose and lignin using ionic liquids: a process for recovery from paper-based waste. J Chem Technol Biotechnol. 2009;84:1818–27.
33. Erdmenger T, Haensch C, Hoogenboom R, Shubert US. Homogeneous tritylation of cellulose in 1-butyl-3-methylimidazolium chloride. Macromol Biosci. 2007;7:440–5.
34. Fukumoto K, Yoshizawa M, Ohno H. Room temperature ionic liquids from 20 natural amino acids. J Am Chem Soc. 2005;127:2398–9.
35. Kagimoto J, Noguchi K, Murata K, Fukumoto K, Nakamura N, Ohno H. Polar and low viscosity ionic liquid mixtures from amino acids. Chem Lett. 2008;37:1026–7.
36. Ohno H, Fukumoto K. Amino acid ionic liquids. Acc Chem Res. 2007;40:1122–9.
37. Fukumoto K, Ohno H. LCST-type phase changes of a mixture of water and ionic liquids derived from amino acids. Angew Chem Int Ed. 2007;46:1852–5.
38. Kagimoto J, Taguchi S, Fukumoto K, Ohno H. Hydrophobic and low-density amino acid ionic liquids. J Mol Liq. 2010;153:133–8.
39. Ohira K, Yoshida K, Hayase S, Itoh T. Amino acid ionic liquid as an efficient cosolvent of dimethyl sulfoxide to realize cellulose dissolution at room temperature. Chem Lett. 2012;41:987–9.
40. Hirao M, Sugimoto H, Ohno H. Preparation of novel room-temperature molten salts by neutralization of amines. J Electrochem Soc. 2000;147:4168–72.
41. King AWT, Asikkala J, Mutikainen I, Jarvi P, Kilpelainen I. Distillable acid-base conjugate ionic liquids for cellulose dissolution and processing. Angew Chem Int Ed. 2011;50:6301–5.
42. Mazza M, Catana DA, Garcia CV, Cecutti C. Influence of water on the dissolution of cellulose in selected ionic liquids. Cellulose. 2009;16:207–15.
43. Gericke M, Liebert T, Seoud AE, Heinze T. Tailored media for homogeneous cellulose chemistry: ionic liquid/co-solvent mixtures. Macromol Mater Eng. 2011;296:483–93.

44. Hauru LKJ, Hummel M, King AWT, Kilpeläinen I, Sixta H. Role of solvent parameters in the regeneration of cellulose from ionic liquid solutions. Biomacromolecules. 2012;13:2896–905.
45. Cammarata L, Kazarian SG, Salter PA, Welton T. Molecular states of water in room temperature ionic liquids. Phys Chem Chem Phys. 2001;3:5192–200.
46. Troshenkova SV, Sashina ES, Novoselov NP, Arndt K-F, Jankowsky S. Structure of ionic liquids on the basis of imidazole and their mixtures with water. Russ J Gen Chem. 2010;80:106–11.
47. Abe M, Fukaya Y, Ohno H. Fast and facile dissolution of cellulose with tetrabutylphosphonium hydroxide containing 40 wt% water. Chem Commun. 2012;48:1808–10.
48. Fort DA, Remsing RC, Swatloski RP, Moyna P, Moyna G, Rogers RD. Can ionic liquids dissolve wood? Processing and analysis of lignocellulosic materials with 1-n-butyl-3-methylimidazolium chloride. Green Chem. 2007;9:63–9.
49. Kilpeläinen I, Xie H, King A, Granstrom M, Heikkinen S, Argyropoulos DS. Dissolution of wood in ionic liquids. J Agric Food Chem. 2007;55:9142–8.
50. Wang X, Li H, Cao Y, Tang Q. Cellulose extraction from wood chip in an ionic liquid 1-allyl-3-methylimidazolium chloride (AmimCl). Bioresour Technol. 2011;102:7959–65.
51. Miyafuji H, Suzuki N. Observation by light microscope of sugi (Cryptomeria japonica) treated with the ionic liquid 1-ethyl-3-methylimidazolium chloride. J Wood Sci. 2011;57:459–61.
52. Sun N, Rahman M, Qin Y, Maxim ML, Rodriguez H, Rogers RD. Complete dissolution and partial delignification of wood in the ionic liquid 1-ethyl-3-methylimidazolium acetate. Green Chem. 2009;11:646–55.
53. Casas A, Oliet M, Alonso MV, Rodriguez F. Dissolution of Pinus radiata and Eucalyptus globulus woods in ionic liquids under microwave radiation: lignin regeneration and characterization. Sep Purif Technol. 2012;97:115–22.
54. Li W, Sun N, Stoner B, Jiang X, Lu X, Rogers RD. Rapid dissolution of lignocellulosic biomass in ionic liquids using temperatures above the glass transition of lignin. Green Chem. 2011;13:2038–47.
55. Miyafuji H, Miyata K, Saka S, Ueda F, Mori M. Reaction behavior of wood in an ionic liquid, 1-ethyl-3-methylimidazolium chloride. J Wood Sci. 2009;55:215–19.
56. D'Andola G, Szarvas L, Massonne K, Stegmann V. (BASF), Ionic liquids for solubilizing polymers. 2008, WO 043837.
57. Pu Y, Jiang N, Ragauskas AJ. Ionic liquid as a green solvent for lignin. J Wood Chem Technol. 2007;27:23–33.
58. Lee SH, Doherty TV, Linhardt JS. Ionic liquid-mediated selective extraction of lignin from wood leading to enhanced enzymatic cellulose hydrolysis. Biotechnol Bioeng. 2009;102:1368–76.
59. Tan SSY, MacFarlane DR, Upfal J, Edye LA, Doherty WOS, Patti AF, Pringle JM, Scott JL. Extraction of lignin from lignocellulose at atmospheric pressure using alkylbenzenesulfonate ionic liquid. Green Chem. 2009;11:339–45.
60. Pinkert A, Goeke DF, Marsh KN, Pang S. Extracting wood lignin without dissolving or degrading cellulose: investigations on the use of food additive-derived ionic liquids. Green Chem. 2011;13:3124–36.
61. Fu D, Mazza G, Tamaki Y. Lignin extraction from straw by ionic liquids and enzymatic hydrolysis of the cellulosic residues. J Agric Food Chem. 2010;58:2915–22.
62. Anugwom I, Mäki-Arvela P, Virtanen P, Willför S, Sjöholm R, Mikkola J-P. Selective extraction of hemicelluloses from spruce using switchable ionic liquids. Carbohydr Polym. 2012;87:2005–11.
63. Xu C, Leppänen A-S, Eklund P, Holmlund P, Sjöholm R, Sundberg K, Willför S. Acetylation and characterization of spruce (Picea abies) galactoglucomannans. Carbohydr Res. 2010;345:810–16.
64. Usuki T, Yasuda N, Yoshizawa-Fujita M, Rikukawa M. Extraction and isolation of shikimic acid from Ginkgo biloba leaves utilizing an ionic liquid that dissolves cellulose. Chem Commun. 2011;47:10560–2.

65. Farina V, Brown JD. Tamiflu: the supply problem. Angew Chem Int Ed. 2006;45:7330–4.
66. Li Q, He Y-C, Xian M, Jun G, Xu X, Yang J-M, Li L-Z. Improving enzymatic hydrolysis of wheat straw using ionic liquid 1-ethyl-3-methyl imidazolium diethyl phosphate pretreatment. Bioresour Technol. 2009;100:3570–5.
67. Tan HT, Lee KT. Understanding the impact of ionic liquid pretreatment on biomass and enzymatic hydrolysis. Chem Eng J. 2012;183:448–58.
68. Uju N, Shoda Y, Nakamoto A, Goto M, Tokuhara W, Noritake Y, Katahira S, Ishida N, Nakashima K, Ogino C, Kamiya N. Short time ionic liquids pretreatment on lignocellulosic biomass to enhance enzymatic saccharification. Bioresour Technol. 2012;103:446–52.
69. Bahcegul E, Apaydin S, Haykir NI, Tatli E, Bakir U. Different ionic liquids favor different lignocellulosic biomass particle sizes during pretreatment to function efficiently. Green Chem. 2012;14:1896–903.
70. Abe M, Fukaya Y, Ohno H. Extraction of polysaccharides from bran with phosphonate or phosphinate-derived ionic liquids under short mixing time and low temperature. Green Chem. 2010;12:1274–80.
71. Padmanabhan S, Kim M, Blanch HW, Prausnitz JM. Solubility and rate of dissolution for Miscanthus in hydrophilic ionic liquids. Fluid Phase Equilib. 2011;309:89–96.

Chapter 3
Choline Chloride-Derived ILs for Activation and Conversion of Biomass

Karine De Oliveira Vigier and François Jérôme

Abstract The progressive introduction of biomass in chemical processes has dramatically changed the way how we design a catalytic process. Among different strategies, assisted catalysis is expected to play a pivotal role in the future. In this context, ChCl-derived ionic liquids and deep eutectic solvents has recently emerged as promising solvents to assist a conventional catalyst in the selective conversion of biomass. In particular, their ability to disrupt the hydrogen bond network of biopolymers, their ability to stabilize polar chemicals and their low miscibility with common low boiling point solvents open a promising route for the conversion of biomass in a more sustainable way. Beside their low price and low ecological footprint, we wish to demonstrate here that these neoteric solvents have processing advantages that no other solvent can provide in the field of biomass.

Keywords Catalysis • Bioinspired ionic liquids • Deep eutectic solvents • Biomass • Carbohydrates

Abbreviations

BMIM	1-butyl 3-methyl imidazolium
ChCl	Choline chloride
DMSO	Dimethylsulfoxide
HMF	5-hydroxymethylfurfural
IL	Ionic liquid

K. De Oliveira Vigier • F. Jérôme (✉)
Institut de Chimie des Milieux et Matériaux de Poitiers (IC2MP)
CNRS-University of Poitiers, ENSIP, 1 rue Marcel Doré, 86022 Poitiers, France
e-mail: francois.jerome@univ-poitiers.fr

Z. Fang et al. (eds.), *Production of Biofuels and Chemicals with Ionic Liquids*,
Biofuels and Biorefineries 1, DOI 10.1007/978-94-007-7711-8_3,

3.1 Introduction

The progressive introduction of renewably-sourced raw materials in chemical processes has dramatically changed the way how we design a catalytic reaction and catalysis is now facing to new technological and scientific challenges in this area. Beside the necessity to find innovative ways capable of selectively activating these renewable raw materials, modern catalysis has also to take into account resource management (*i.e.* carbon, water and metals) to ensure the sustainability of these processes. If during several years catalysis aimed at building new molecules, catalysis has now to integrate the notion of deconstruction (*e.g.* disassembling of (bio)polymers). Response to all these constraints is however not self-satisfied anymore and catalysis also has to provide chemicals with similar prices and even superior performances than chemicals derived from fossil reserves in order to favour their emergence on the market.

The progressive introduction of biomass, especially renewable polyols such as cellulose, hemicelluloses, monomeric carbohydrates and glycerol, in chemical processes is a clear illustration of this fundamental change that is now operating catalysis. In particular, due to the complex structure and high oxygen content of biomass, catalysis is now facing to new fundamental questions that are currently hampering the industrial emergence of bio-based derivatives such as (1) how to control the regioselectivity of reaction since the presence of numerous hydroxyl groups (and different linkages) can lead to the formation of many side products, (2) how to overcome the low accessibility of biopolymers to catalyst, a major bottleneck in the deconstruction of biomass, (3) how to activate biomass without degrading carbohydrates, (4) what is the effect of water, a contaminant of biomass, on catalyst activity, selectivity, stability, and (5) how to overcome the low solubility of biomass. The specialized literature (academic and industrial) and prospective reports from different institutions and governments estimate that more than 10 years of fundamental researches are still needed to achieve mature industrial processes based on the use of biomass.

Faced with the introduction of biomass in chemical processes, several strategies are under investigation. The first one consists in a direct transfer of actual catalytic technologies based on fossil carbon to renewable carbon. This approach is for instance efficient from vegetable oils and actually explains the large number of publications/patents dedicated to this raw material although it represents less than 5 % of the worldwide production of biomass [1]. Fatty derivatives indeed have structures close to those of hydrocarbons, thus allowing a possible rapid transfer of catalytic technology with minimal cost investments. On the other hand, glycerol, the main co-product of vegetable oils, can be used as a C3 chemical to enter the propene platform [2]. However, this approach can hardly be transposed to ligno-cellulosic biomass (95 % of the worldwide production of biomass!!) mainly because current catalytic systems are not adapted to these oxygenated raw materials that exhibit very complex structures. In this context, a second strategy is under investigation and consists in designing novel catalytic surfaces capable of

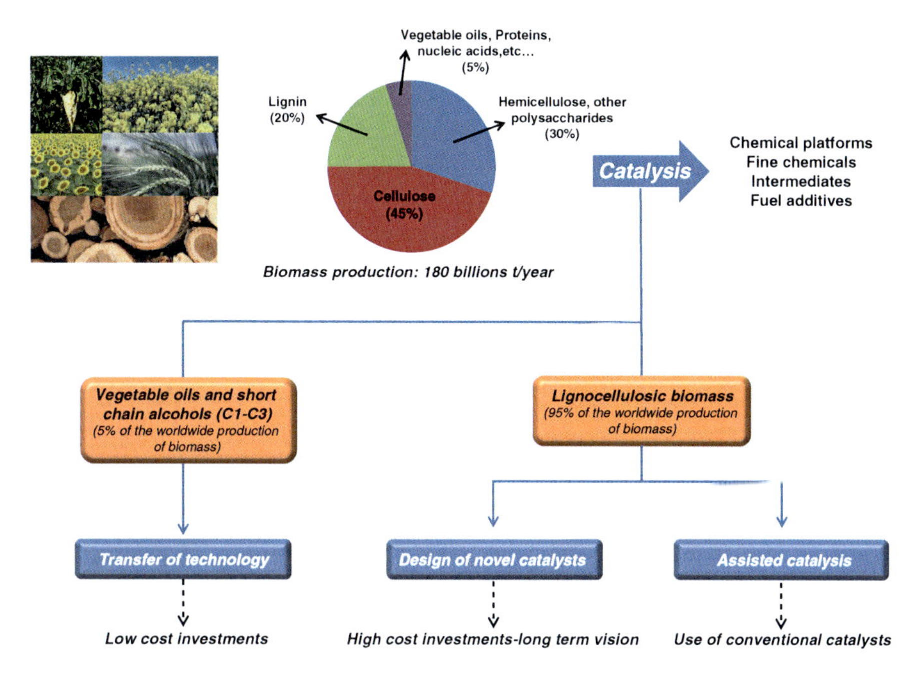

Scheme 3.1 Heterogeneous catalysis applied to biomass processing

selectively activating biopolymers [3]. This long term vision is necessary but clearly requires important cost investments. Indeed, as compared to homogeneous catalysis for which all elementary steps of the catalytic cycles are known at a molecular level, it is not the case for heterogeneous catalysis for which the design of novel solid catalysts is still empirical mostly due to the difficulty to have information on the catalytic sites at an atomistic level. Assisted catalysis is another concept which is now gaining more and more interest in the field of biomass processing. The idea consists in finding innovative ways capable of assisting or driving a conventional catalyst in the selective conversion biomass. For instance, physical methods such as ultrasound or ball-milling are already known techniques to help solid catalysts in the conversion of recalcitrant substrates such as cellulose or lignocellulose. From 2000, novel and innovative media such as bio-inspired ionic liquids and deep eutectic solvents have emerged in the current literature. These new media have processing advantages (large dissolution of polyols, insolubility with many organic solvents, tunable electrochemical window, tunable acidity, tolerance to water, etc....) that no other solvent can provide. Their abilities to deactivate water, to stabilize or destabilize reaction intermediates, to disrupt hydrogen bond networks now open efficient tools for assisting a catalyst in the selective deconstruction and conversion of structurally complex raw materials such as biopolymers to more value added chemicals (Scheme 3.1).

In this book chapter, we report the most recent advances made in the field of catalytic conversion of biomass assisted by choline-derived solvent. In particular, their ability to decrease the crystallinity of cellulose, to assist the deconstruction of biopolymers or to promote the conversion of carbohydrates to higher value added chemicals is discussed. Additionally, at the end of the manuscript, we will discuss the contribution of these neoteric solvents for the purification of bio-based chemicals such as biodiesel and furanic esters.

3.2 Decrease of the Cellulose Crystallinity in ChCl-Derived ILs

3.2.1 Dissolution of Cellulose in Non-derivatizing Media

Cellulose is a highly valuable biopolymer of glucose from which chemical platforms, intermediates, ethanol and fuel additives can be then produced [4, 5]. All these processes initially imply the catalytic deconstruction of cellulose to glucose. The high crystallinity of cellulose is a serious bottleneck which is at the origin of the low accessibility of cellulose to (bio)catalysts. Hence, in many cases, harsh conditions of pressure and temperature are required for the deconstruction of cellulose making the control of the reaction selectivity very difficult. To overcome this issue, cellulose is generally subjected to a pre-treatment process prior to catalytic deconstruction. This pre-treatment aims at favoring a better accessibility of the cellulosic backbone to catalyst by reducing its crystallinity or particle size or degree of polymerization for instance. In this context, much effort has been recently devoted to the search of innovative media capable of dissolving and thus disrupting the supramolecular organization of cellulose. Dissolution of cellulose in a non-derivatizing solvent is an interesting approach that allows a change of the cellulose structure from a highly crystalline to a low crystalline structure, a key parameter in the subsequent catalytic hydrolysis of cellulose to glucose. After the dissolution process, cellulose is generally recovered, by precipitation, upon addition of an anti solvent such as ethanol or water. Historically, mixtures of DMSO/LiCl or DMA/LiCl (among other combinations) and more recently *N*-methylmorpholine-*N*-oxide (Lyocell process) have been used for the dissolution/decrystallization of cellulose [6]. Although these systems ensure a drastic decrease in the crystallinity index of cellulose, their recycling is difficult and rather expensive. Recently, ionic liquids (ILs) have received considerable attentions because of their ability to dissolve and thus to decrease the crystallinity index of cellulose (Scheme 3.2).[1]

Dissolution of cellulose in ILs has been firstly demonstrated at the beginning of the twentieth century in particular using ethyl ammonium nitrate [8, 9]. Nowadays, the use of ionic liquids for the dissolution/decrystallization of cellulose is now witnessing a sort of renaissance with the recent emergence of room temperature

[1] As an example see pioneer work of [7].

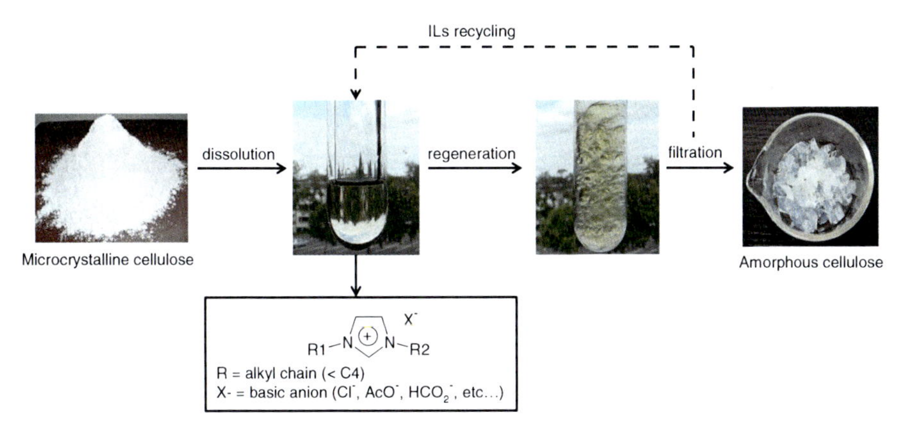

Scheme 3.2 Decrystallization of the cellulosic fraction of lignocellulosic biomass by dissolution in ILs

ILs. To date, plenty of works have been recently reported in this field of chemistry and this topic is too large to be summarized here. Complementary information to this section can be found in excellent recent reviews.[2]

Analysis of the specialized literature reveals that two parameters mainly govern the dissolution of cellulose in ILs (1) the ability of ILs to disrupt the extensive hydrogen-bond network of cellulose and (2) hydrophobic interactions. In ILs, dissolution rate of cellulose closely depends on the temperature, time of heating and molecular weight of ILs.[3]

Anion of the ILs plays an important role in the dissolution process by inducing polar-interaction with the hydroxyl groups of cellulose thus weakening the hydrogen bond network of cellulose. To date, chloride is one of the most efficient anion but its exact role is still subject to controversy in the current literature. Other anions such as acetate, formate or phosphate have been also proven to be effective. More generally, anion with a basic character seems to be more favorable for the dissolution of cellulose. The cation composing the ILs plays also a major role in the dissolution process. Due to intra- and intermolecular hydrogen bonding, cellulose is composed of flat ribbons with sides that differ markedly in their polarity. Hence, amphiphilic cations are generally required in order to ensure an efficient dissolution of cellulose. The size of the cation is also a parameter that is taken into account in few literatures and an optimal size should be found in order to favor the diffusion of ILs within the cellulose microfibrils. In this context, the imidazolium moiety closely meets all these requirements.

Although elucidation of the exact mechanism governing the dissolution of cellulose in ILs is still not really clear, combination of an amphiphilic cation with a basic anion seems to be a good compromise. It is more or less accepted that the

[2] As an example see pioneer work of Richard et al. [10].

[3] Lindman et al. [11] and references cited therein.

cation has the role to slide open the cellulose fibrils and to transport the anion within the cellulose backbone where it interacts with the hydrogen bond network. To date, 1-butyl-3-methyl-imidazolium chloride and 1-ethyl-3-methyl-imidazolium acetate are considered as the best ILs for the dissolution of cellulose. Despite the remarkable ability of these ILs to dissolve cellulose, their industrial emergence is unfortunately hampered by their high cost, reactivity, toxicity and high viscosity. Additionally, RTILs are highly hygroscopic and presence of water, even in a trace amount, has a detrimental effect on the dissolution of cellulose. Hence, room temperature ILs are nowadays only regarded as excellent models to understand the mechanism governing the dissolution/decrystallization of cellulose.

3.2.2 Regeneration of Cellulose

Temperature plays an important role on the dissolution rate of cellulose. Because a change of the cellulose structure from a highly crystalline to a low crystalline form is not thermodynamically favorable, dissolution process of cellulose generally occurs at relatively high temperature. Under these conditions, cellulose and, in some cases, the ILs, are partly degraded making the long term viability of these systems a serious limitation. In this context, assistance of microwave has been explored to accelerate the dissolution process and promising results have been reported [12]. Once dissolved, cellulose is regenerated by precipitation upon addition of an antisolvent such as ethanol, water or acetone. The regeneration of cellulose from ILs is a very important step and should be closely controlled. Indeed, owing to its amphiphilic nature, cellulose is known for its ability to encapsulate a wide range of organic substrates including ILs. The presence of residual ILs in regenerated cellulose is problematic not only because the anion may lead to a denaturation of enzymes or poisoning of acid sites during the subsequent (bio) catalytic depolymerization of cellulose but also because ILs are relatively expensive and the entire amount of ILs need to be recovered in order to design a viable process. Recent studies have shown that regeneration of cellulose from [BMIM]Cl at temperature around 60 °C or assistance of ultrasound allows to recover nearly 99 % of the ILs offering a suitable route to limit the contamination of cellulose.[4]

3.2.3 Towards Bio-inspired ILs

As smartly recently stated by M. Francisco et al. [14] *finding an eco-efficient solvent for the dissolution of cellulose and more largely lignocellulosic biomass is becoming the Achille's heel of renewable chemicals and biofuels processing.*

[4] As a recent selected example see [13].

In this context, few groups have attempted the design of bio-inspired ILs from choline chloride (ChCl) with the aim of dissolving (and thus decreasing the crystallinity index of cellulose) in a more sustainable way. ChCl is a very cheap (<2€/kg), biodegradable and non-toxic quaternary ammonium salt which can be either extracted from biomass or readily synthesized from fossil reserves (million metric tons per year) through a very high atom economy process. Its ionic structure makes of this organic salt a suitable candidate for the design of safer solvents that are particularly promising for biomass processing. In this context, two strategies are employed to produce media from ChCl (1) a chloride metathesis to produce the so-called bio-inspired ILs or (2) its combination with safe hydrogen bond donors such as urea, renewable carboxylic acids (*e.g.* oxalic, citric, succinic or aminoacids) or renewable polyols (*e.g.* glycerol, carbohydrates) to produce a deep eutectic solvent (DES). More information regarding DES is provided later in the chapter. To date the number of examples involving bio-renewable ChCl-derived solvents for the dissolution of cellulose and more largely lignocellulosic biomass is rather scarce mostly due to the novelty of these systems.

ChCl being a solid with a high melting point, one of the main strategies reported in the current literature consists in properly exchanging the chloride anion by acetate or an amino acid affording ILs that are liquid at a temperature below than 70 °C. All these so-called bio-inspired ILs were conveniently prepared by neutralization of the commercially available choline hydroxide derivative with the corresponding acid (yield >95 %). Main advantages of these ILs stem from their convenient synthesis, biodegradability, low toxicity and low price. To date, these bio-inspired ILs were essentially used in various catalytic reactions such as aldol [15] and Knoevenagel [16] reactions for instance but their use for biomass processing remained scarce.

In 2010, C. S. Pereira and his co-workers reported the efficient use of cholinium ethanoate and lactate for the dissolution of refined cork, an insoluble residue from the cork manufactures, composed of ~20 wt% of polysaccharides, ~30 wt% of poly (phenolics) ("lignin-like") and ~50 wt% suberine [17]. Ability of choline-derived ILs to dissolve refined cork has been investigated at 100 °C. The residue (not soluble) was then analyzed by ATR-FTIR to identify which polymer of refined cork has been extracted. Choline ethanoate was able to dissolve a larger quantity of refined cork than the reference 1,3-dialkylimidazolium ionic liquid, especially of the aromatic suberin component (Table 3.1).

Other anions such as butanoate, hexanoate, methylpropanoate and lactate were also tested. Among them, lactate was found the less efficient (extraction efficiency of 20.7 and 39.7 % for lactate and ethanoate, respectively). An increase of the chain length of the anion led to an improvement of the extraction efficiency and best results were obtained with the hexanoate anion for which the extraction efficiency reached 64.9 %. The ability of choline-derived ILs to dissolve refined cork follow the pK_a value of the conjugate acid of the anion *i.e.* an increase of the basicity led to an increase of the extraction efficiency. ATR-FTIR analyses revealed a drastic reduction in the aliphatic and aromatic bands of refined cork after extraction suggesting that tested choline-derived ILs mostly extract suberin.

Table 3.1 Dissolution of cork biopolymers in choline-derived ILs

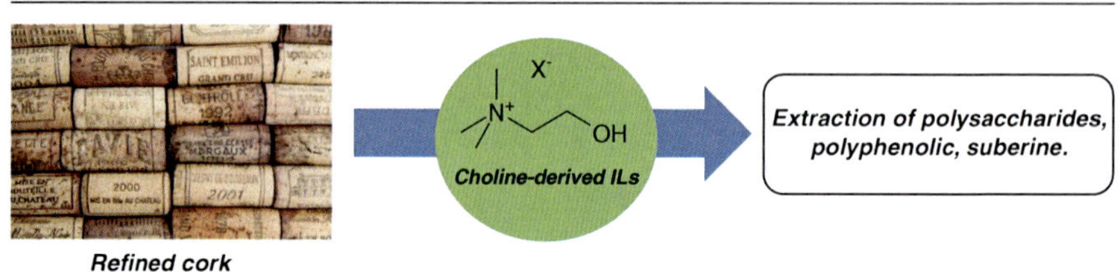

Entry	Anion (X^-)	Extraction efficiency (%)	pK_a
1	$CH_3CO_2^-$	39.7	4.76
2	$CH_3(CH_2)_2CO_2^-$	44.1	4.84
3	$(CH_3)_2CHCO_2^-$	55.1	4.83
4	$C_5H_{11}CO_2^-$	64.9	4.85
5	Lactate	20.7	3.86

In 2012, Zhang et al. investigated the dissolution of microcrystalline cellulose (PH AVICEL 106) in choline acetate [18]. Although choline acetate ([Ch]OAc) was not capable of dissolving microcrystalline cellulose in a large extent (solubility <0.2 and ~0.5 wt% after 5 min and 12 h, respectively), it is noteworthy that microcrystalline cellulose started to swell after immersion in [Ch]OAc for 12 h at 110 °C. After filtration, cellulose looked like a flocky precipitate rather than a powder, suggesting that this media does affect the supramolecular organization of cellulose. To ensure a complete dissolution of cellulose, effect of additives was investigated. Among various tested additives, it was shown that addition of 5–15 wt% of tributylmethyl ammonium chloride ([TBMA]Cl) in [Ch]OAc dramatically enhanced the dissolution of microcrystalline cellulose. In particular, in the presence of 15 wt% of ([TBMA] Cl), 6 wt% of cellulose was dissolved within only 10 min at 110 °C. For comparison, when the dissolution of microcrystalline cellulose was performed in neat 1-butyl-3-methylimidazolium chloride (commonly used for the dissolution of cellulose), only 4 wt% of MCC were dissolved after 8 h of reaction, further demonstrating the effectiveness of the [Ch]OAc/[TBMA]Cl mixture.

As commonly performed in the case of imidazolium-based ILs, cellulose was then regenerated by addition of ethanol and recovered. XRD analyses performed on regenerated cellulose revealed that the cellulose structure was successfully changed from a high to a low crystalline structure indicating that the [Ch]OAc/[TBMA]Cl mixture was capable of disrupting the hydrogen bond network of cellulose. During the dissolution process, the glucose units were not damaged (checked by infra-red) while viscosimetry analyses revealed that the degree of polymerization of cellulose remained unchanged before and after the dissolution process. After washing of regenerated cellulose, no nitrogen was detected by elemental analysis suggesting that the [Ch]OAc/[TBMA]Cl mixture can be conveniently separated from cellulose.

After filtration of regenerated cellulose and removal of ethanol, the [Ch]OAc/[TBMA]Cl mixture was recycled three times without appreciable decrease of its dissolution abilities. After the third run, the ability of the [Ch]OAc/[TBMA]Cl

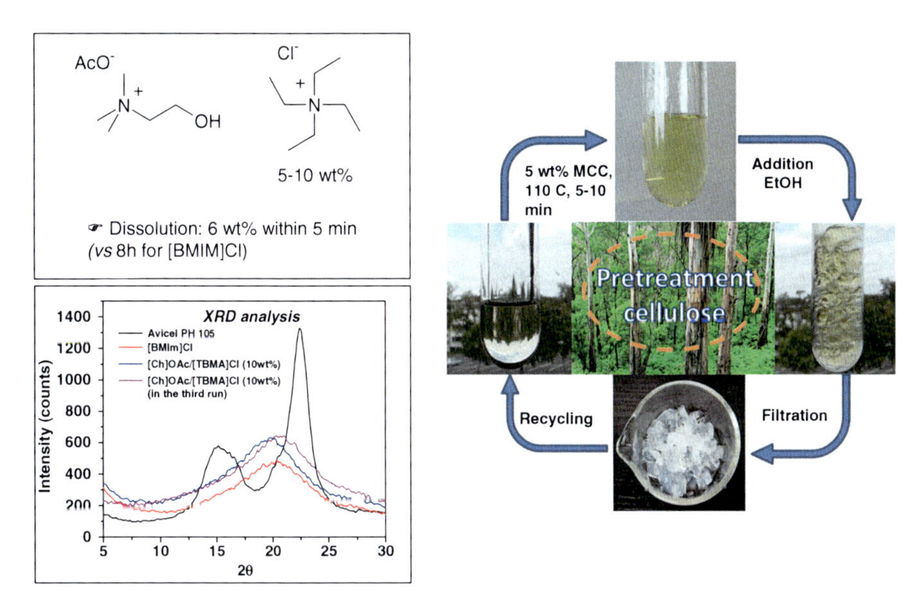

Scheme 3.3 Dissolution/regeneration of microcrystalline cellulose in Choline acetate

mixture to dissolve cellulose however started to drop mainly due to the accumulation of impurities in the solution that hampered the long term recycling of this system. A similar trend is observed with imidazolium-based ILs. The whole dissolution/regeneration process is summarized in Scheme 3.3.

In the same year, Zong and his co-workers reported the design of ILs by combining cholinium as a cation and amino acids as anions affording the so-called [Ch][AA] with AA = glycine, alanine, serine, proline, among many others. Liu et al. [19] Ability of these bio-inspired ILs to dissolve biopolymers such as lignin, xylan (a model of hemicellulose) and cellulose was investigated. All prepared [Ch][AA] were fully characterized in term of viscosity, stability (TGA analysis), melting point (DSC analysis), alkalinity and optical rotation in order to rationalize the efficiency of [Ch][AA] in the dissolution of biopolymers. In a first approximation, authors observed that [Ch][AA] with a high alkalinity and low viscosity are more favorable for the dissolution of lignin. In particular, [Ch][glycine] was found to be the best ILs with a dissolution of up to 220 mg of lignin per gram of ILs at 90 °C. Although alkalinity and viscosity of [Ch][AA] also exerted an influence on the dissolution of xylan, this biopolymer was found much less soluble than lignin. Cellulose, a recalcitrant biopolymer, was however insoluble.

Despite the low solubility of xylan and cellulose in [Ch][glycine], it can be used for the pre-treatment of rice straw. In particular, a pre-treatment of rice straw in [Ch][glycine] at 90 °C for 24 h prior to enzymatic hydrolysis, led to an enhancement of the glucose production. For instance, after pre-treatment of rice straw in [Ch][glycine], the concentration of glucose obtained after enzymatic hydrolysis was improved from 0.31 to 2.05 g L^{-1} which was attributed to the ability of the [Ch]

Table 3.2 Dissolution of biopolymers in [Ch][AA] ILs

Entry	ILs	T_g/T_d (°C)[a]	Viscosity (mPa/s)	pH (5 mM)	Lignin (mg/g)	Xylan (mg/g)	Cellulose (mg/g)
1	[Ch][Gly]	−61/150	121	10.3	220	76	<5
2	[Ch][Ala]	−56/159	163	10.2	180	77	<5
3	[Ch][Ser]	−55/182	402	9.8	170	70	<5
4	[Ch][Thr]	−39/172	454	9.8	160	85	<5
5	[Ch][Val]	−74/177	372	10.3	70	15	<5
6	[Ch][Leu]	−47/175	476	10.2	150	40	<5
7	[Ch][Ile]	−47/175	480	10.3	170	40	<5
8	[Ch][Met]	−61/178	330	10.1	150	75	<5
9	[Ch][Phe]	−60/160	520	9.7	140	65	<5
10	[Ch][Trp]	−12/174	5,640	10.2	90	10	<5
11	[Ch][Pro]	−44/163	500	10.7	170	75	<5
12	[Ch][Asp]	−22/202	2,060	6.8	< 10	< 1	<1
13	[Ch][Glu]	−18/202	2,308	6.7	26	< 1	<5
14	[Ch][Asn]	−14/187	1,903	9.5	16	5	<5
15	[Ch][Gln]	−40/203	2,589	8.0	50	5	<5
16	[Ch][Lys]	−48/165	460	10.4	140	65	<5
17	[Ch][Hys]	−40/171	980	10.0	140	35	<5
18	[Ch][Arg]	−10/163	1,002	11.3	110	25	<5

[a] T_g and T_d refer to glass transition and decomposition temperature, respectively

[glycine] to partly dissolve lignin, thus making more accessible the hemicellulosic and cellulosic fraction to enzymes. Optimization has been recently reported by same authors [20] (Table 3.2).

In 2012, Itoh and co-workers reported the use of what they have called "ionic liquids inspired by Nature" for the dissolution of cellulose [21]. Although this work does not deal with the use of choline, it opens key data for improving the ability of ChCl-derived ILs to dissolve cellulose. After examination of the protein sequences of several cellulase (enzymes responsible for the hydrolysis of cellulose), authors hypothesized that amino-acids might be suitable anions to disrupt the hydrogen bond network of cellulose and thus enable its dissolution. As a cation, authors have highlighted the superior performances of *N*,*N*-diethyl-*N*-(2-methoxyethyl)-*N*-methylammonium, a cation with a structure close to that of an etherified cholinium. In particular, after combination of this cation with amino-acids such as tryptophan,

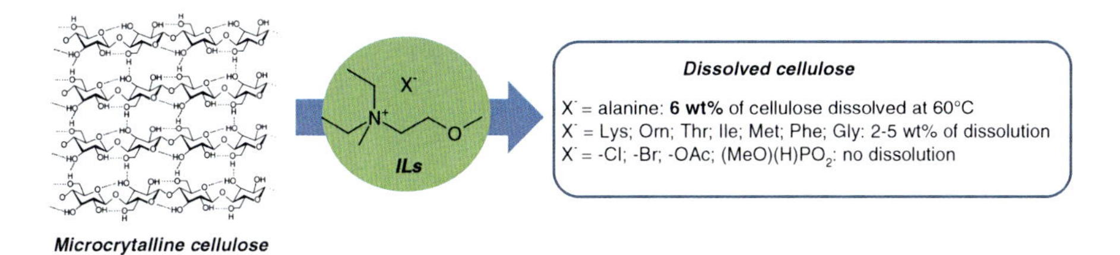

Scheme 3.4 Dissolution of microcrystalline cellulose in amino acid derived ILs

alanine, cysteine, lysine, among many others, cellulose was efficiently dissolved at 100 °C. Among tested amino-acids, alanine was found to be the best achieving up to 6 wt% of dissolution of cellulose at 60 °C. All other tested anions such as Cl^-, AcO^-, HO^- were found inefficient further supporting the pivotal role played by amino acid in the dissolution of cellulose. More importantly, these amino acid-derived ILs were tolerant to the presence of up to 2 wt% of water which is an important point considering that biomass always contained water. At higher temperature (150 °C), the dissolution rate of cellulose as well as the solubility of cellulose was greatly improved but a degradation of ILs was noticed at such temperature and authors recommended to not proceeding dissolution experiments at a temperature higher than 120 °C. Once dissolved, cellulose can be regenerated after addition of water as an anti-solvent. Precipitated cellulose was of the Type II supporting that such ILs are able to induce a change of the crystalline structure of cellulose. Impact of this change of crystallinity on the recalcitrance of cellulose to deconstruction was then evaluated in the enzymatic hydrolysis of cellulose. After regeneration, 88 % of cellulose was converted after 10 h of reaction at 50 °C and pH5 which has to be compared to the 40 % obtained without pre-treatment in ILs. In our views, this work is of prime importance within the scope of this chapter since it indirectly suggests that combination of amino acid with an etherified cholinium cation should provide competitive bio-inspired ILs for the dissolution of cellulose. Such aspect is the topic of current investigations in our group (Scheme 3.4).

3.2.4 Deep Eutectic Solvents

Recently, P. Abbott and co-workers have introduced the concept of Deep Eutectic Solvents (DES). A DES is a fluid generally composed of two or three cheap and safe components which are capable of associating each others, often through hydrogen bond interactions, to form a eutectic mixture. The resulting DES is characterized by a melting point lower than that of each individual component. Generally, DESs are characterized by a very large depression of freezing point and are liquid at temperatures lower than 150 °C [22]. Note that most of them are liquid between room temperature and 70 °C. In most cases, a DES is obtained by mixing a quaternary

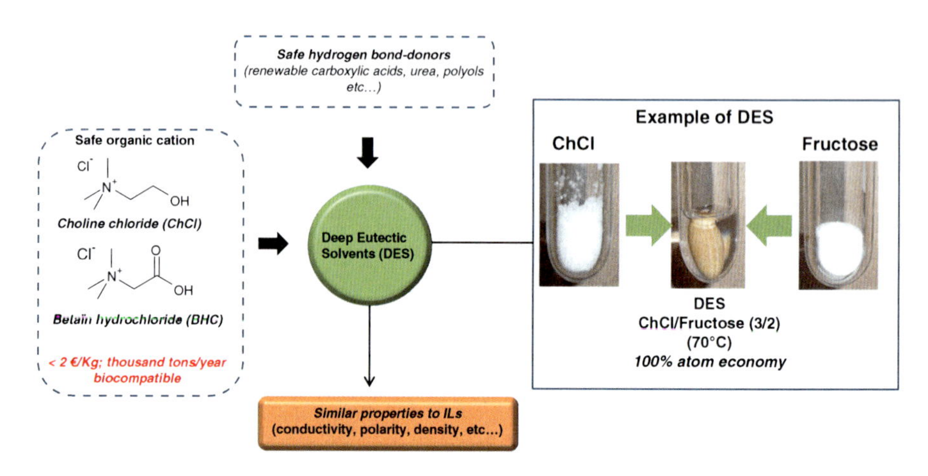

Scheme 3.5 Deep Eutectic Solvents

ammonium salt with metal salts or a hydrogen bond donor (HBD) that has the ability to complex the halide anion of the quaternary ammonium salt.

Owing to its low cost, biodegradabity and low toxicity, choline chloride (ChCl) and more recently glycine betaine (zwitterionic or protonic form) have been recently proposed as an organic salt to produce eutectic mixtures generally in combination with cheap and safe HBDs such as urea, renewable carboxylic acids (*e.g.* oxalic, citric, succinic or amino acids) or renewable polyols (*e.g.* glycerol, carbohydrates). As compared to the traditional ionic liquids (ILs), DESs derived from ChCl gather many advantages such as (1) low price, (2) 100 % atom-economy synthesis (no purification is required), and (3) most of them are biodegradable [23], biocompatible [24] and non-toxic [25] reinforcing the sustainability of these media. Physicochemical properties of DESs (density, viscosity, refractive index, conductivity, surface tension, etc.) are very close to those of common ILs. Thereby, they have the potential to advantageously replace ILs in many applications such as metal and oxides dissolution, catalysis, electrochemistry and material preparations. Additionally, through hydrogen bond interaction, DESs have the unique ability to stabilize and thus to lower the reactivity of water, opening the route to chemical transformations that are normally not feasible in hygroscopic solvents [26] (Scheme 3.5).

In 2007, Abbott and co-workers defined DESs using the general formula $R_1R_2R_3R_4N^+,X^-Y^-$ [27]:

Type I DES Y = MClx, M = Zn, Sn, Fe, Al, Ga
Type II DES Y = MClx.yH_2O, M = Cr, Co, Cu, Ni, Fe
Type III DES Y = R_5 Z with Z = –$CONH_2$, –COOH, –OH

Note that the same group also defined a fourth type of DES which is composed of metal chlorides (e.g. $ZnCl_2$) mixed with different HBDs such as urea, ethylene glycol, acetamide or hexanediol (type IV DES).

Similar to the case of ChCl-derived bio-inspired ILs, the use of DES for dissolution of cellulose has been scarcely reported mostly due to the novelty of these systems. Although based on the state of the art, protic groups such as –OH or –COOH are clearly not favorable for the dissolution of cellulose, their involvement in the formation of a DES drastically reduces their protic nature (the –OH group being involved in hydrogen bond interaction) thus offering a better chance to achieve the dissolution of cellulose.

In 2012, Georgia Tech Research Corporation has patented the dissolution of microcrystalline cellulose (AVICEL) in various DESs made of ChCl and betaïne monohydrate [28]. Neat DES made of ChCl and urea (or malonic acid or formamide) were not able to dissolve microcrystalline cellulose. When DESs were diluted with a basic solution (NaOH or NaOAc) together with a prolonged incubation time, a swelling of cellulose was however observed. By means of XRD analyses, a decrease of the crystallinity index of cellulose of 15–20 % was noticed suggesting that these systems can partly interact with the hydrogen bond network of cellulose. In agreement with previous results, regenerated cellulose was less recalcitrant to hydrolysis after pretreatment in basified ChCl-derived DES. Although neat DESs do not dissolve cellulose, one should however mention that their combination with a basic solution allowed avoiding the large amount of base traditionally required for the dissolution of cellulose. Next, authors highlighted the possible formation of DES from betaïne monohydrate and urea. As compared to ChCl, betaïne monohydrate is more attractive due to its lower cost and its direct availability from biomass (co-product of the sugar beet industry). DESs are made with more difficulty from betaïne than from ChCl and only urea was found eligible as a hydrogen bond donor in such case. Such betaïne derived DESs is however highly viscous. Following the same strategy than that used from ChCl authors found that a pretreatment of cellulose in the betaïne/urea DES led to a decrease of 15 % in the crystallinity index of cellulose. Although these systems have allowed the crystallinity index of cellulose to be slightly decreased, DESs are however not capable of dissolving cellulose presumably because DES components are already involved in hydrogen bond interaction making difficult their interaction with the hydrogen bond network of cellulose. Additionally, removal of DESs from regenerated cellulose is not an easy task and extensive washing are required.

In the same year, M. Francisco and co-workers investigated a series of 26 different DESs in the dissolution of cellulose, lignin and starch [14]. Since selected mixtures exhibited no melting point by differential scanning calorimetry (only glass transition), such mixtures were more considered as low transition temperature mixing (LTTM) rather than real DESs. All solubility measurements were determined by using the cloud point method within a range of temperature of 60 and 100 °C. This method consists in the progressive addition of a biopolymer in LTTMs up to the observation of a turbid solution. Among them, LTTMS made of lactic acid-ChCl were found particularly efficient for lignin dissolution. A clear solubility enhancement of lignin was even observed with an increase of the lactic acid content. Reversely, LTTMs made of malic acid were found more efficient for the dissolution of starch than for the dissolution of lignin. Among tested melts, malic acid-proline melt efficiently dissolved starch and dissolution ability can be

Table 3.3 Solubility of lignin, starch and cellulose in various LTTMs

			Biopolymer solubility (wt%)		
Entry	DES	T (°C)	Lignin	Starch	Cellulose
1	Lactic acid/proline (2/1)	60	7.56	0.00	0.00
2	Lactic acid/betaïne (2/1)	60	12.03	0.00	0.00
3	Lactic acid/ChCl (3/1)	60	4.55	0.00	0.00
4	Lactic acid/ChCl (2/1)	60	5.38	0.00	0.00
5	Lactic acid/ChCl (5/1)	60	7.77	0.00	0.00
6	Lactic acid/ChCl (10/1)	60	11.82	0.13	0.00
7	Lactic acid/Hystidine (9/1)	60	11.88	0.13	0.00
8	Lactic acid/Glycine (9/1)	60	8.77	0.00	0.00
9	Lactic acid/Alanine (9/1)		8.47	0.26	0.00
10	Malic acid/Alanine (1/1)	100	1.75	0.59	0.11
11	Malic acid/Betaïne (1/1)	100	0.00	0.81	0.00
12	Malic acid/ChCl (1/1)	100	3.40	7.10	0.00
13	Malic acid/Glycine (1/1)	100	1.46	7.65	0.14
14	Malic acid/Proline (1/1)	100	0.00	0.00	0.00
15	Malic acid/Proline (1/2)	100	6.09	0.32	0.24
16	Malic acid/Proline (1/3)	100	14.90	5.90	0.78
17	Malic acid/Hystidine (2/1)	85	0.00	0.00	0.00
18	Malic acid/Nicotinic acid (9/1)	85	0.00	0.00	0.00
19	Oxalic acid/Betaïne (1/1)[a]	60	0.66	0.00	0.00
20	Oxalic acid/Proline (1/1)[a]	60	1.25	0.00	0.00
21	Oxalic acid/ChCl (1/1)[a]	60	3.62	2.50	0.00
22	Oxalic acid/Glycine (3/1)[a]	85	0.28	0.00	0.00
23	Oxalic acid/Nicotinic acid (9/1)[a]	60	0.00	2.83	0.00
24	Oxalic acid/Hystidine (9/1)[a]	60	0.00	0.00	0.25
25	Oxalic acid/ChCl (1/1)[b]	60	0.00	0.15	0.00
26	Oxalic acid/Proline (1/1)[b]	60	0.00	0.15	0.00

[a]Dihydrate oxalic acid
[b]Anhydrous oxalic acid

improved by increasing of the proline ratio. Note that no clear rationalization was proposed and the search of LTTMs for the dissolution of lignin or starch still remains empirical.

In agreement with the above-described work patented by Georgia Tech Research Corporation, no significant dissolution of cellulose was observed in all tested LTTMs. Nevertheless, in the typical case of LTTMs derived from proline, turbid solution was observed with cellulose and no evidence of solid particle was further detected further supporting the superior ability of aminoacid for interacting with cellulose as described above.

Having all these results in hand, authors next checked the ability of LTTMs for the delignification of wheat straw. Using a Lactic acid-ChCl melt (2/1), 2 wt% of lignin was extracted after incubation overnight at 60 °C. Although no solubility data was provided, authors claimed that the solubility of wheat straw can be improved using a malic acid/proline melt (1/3) which is consistent with results presented in Table 3.3.

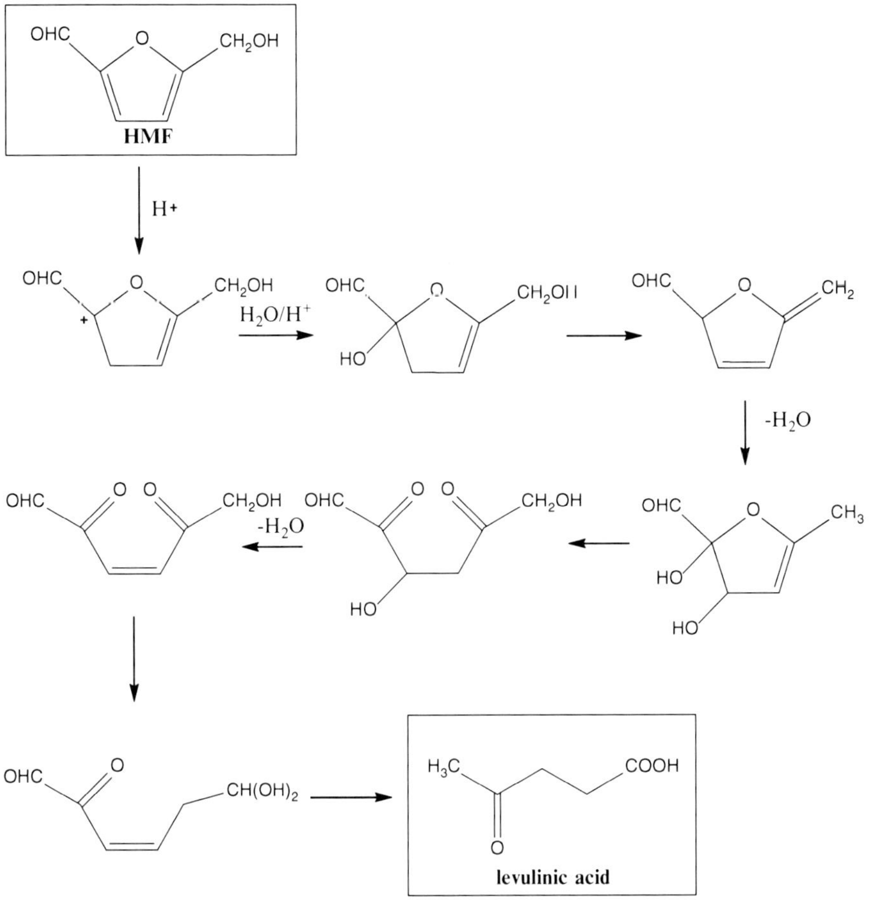

Scheme 3.6 Acid-catalyzed rehydration of HMF to levulinic and formic acids

3.3 Conversion of Carbohydrates to 5-Hydroxymethylfurfural (HMF)

The synthesis of HMF is nowadays one of the most investigated reactions from biomass. HMF is indeed considered as a chemical platforms from which new generations of biofuels (ex: dimethylfurane) and a wide range of intermediates, monomers and many other fine chemicals can be then produced [29, 30]. This old reaction is now witnessing a sort of renaissance due to the scarcity of oils. HMF is produced through a triple acid-catalyzed dehydration of hexoses. In this reaction the nature of the solvent plays a pivotal role by ensuring (i) the dissolution of carbohydrates, including biopolymers (ii) the dilution of released water, thus limiting side reactions such as the rehydration of HMF to levulinic and formic acids (Scheme 3.6) and (iii) determine the choice of the work-up procedure. Obviously, the solvent

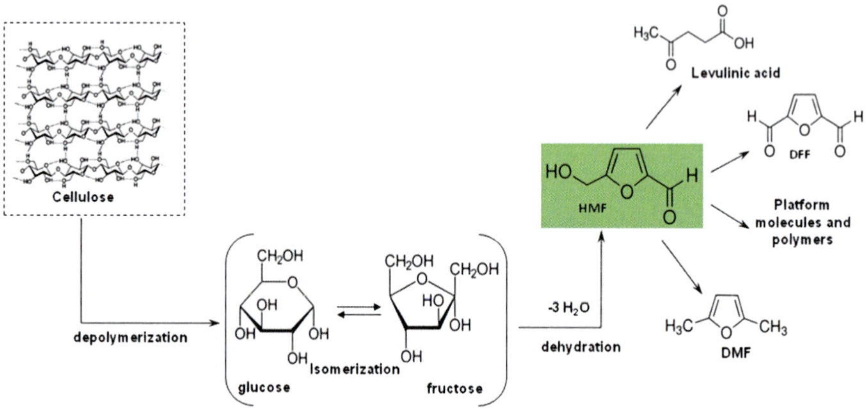

Scheme 3.7 Acid-catalyzed conversion of carbohydrates to HMF

should be also inert. In the current literature, several solvents have been proposed. Among them, dimethylsulfoxide (DMSO), water, mixtures of water and organic compounds, and ionic liquids have been particularly investigated.

In DMSO, yields of HMF greater than 85 % were obtained but the extraction of HMF from DMSO still remains rather complex and thus expensive. Additionally, under acidic conditions, DMSO may be decomposed leading to the formation of toxic products decreasing the sustainability/attractiveness of the process. In water, yields of HMF are rather low due to the side acid-catalyzed rehydration of HMF to levulinic and formic acids (Scheme 3.6). For this reason, water is often used in combination with organic solvents. This strategy affords higher yields but one should notice that reported yields are still lower than in the presence of DMSO.

In recent years, ILs have received considerable attention for this reaction. Yields of HMF obtained in ILs are comparable to those obtained in DMSO while HMF can be conveniently recovered by liquid-liquid phase extraction using for instance methylisobutylketone (MIBK), tetrahydrofurane or butanol. Note that the extraction of HMF from ILs can be carried out in a continuous mode, thus allowing side reactions involving HMF to be limited at the same time. It should be noted that same strategy was employed in water (biphasic system) in order to limit the rehydration of HMF. In such case, sodium chloride is generally used in order to facilitate the extraction of HMF (salt-out effect). Unfortunately, the price and toxicity of ILs together with problems linked to their long term recycling are currently hampering their utilization at an industrial scale. Clearly, the industrial emergence of HMF requires chemists to urgently develop innovative processes to produce HMF in a more sustainable way from biomass. In this context, the above-described DESs have received more and more attention for the conversion of hexoses to HMF. In the next section, we present the most recent innovative works reported in this field of chemistry. Although this topic has emerged very recently, we wish to demonstrate here that DESs have the potential to produce HMF in a more rational way than in conventional solvents (Scheme 3.7).

In 2008, B. Han and co-workers have reported that HMF can be produced in acidic ChCl-derived DESs [31]. In particular, in a melt composed of ChCl and citric acid (a cheap and renewable carboxylic acid), authors have shown that fructose can be converted to HMF with more than 76 % yield (at 80 °C, 1 h, ratio DES/fructose = 5). Other DESs made of renewably sourced carboxylic acids such as oxalic and malonic acids have been also successfully used. Owing to the low solubility of HMF in the ChCl/citric acid DES, the process can be performed in a biphasic system using ethyl acetate as an extraction solvent. Like in the case of ILs, continuous extraction of HMF not only facilitated the isolation of HMF but also allowed the selectivity to HMF to be increased. Indeed, under such conditions, HMF was obtained with a yield as high as 91 %. After removal of the ethyl acetate phase containing HMF, the ChCl/citric acid DES was recycled. A slight decrease of the HMF yield was observed upon recycling experiments mainly due to the accumulation of water in the reaction media that affects the selectivity of the process (presumable acid catalyzed rehydration of HMF to levulinic and formic acid). After drying of the used ChCl/citric acid eutectic mixture, the initial yield of HMF was recovered further demonstrating (i) the negative effect of water on the HMF selectivity and (ii) the stability of the ChCl/citric acid DES under reported conditions.

Next, the same group has transposed this work to the tandem hydrolysis/dehydration of inulin, a biopolymer of fructose, to HMF in the presence of acidic DESs such as ChCl/citric acid monohydrate or ChCl/oxalic acid dehydrate [35]. They have shown that, at 70 °C, the solubility of inulin was 150 and 28 $mg.g^{-1}$ in ChCl/oxalic acid and ChCl/citric acid, respectively. The tandem hydrolysis/dehydration of inulin (81.0 mg, 0.5 mmol fructose units) to HMF was performed at a temperature within a range of 50 and 90 °C. At 80 °C, the maximum yield of HMF was 55 % in both DESs within 2 h. Note that the conversion rate of fructose to HMF is higher in the ChCl/oxalic acid DES than in the ChCl/citric acid one due to a difference of acid strength. The authors have also demonstrated that addition of a suitable amount of water as soon as the beginning of the reaction has a positive effect on the formation of HMF. In particular, authors have shown that in ChCl/oxalic acid and ChCl/citric acid DESs water does not affect the selectivity of the reaction as soon as the water/fructose units molar ratio remained lower than 31. When the water content was increased, secondary reactions such as rehydration of HMF became the dominant reactions. The reaction temperature is of prime importance in this process and closely governs the selectivity of each elementary step. The hydrolysis of inulin to fructose was optimal at 50 °C while 80 °C was found to be necessary to dehydrate fructose to HMF. In this context, the one pot process was carried out in two steps involving (i) hydrolysis of inulin at 50 °C for 2 h and (ii) heating of the solution to 80 °C for another 2 h in order to dehydrate *in-situ* produced fructose to HMF. Using this procedure, the selectivity to HMF (65 %) was found to be higher. For the same reasons to those described above, the HMF production rate was found to be higher in the ChCl/oxalic acid DES than in the ChCl/citric acid one. To further increase the yield of HMF, authors have attempted the reaction in a biphasic system using acetyl acetate as an extraction solvent. In agreement with previous reports, the yield of HMF was increased from 57 to 64 %

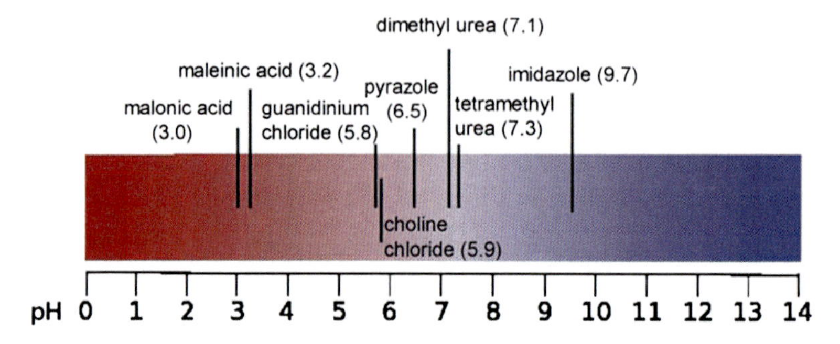

Scheme 3.8 pH values of different melts made of fructose as a hydrogen bond donor

and the extraction was found to be fully selective to HMF. After elimination of the ethyl acetate phase containing HMF, the possible recycling of the DES was investigated. Although water was unavoidably accumulated in the DES phase, at least six runs were performed without appreciable decrease of the HMF yield.

In 2009, König and co-workers investigated the production of HMF from a melt composed of D-fructose (40 wt%) and *N,N′*-dimethyl urea (DMU) heated at 110 °C for 2 h in the presence of $CrCl_2$ and $CrCl_3$ (10 mol% each). Unfortunately, even in biphasic conditions, (using ethyl acetate as an extraction solvent) low yields of HMF were obtained (6 and 2 % with $CrCl_2$ and $CrCl_3$, respectively). Other catalysts such as $FeCl_3$ and $AlCl_3$ gave similar results. Only Amberlyst 15, a sulfonated ion-exchange resin, provided a 27 % yield of HMF in this system. Authors have next investigated the production of HMF using different urea derivatives (urea, DMU and *N,N′*-tetramethyl urea (TMU)). Reactions were performed in the presence of $FeCl_3$ (10 mol%) and heated at 100 °C for 1 h. Conversely to urea and DMU, TMU-based melt gave HMF with an excellent yield of 89 %. These results tend to show that the presence of –NH– groups on urea and DMU are detrimental for the selectivity of the reaction. Although high yields have been obtained with TMU, the toxicity and problem of separation arising from the use of TMU represent two serious limitations. It is noteworthy that these results are in accordance with a previous work of B. Han and co-workers who have reported that basic DESs (ChCl/urea) in combination with Lewis acid ($ZnCl_2$, $CrCl_3$) or ChCl/metal chlorides-based DES were poorly efficient in the dehydration of fructose to HMF [32]. König et al. have next developed novel carbohydrate-derived melts (based on choline chloride as a hydrogen bond acceptor) with low melting point, low viscosity, low toxicity and high sugar content. The pH values of the different melts are presented in Scheme 3.8. Production of HMF in these melts was tested using a content of fructose of 40 wt% and Amberlyst 15 or $FeCl_3$ as catalysts. Reactions were carried out at 100 °C for 1 h. Among all tested melts, only ChCl/fructose DES has allowed the production of 25 and 40 % yield of HMF in the presence of A15 and $FeCl_3$, respectively. When no ChCl was employed, levulinic acid (20 %) was detected instead of HMF further supporting the key role played by ChCl.

Table 3.4 Acid-catalyzed dehydration of carbohydrates to HMF in various melts

Carbohydrates *(fructose, glucose, sucrose, inulin)* — melts / acid catalyst, $- 3\ H_2O$ → **HMF**

	Yield of HMF (%) from			
Catalysts	D-fructose[a]	Inulin[d]	Sucrose[c]	D-glucose[b]
Montmorillonite	49	7	35	7
Amberlyst 15	40	54	27	9
pTsOH	67	57	25	15
$Sc(OTf)_3$	55	44	28	9
$CrCl_2$	40	36	62	45
$CrCl_3$	60	46	43	31
$ZnCl_2$	8	3	6	6
$FeCl_3$	59	54	27	15

Amberlyst 15:50 mg
[a]400 mg fructose, 600 mg ChCl, 100 °C, 0.5 h
[b]400 mg glucose 600 mg ChCl, 110 °C, 0.5 h
[c]500 mg sucrose, 500 mg ChCl, 100 °C, 1 h
[d]500 mg inulin 500 mg ChCl, 90 °C, 1 h

Based on these results, authors have screened the activity/selectivity of several homogeneous, heterogeneous, Bronsted and Lewis acid catalysts in ChCl/carbohydrate melts. Results are reported in Table 3.4. Except in the case of $ZnCl_2$, it was found that, for all other tested catalysts (Table 3.4), HMF can be obtained with 40–60 % yields from melts composed of ChCl and fructose or inulin. Tested solid catalysts such as Montmorillonite and Amberlyst 15 were also capable of promoting the dehydration of fructose to HMF (40–49 % yield). Conversely to the case of fructose, Montmorillonite was found however poorly active from inulin which was ascribed to the low efficiency of Montmorillonite in the catalytic hydrolysis of inulin to fructose, a pre-requisite step prior formation of HMF.

Conversion of glucose to HMF is more challenging and requires first an isomerization step to fructose before dehydration to HMF. In accordance with a previous work of Zhao et al. performed in imidazolium-based ILs [33], chromium-based catalysts were found the most efficient catalysts in tested DES affording HMF with 31–45 % yield and 43–62 % yield from glucose and sucrose, respectively.

In 2012, K. De Oliveira Vigier et al. reported that betaïne hydrochloride (BHC), a co-product of the sugar beet industry, can be used as a renewably sourced Brönsted acid in combination with ChCl and water for the production of HMF from fructose and inulin [34] In a ternary mixture ChCl/BHC/water (10/0.5/2), HMF was produced with 63 % yield (at 130 °C from 40 wt% of fructose). As observed by B. Han and co-workers, when the reaction was performed in a biphasic system using methylisobutylketone (MIBK) as an extraction solvent, HMF was recovered with a purity higher than 95 % (isolated yield of 84 % from 10 wt% of fructose) further demonstrating that these systems similarly behave to the

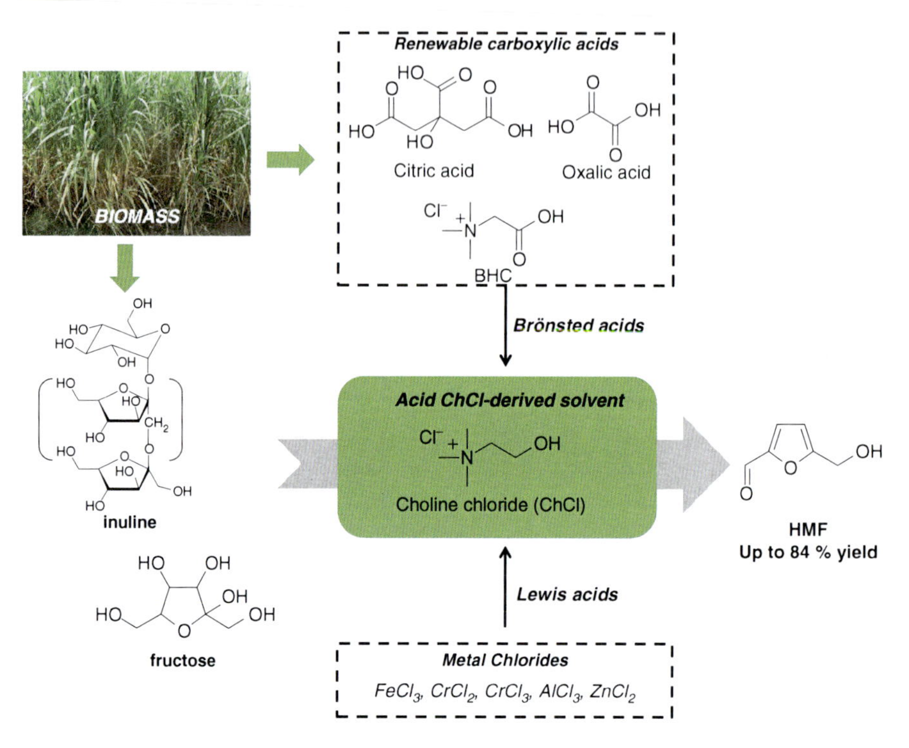

Scheme 3.9 Acid-catalyzed dehydration of inulin and fructose in ChCl-derived DES

traditional imidazolium-derived ILs (Scheme 3.9). After recovery of the MIBK phase containing HMF, the ChCl/BHC/water system was successfully recycled seven times. In same work, authors have shown that the dehydration of fructose to HMF can also conveniently take place in a BHC/glycerol (1:1) mixture, a DES exclusively made of renewably-sourced chemicals. Although 51 % yield of HMF was successfully obtained at 110 °C (from 10 wt% of fructose), extraction of HMF from the BHC/glycerol medium remained very difficult due to the very high solubility of HMF in this mixture.

Later, F. Liu et al. have shown that, under exposure to CO_2, fructose and inulin can be converted to HMF in a ChCl/fructose DES [35]. The pK_a of carbonic acid is low enough to catalyze the dehydration of fructose to HMF as previously shown by B. Han [36]. In this work, CO_2 initially reacted with water contained in fructose and in the DES resulting in the formation of carbonic acid in a sufficient amount to initiate the dehydration of fructose to HMF. After 90 min of reaction at 120 °C under 4 MPa of CO_2, a yield of 74 % of HMF was obtained from 20 wt% of fructose. Due to the high selectivity of the reaction and the insolubility of ChCl in MIBK, it was also possible to selectively extract HMF which was recovered with a purity of 98 %. Remarkably, the present system was tolerant to very high loading of fructose whereas most of reported solvents suffer from a low selectivity to HMF

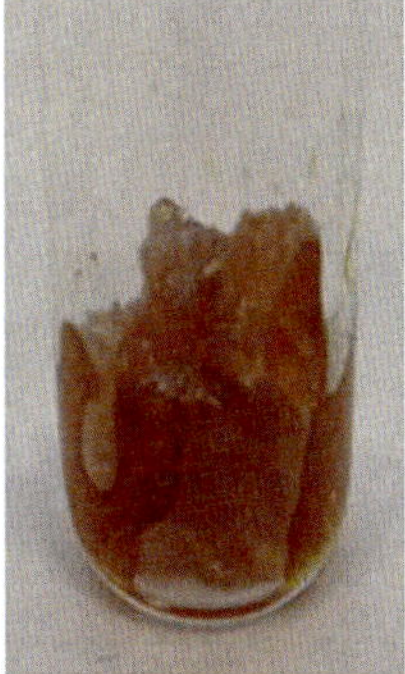

Scheme 3.10 Pictures of various ChCl/HMF mixtures at room temperature

when the fructose loading was higher than 20 wt%. For instance, fructose with a loading of 100 wt% was successfully dehydrated to HMF in a ChCl/CO_2 system without appreciable decrease of the HMF yield (66 %). The authors have ascribed the tolerance of such system to high loading of fructose to strong interaction between produced HMF and ChCl resulting in the stabilization of HMF in the reaction media. As shown in Scheme 3.10, when neat HMF and ChCl were mixed together a melt was readily obtained at a fructose content higher than 60 wt%. Such melt might be responsible for the surprising stability of HMF in such system when using high loading of fructose. It is indeed known that when a chemical is engaged in the formation of a DES its reactivity is drastically reduced.

3.4 Biodiesel from Soybean Oil

3.4.1 Synthesis of Fatty Acid Methyl Esters in the Presence of DES

Biodiesel is a renewable biofuel made from oils or fats that can be used directly in the diesel engine [37]. Biodiesel is biodegradable, non-toxic and generates less pollutant emissions than "conventional" diesel [38] There are four main routes to produce biodiesel, direct use and blending of raw oils, micro-emulsions, thermal cracking and transesterification. The transesterification reaction is the most common method to synthesize biodiesel. Much effort has been paid to decrease the total cost of biodiesel since it is significantly more expensive than fossil-derived diesel [39]. Several studies have focused on the choice of the raw materials since it is the main cost contributing factor [40]. Thus waste cooking oils and industrial oils such as sludge palm oil (SPO) and acidic crude palm oil (ACPO) were used in the biodiesel production. During methanolysis of oils, glycerol is released. Due to high difference of polarity with biodiesel, released glycerol and excess of methanol can be removed by phase decantation. However, complete removal of glycerol from biodiesel is not an easy task and postpurification processes are necessary in order to reach the required ASTM specifications.

ILs can also be used for the production of biodiesel since it can act both as catalyst and solvent. However, their cost and their complex synthesis are not competitive. In this context, few groups have attempted the use of DES for the removal of glycerol from biodiesel at the end of the methanolysis process.

Hayyan et al. have shown that DES can be used as a catalyst for esterification of free fatty acids (FFA) contained in vegetable oils. Hayyan et al. [40] They have used phosphonium-based DES (P-DES, made of *p*-toluenesulfonic acid monohydrate and alkyltriphenylphosphonium) in the pre-treatment of low grade oils. The authors have studied two catalyzed reactions (esterification and transesterification) to produce biodiesel from Low grade crude palm oil (LGCPO), an agro-industrial raw material generated from oil palm mills. LGCPO was only considered for biodiesel production by very few studies. The esterification was performed in the presence of P-DES catalyst and methanol. Pre-treatment of LGCPO *via* esterification is necessary for conversion of high FFA to fatty acid methyl ester (FAME) since FFA causes a poisoning of basic sites used in biodiesel production. After treatment with the DES, only 0.88 % of FFA remained in LGPCO. Additionally, the DES can be recycled three times which represent a considerable advantage as compared to *p*-toluenesulfonic acid traditionally used in such case. The authors have also performed the transesterification of pre-treated LGPCO to produce FAME in the presence of P-DES. At the end of the reaction, it was shown that the P-DES was completely removed from the biodiesel since no P and K were detected.

Zhao et al. [41] have studied the enzymatic preparation of biodiesel from soybean oil using a ChCl/Glycerol DES (1:2 molar ratio). The transesterification reaction was performed in a mixture of DES and methanol. Different enzymes were tested. Reaction was heated at 50 °C in an oil bath. Authors have demonstrated that the highest triglycerides conversion was 88 %. This high conversion rate was obtained in the presence of Novozym 435/mL, 0.2 % (v/v) of water; 50 °C and 24 h with a volumic ratio of 7:3 of DES/methanol. DES is biocompatible with lipase. Moreover some authors have shown that DES can be used to reduce the purification cost of biodiesel through transesterification. This work has demonstrated that this DES is biocompatible with enzymes, which is of interest for the valorization of biomass.

3.4.2 Extraction with DES

The possibility of using DES to extract glycerol was successfully demonstrated on palm oil-derived biodiesel using KOH as a basic catalyst. Authors have used a DES composed of ChCl and glycerine [42] as a solvent to extract residual glycerine contained in biodiesel. They have shown that the best DES/biodiesel ratio is inversely proportional with the percentage of extracted glycerine. The DES composition is also of prime importance. DESs with low content of glycerol are generally preferred to extract residual glycerine from biodiesel. Generally, the composition of starting ChCl/glycerol eutectic mixture is adjusted in order to have, after extraction of the residual glycerol from biodiesel, an ChCl/glycerine close to the ideal composition of the DES. Best separation was achieved using a DES:biodiesel and a ChCl/glycerine DES molar composition of 1:1. More importantly, at the end of the reaction, ChCl can be recovered by precipitation and reused in combination with glycerol. Shabaz et al. [43] have studied the removal of glycerol from palm oil-derived biodiesel using phosphonium-based salt with different hydrogen bond donors (HBD). Novel DESs based on methyltriphenylphosphonium bromide as salts and glycerin, ethylene glycol, and triethyleneglycol as hydrogen bond donor were prepared. Glycerol based DESs were not highly efficient to remove residual glycerine contained in biodiesel. Only DESs composed of a 1:2 ChCl/glycerol molar ratio while respecting a DES:biodiesel molar ratio of 2:1, 2.5:1, and 3:1 were found to be efficient. However, DESs made of ethylene glycol or triethylene glycol were found to be more efficient in removing residual glycerol from biodiesel. The optimum DES/biodiesel molar ratios using ethylene glycol or triethylene glycol were 0.75:1.

The authors have also demonstrated that the residual catalyst KOH used in the transesterification of oils can be removed from the reaction media using DES based on choline chloride or methyltriphenylphosphonium bromide (MTPB) salts [44]. In such case, glycerol, ethylene glycol 2,2,2-trifluoroacetamide and triethylene glycol were used as hydrogen bond donors. An increase of the DES/biodiesel and ChCl/HBD led to a higher KOH extraction efficiency. For instance, the ChCl/glycerol and MTPB/glycerol DESs allowed removal of 98.5 and 94.6 % respectively of KOH from palm oil-based biodiesel.

Very recently, Pablo Dominguez de Maria and coworkers have studied the (trans)esterification of HMF with different acyl donors (ethyl actetate, ethyl hexanoate, dimethyl carbonate, soybean oil, propionic acid, hexanoic acid, lauric acid) in the presence of a biocatalyst [45]. Under solvent free conditions, the yields of HMF esters were higher than 80 % after 24 h of reaction at 40 °C. Although no solvent was used during the (trans) esterification reaction, the selective separation of the unreacted HMF from HMF esters is necessary since the reaction was not complete. In this context, ChCl-based DESs were used for the selective extraction of HMF from HMF esters. Following this approach, more than 90 % of HMF esters, along with a purity higher than 99 %, were recovered after selective extraction of residual HMF by the DES. Investigated DESs were composed of ChCl and either glycerol or xylitol or urea. Regardless of the nature of the DES, the separation was always very selective and the optimal DES/reaction mixture volume ratio was found to be 1. It is noteworthy that this study confirms the previous work of the same authors where alcohol-esters mixtures were efficiently separated using DES [46] (Scheme 3.11).

Scheme 3.11 Separation of HMF and HMF esters using DES as an extraction solvent

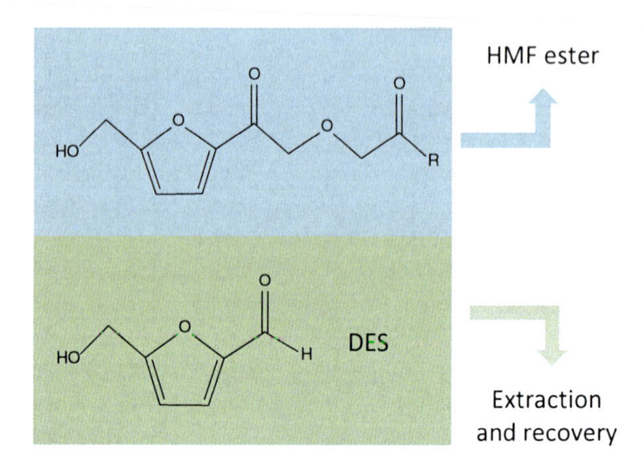

3.5 Conclusion

From 2000, new generation of solvents so-called bio-inspired ILs and DES derived from ChCl and glycine betaïne have emerged as promising candidates for biomass processing. More than a sustainable alternative to the traditional imidazolium-derived ILs (low price, low ecological footprint), these neoteric solvents have clearly processing advantages that no other solvent can provide in the field of biomass. In particular, their tunable viscosity, their ability to dissolve carbohydrates and related biopolymers, their ability to chemically stabilize polar molecules and their immiscibility with commonly used low boiling point solvents has open the route to the design of eco-efficient processes.

In the field of biopolymer dissolution, ILs derived from ChCl are quite efficient especially when combined with a basic anion derived from amino-acid. Although these systems can dissolve various biopolymers such as lignin, starch or suberine, their ability to dissolve cellulose is unfortunately more problematic mainly due to the presence of –OH group on the cholinium cation which is clearly not favorable for the dissolution of cellulose. Addition of additives drastically improves the ability of ChCl-derived ILs to dissolve cellulose but at the expense of the sustainability of the process. One should comment that recent reported works on cations exhibiting close structure to ChCl has shown the way how to design more efficient system from ChCl. In particular, the etherification of the –OH group of choline should be an attractive way. The direct etherification of cholinium cation with short chain alcohols is however quite difficult to be performed under competitive route and innovation in this direction is required in order to widen the scope and use of these systems.

Like ChCl-derived ILs, DESs are capable of dissolving various biopolymers except cellulose presumably because their formation results in the auto-association of two components through hydrogen bond interaction, thus preventing an efficient interaction of these systems with the hydrogen bond network of cellulose. DESs are however much more efficient in the conversion of monomeric carbohydrates such as fructose or glucose. In particular, the ability of ChCl to produce DES with monomeric carbohydrates or low molecular weight biopolymers such as inulin have allowed the production of HMF in a more competitive way than using conventional solvents. Additionally, the ability of ChCl to stabilize hydrogen bond donor such as HMF provides catalytic processes that are tolerant to high loading of fructose, a main drawback encountered with other solvents.

We are fully convinced that ChCl or glycine betaine-derived ILs and DES do have the potential to open new horizons in the field of catalysis applied to biomass. Although promising results have been reported, this approach is not mature yet for use on a large scale and few issues need to be overcome such as the relative instability of ChCl at temperature higher than 120 °C, in some applications their viscosity and, as mentioned above, the necessity to find sustainable routes for the chemical functionalization of the cholinium cation with the aim of widen their use.

References

1. Biermann U, Bornscheuer U, Meier MAR, Metzger JO, Schfer HF. Oils and fats as renewable raw materials in chemistry. Angew Chem Int Ed. 2011;50(17):3854–71.
2. Jérôme F, Barrault J. Selective conversion of glycerol into functional monomers via catalytic processes. In: Mathers RT, Meir MAR, editors. Green polymerization methods. Weinheim: Wiley; 2009.
3. Corma A, Iborra S, Velty A. Chemical routes for the transformation of biomass into chemicals. Chem Rev. 2007;107(6):2411–502.
4. Van de Vyver S, Peng L, Geboers J, Schepers H, De Clippel F, Gommes CJ, Goderis B, Jacobs PA, Sels BS. Sulfonated silica/carbon nanocomposites as novel catalysts for hydrolysis of cellulose to glucose. Green Chem. 2010;12:1560–3.
5. Serrano-Ruiz JC, Dumesic JA. Catalytic routes for the conversion of biomass into liquid hydrocarbon transportation fuels. Energy Environ Sci. 2011;4:83–99.
6. Pinkert A, Marsh KN, Pang S. Reflections on the solubility of cellulose. Ind Eng Chem Res. 2010;49:11121–30.
7. Richard Swatloski P, Spear SK, Holbrey JD, Rogers RD. Dissolution of cellulose with ionic liquids. J Am Chem Soc. 2002;124:4974–5.
8. Walden P. Über die Molekulargrösse und elektrische Leitfähigkeit einiger gesehmolzenen Salze. Bull Acad Impér Sci St Petersburg. 1914;1800:405–22.
9. Graenacher C. Cellulose dissolution. US Patent 1943176; 1934.
10. Tadesse H, Luque R. Advances on biomass pretreatment using ionic liquids: an overview. Energy Environ Sci. 2011;4:3913–29.
11. Lindman B, Karlström G, Stigsson L. On the mechanism of dissolution of cellulose. J Mol Liq. 2010;156:76–81.
12. Mikkola J, Kirilin A, Tuuf J, Pranovich A, Holmbom B, Kustov LM, Murzin DY, Salmi T. Ultrasound enhancement of cellulose processing in ionic liquids: from dissolution towards functionalization. Green Chem. 2007;9:1229–37.
13. Gupta KM, Hu Z, Jiang J. Molecular insight into cellulose regeneration from a cellulose/ionic liquid mixture: effects of water concentration and temperature. RSC Adv. 2013;3:4425–33.
14. Francisco M, Van Den Bruinhorst A, Kroon MC. New natural and renewable low transition temperature mixtures (LTTMs): screening as solvents for lignocellulosic biomass processing. Green Chem. 2012;14:2153–7.

15. Hu S, Jiang T, Zhang Z, Zhu A, Han B, Song J, Xie Y, Li W. Functional ionic liquid from biorenewable materials: synthesis and application as a catalyst in direct aldol reaction. Tetrahedron Lett. 2007;48:5613–17.
16. Moriel P, Garcia-Suarz EJ, Martinez M, Garcia AB, Montes-Moran MA, Calvino-Casilda V, Banares MA. Synthesis, characterization and catalytic activity of ionic liquids based on bioresources. Tetrahedron Lett. 2010;51:4877.
17. Garcia H, Ferreira R, Petkovic M, Ferguson JL, Leitao MC, Nimal Gunaratne HQ, Seddon KR, Nebelo LPN, Pereira CS. Dissolution of cork biopolymers in biocompatible ionic liquids. Green Chem. 2010;12:367–9.
18. Zhang Q, Benoit M, De Oliveira Vigier K, Barrault J, Jérôme F. Green and inexpensive choline-derived solvents for cellulose decrystallization. Chem Eur J. 2012;18(4):1043–6.
19. Liu QP, Hou X-D, Li N, Zong M-H. Ionic liquids from renewable biomaterials: synthesis, characterization and application in the pre-treatment of biomass. Green Chem. 2012;14:304–7.
20. Hou X-D, Smith TJ, Li N, Zong MH. Novel renewable ionic liquids as highly effective solvents for pretreatment of rice straw biomass by selective removal of lignin. Biotechnol Bioeng. 2012;109(10):2484–93.
21. Ohira K, Abe Y, Kawatsura M, Suzuki K, Mizuno M, Amano Y, Itoh T. Design of cellulose dissolving ionic liquids inspired by nature. ChemSusChem. 2012;5:388–91.
22. Zhang Q, De Oliveira Vigier K, Royer S, Jérôme F. Deep eutectic solvents: syntheses, properties and applications. Chem Soc Rev. 2012;41:7108–46.
23. Yu Y, Lu X, Zhou Q, Dong K, Yao H, Zhang S. Biodegradable naphthenic acid ionic liquids: synthesis, characterization, and quantitative structure–biodegradation relationship. Chem Eur J. 2008;14:11174–82.
24. Weaver KD, Kim HJ, Sun J, MacFarlane DR, Elliott GD. Cyto-toxicity and biocompatibility of a family of choline phosphate ionic liquids designed for pharmaceutical applications. Green Chem. 2010;12:507–13.
25. Ilgen F, Ott D, Kralish D, Reil C, Palmberger A, König B. Conversion of carbohydrates into 5-hydroxymethylfurfural in highly concentrated low melting mixtures. Green Chem. 2009;11:1948–54.
26. Cammarata L, Kazarian SG, Salter PA, Welton T. Molecular states of water in room temperature ionic liquids. Phys Chem Chem Phys. 2001;3:5192–200.
27. Abbott AP, Barron JC, Ryder KS, Wilson D. Eutectic-based ionic liquids with metal-containing anions and cations. Chem Eur J. 2007;13:6495–501.
28. Hertel MR, Bommarius AS, Realff MJ, Kang Y. Deep eutectic solvent systems and methods. WO 2012145522; 2012.
29. Lichtenthaler FW. Unsaturated *O*- and *N*-heterocycles from carbohydrate feedstocks. Acc Chem Res. 2002;35:728–37.
30. Moreau C, Belgacem MN, Gandini A. Recent catalytic advances in the chemistry of substituted furans from carbohydrates and in the ensuing polymers. Top Catal. 2004;27:11–30.
31. Hu S, Zhang Z, Zhou Y, Han B, Fan H, Li W, Song J, Xie Y. Conversion of fructose to 5-hydroxymethylfurfural using ionic liquids prepared from renewable materials. Green Chem. 2008;10:1280–3.
32. Hu S, Zhang Z, Zhou Y, Song J, Fan H, Han B. Direct conversion of inulin to 5-hydroxymethylfurfural in biorenewable ionic liquids. Green Chem. 2009;11:873–7.
33. Zhao H, Holladay JE, Brown H, Zhang ZC. Metal chlorides in ionic liquid solvents convert sugars to 5-hydroxymethylfurfural. Science. 2007;316:1597–600.
34. De Oliveira Vigier K, Benguerba A, Barrault J, Jérôme F. Conversion of fructose and inulin to 5-hydroxymethylfurfural in sustainable betaine hydrochloride-based media. Green Chem. 2012;14:285–9.
35. Liu F, Barrault J, De Oliveira Vigier K, Jérôme F. Dehydration of highly concentrated solution of fructose to 5-hydroxymethylfurfural in cheap and sustainable choline chloride/CO_2 system. ChemSusChem. 2012;5:1223–6.

36. Li X, Hou M, Han B, Wang X, Zou L. Solubility of CO_2 in a choline chloride + urea eutectic mixture. J Chem Eng Data. 2008;53:548–50.
37. (a) Ma F, Hanna MA. Biodiesel production: a review. Bioresour Technol. 1999;70:1–15; (b) Canacki M. The potential of restaurant waste lipids as biodiesel feedstocks. Bioresour Technol. 2007;98: 183–90.
38. (a) Wang WG, Luons DW, Clark NN, Gautan M, Norton PM. Emissions from nine heavy trucks fueled by diesel and biodiesel blend without engine modification. Environ Sci Technol. 2000;34: 933–39; (b) Oh PP, Lau HLN, Chen J, Chong MF, Choo YM. A review on conventional technologies and emerging process intensification (PI) methods for biodiesel production. Renew Sustain Energy Rev. 2012;16:5131–45.
39. Canacki M, Gerpen JV. Biodiesel production from oils and fats with high free fatty acids. Trans ASAE. 2001;44:1429–36.
40. Hayyan A, Hashim MA, Mjalli FS, Hayyan M, AlNashef IM. A novel phosphonium-based deep eutectic catalyst for biodiesel production from industrial low grade crude palm oil. Chem Eng Sci. 2013;92:81–8.
41. Zhao H, Zhang C, Crittle TD. Choline-based deep eutectic solvents for enzymatic preparation of biodiesel from soybean oil. J Mol Catal. 2013;85–86:243–7.
42. Hayyan M, Mjalli FS, Hashim MA, Alnashef IM. A novel technique for separating glycerine from palm oil-based biodiesel using ionic liquids. Fuel Process Technol. 2010;91:116–20.
43. Shahbaz K, Mjalli FS, Hashim MA, AlNashef IM. Using deep eutectic solvents based on methyl triphenyl phosphunium bromide for the removal of glycerol from palm-oil-based biodiesel. Energy Fuels. 2011;25:2671–8.
44. Shahbaz K, Mjalli FS, Hashim MA, AlNashef IM. Eutectic solvents for the removal of residual palm oil-based biodiesel catalyst separation and purification technology. Sep Purif Technol. 2011;81:216–22.
45. Krystof M, Perez-sanchez M, Dominguez de Maria P. Lipase-catalyzed (trans)esterification of 5-hydroxymethylfurfural and separation from HMF esters using deep-eutectic solvents. ChemSusChem. 2013. doi:10.1002/cssc.201200931.
46. Maugeri Z, Leitner W, Dominguez de Maria P. Practical separation of alcohol–ester mixtures using deep-eutectic-solvents. Tetrahedron Lett. 2012;53:6968–71.

Part II
Dissolution and Derivation of Cellulose and Fractionation of Lignocellulosic Materials with Ionic Liquids

Chapter 4
Design of Ionic Liquids for Cellulose Dissolution

Toshiyuki Itoh

Abstract Cellulose consists of linear glucose polymer chains that form a very tight hydrogen-bonded supramolecular structure making it highly resistant to enzymatic degradation. The ionic liquid, 1-butyl-3-methylimidazolium chloride ([C_4mim]Cl), has been found to dissolve cellulose and the regenerated cellulose from the IL solution is less crystalline. To design ionic liquids that dissolve cellulose, Kamlet-Abboud-Taft β-values can be used as a solvent indicator. Amino acid anions have strong interactions between hydroxyl groups in the cellulose molecule: N, N-diethyl,N-methyl,N-(2-methoxy)ethylammonium alanate ([$N_{221(ME)}$][Ala]) thus they are studied in this chapter for cellulose dissolution. Addition of an anti-solvent like water or ethanol to the cellulose/IL solution caused precipitation of cellulose dissolved and the structure of the regenerated cellulose to change to a disordered form. Crystal form of the regenerated cellulose depends on the dissolution solvent; the disordered chain region seems to increase in the order of [$N_{221(ME)}$][Ala] < [C_2mim][OAc] < [C_2mim][$(EtO)_2PO_2$] < [C_2mim]Cl. On the other hand, the order of degree of polymerization of the cellulose is [$N_{221(ME)}$][Ala] > [C_2mim][OAc] > [C_2mim][$(EtO)_2PO_2$] > [C_2mim]Cl. Treatment with [$N_{221(ME)}$][Ala] is therefore much more suitable to use in preparing regenerated cellulose fibers than other commonly used ionic liquids.

Keywords Cellulose dissolution • Amino acid ionic liquids • Regenerated cellulose • A mixed solvent

T. Itoh (✉)
Department of Chemistry and Biotechnology, Graduate School of Engineering, Tottori University, 4-101 Koyama Minami, Tottori, Japan
e-mail: titoh@chem.tottori-u.ac.jp

Z. Fang et al. (eds.), *Production of Biofuels and Chemicals with Ionic Liquids*, Biofuels and Biorefineries 1, DOI 10.1007/978-94-007-7711-8_4,

4.1 Introduction

Cellulose is an important renewable resource for production of biocomposites and biofuel alcohols. However, since it consists of linear glucose polymer chains that form a very tight hydrogen-bonded supramolecular structure, cellulose resists enzymatic degradation. There has been growing interest in the development of a means of modifying cellulose structure to an easily digestible form by biodegradation [1]. Multiple hydrogen bonding among cellulose molecules results in the formation of highly ordered crystalline regions [2]. Therefore, cellulose does not dissolve in water and common organic solvents at ambient conditions. The challenge for dissolving cellulose has a long history [3]. The first attempt was reported early in the 1920s and some mixed solvent systems for cellulose dissolution were developed [1, 3]: sodium hydroxide/carbon disulfide (CS_2) [4] and sodium hydroxide/urea [5] are well known as commercial cellulose derivatizing solvents. Rosenau et al. [6] reported using N-methylmorphorine-N-oxide monohydrate (NMMO) as a solvent for direct dissolution of cellulose in an industrial fiber-making process [6]. Combination of a polar molecular solvent with a salt was also reported to dissolve cellulose: N,N-dimethyl acetoamide (DMA) in combination with LiCl [7], a mixture of DMSO and tetrabutylammonium fluoride (TBAF) [8] were thus developed as cellulose dissolution solvents. Recently, Ohno and co-workers reported an interesting cellulose dissolution system of a mixed solvent of tetrabutylphosphonium hydroxide (TBPH) containing 40 wt% water [9]. In all cases, an appropriate combination of organic salts and polar solvents was essential to realizing high dissolution of cellulose. Fischer et al. [10] reported that molten salt hydrates ($LiX{\bullet}nH_2O$; $X = I^-$, NO_3^-, $CH_3CO_2^-$, ClO_4^-) dissolved cellulose [10]. It is now well recognized that very high polarity of the solvent system might be the key to breaking down the cellulose network and dissolving cellulose even if these solvents have no ionic character [3] (Fig. 4.1). However, there is a serious environmental drawback to such traditional solvent systems: they generally require large quantities of hazardous chemicals and high temperatures. From the standpoint of green chemistry, development of a safe and efficient cellulose dissolution process can be anticipated [3].

Ionic liquids (ILs) usually melt below 100 °C and are becoming attractive alternatives to volatile and unstable organic solvents due to their high thermal stability and nearly non-volatility. The most fascinating nature of ILs is their structural diversity. We are able to design their physicochemical properties, including viscosity, polarity, and hydrophobicity. Numerous papers and several reviews on the ILs have been published [11], and it is now widely recognized that ILs are applicable to the media for many types of chemical reactions [11] and even for enzymatic reactions [12]. ILs have consequently show a unique solubility in many inorganic and organic materials, and it is anticipated that ILs might dissolve insoluble compounds, including cellulose, which is impossible with conventional molecular liquids [11].

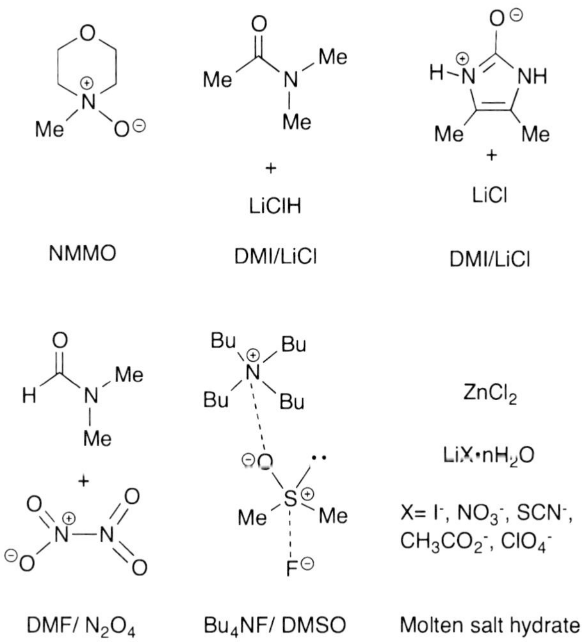

Fig. 4.1 Typical traditional solvent for cellulose dissolution. *NMMO* N-methylmorpholine-N-oxide, *LiCl* lithium chloride, *DMI* N,N-dimethylimidazolidin-2-one, *DMF* N,N-dimethylformamide, *Bu_4NF* tetrabutylammonium fluoride, *DMSO* dimethylsulfoxide

Swatloski et al. [13] reported a breakthrough on this issue using ionic liquid technology: they found that cellulose dissolved in an ionic liquid, 1-butyl-3-methylimidazolium chloride ([C_4mim]Cl), and that the regenerated cellulose from the IL solution was less crystalline [13]. An increased reaction rate of cellulase-mediated hydrolysis was realized when regenerated cellulose from the ionic liquid solution was subjected to the enzymatic reaction resulting from reduced crystallinity [13]. Since then, extensive investigations have been carried out to develop an IL that possesses the capability to dissolve cellulose and reduce its crystallinity [1, 3, 14]. Most reported ILs are imidazolium based or alkyloxyalkyl-substituted ammonium salts with chloride, formate, acetate, propionate, or phosphate as counter anion (Fig. 4.2) [1, 3, 14].

4.2 Designing Ionic Liquids That Dissolve Cellulose

Dissolution of cellulose in an ionic liquid (IL) was first achieved using imidazolium chloride as mentioned before [13]. Since then, many researchers have used chloride salts as cellulose dissolving ILs [1, 3]. However, most chloride salts have serious disadvantages: high T_m, high viscosity, especially when cellulose is dissolved in the IL, and high corrosive nature. Since these drawbacks are particularly critical when chloride salts are applied in industrial use, ILs must be developed that enable

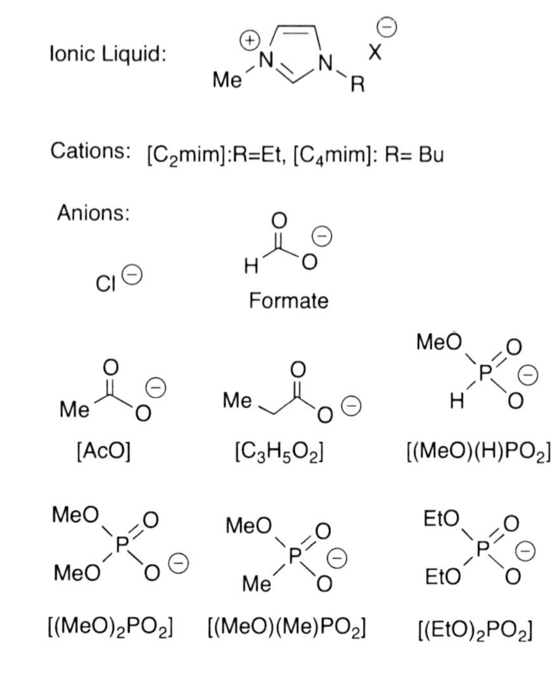

Fig. 4.2 List of imidazolium ionic liquids that show high cellulose dissolution. [AcO] acetate, [$C_3H_5O_2$] propionate, [(MeO)(H)PO_2] methyl phosphonate, [$(MeO)_2PO_2$] dimethyl phosphate, [(MeO)(Me)PO_2] methyl methylphosphonate, [$(EtO)_2PO_2$] diethyl phosphate

efficient processing of cellulose biomass while minimizing the cost of the extra energy.

Cellulose dissolution properties of ILs depend on the hydrogen-bonding characteristics for which the role of the anion is important for loosening hydrogen bonds in the crystalline region of cellulose [1–3]: the hydrogen bond-accepting ability of the anions of ILs seems to be closely linked to the solubility of the cellulose, and the solubility of cellulose in ILs increases almost linearly with the increasing hydrogen bond-accepting ability of the anions [14]. To estimate hydrogen-bond-accepting ability, it is important to know the hydrogen-bonding characteristics. The Kamlet-Abboud-Taft parameters (KAT values) specify three distinct solvent polarities: hydrogen-bonding acidity (α), hydrogen-bonding basicity (β), and dipolarity/polarizability (π^*)[15, 16]. Ohno and co-workers pointed out that the β-values obtained in a solvatochromatic study might be a better indicator of the ability to dissolve cellulose than the pKa values [17, 18]. Hydrogen-bonding basicity is considered to be necessary to dissolve cellulose, because high basicity weakens the inter- and intra-molecular hydrogen bonds in cellulose crystal [14].

Table 4.1 is a list of several ILs that have the capability to dissolve cellulose and their reported KAT values. As shown, β-values depend on the anion which is a good indicator of the cellulose dissolving property.

ILs containing carboxylate anions were reported to show strong hydrogen-bonding basicity: Bonhote et al. [19] reported that [C_2mim][acetate] displays strong hydrogen-bonding basicity [19] and that this salt dissolves cellulose well [8]. Imidazolium salts with carboxylic anions, such as lactate [20] and amino acid

Table 4.1 Cellulose dissolving ability of ionic liquids and their Kamlet-Abboud-Taft (KAT) values

Ionic liquid				KAT value			
Cation	Anion	Cellulose solubility	Viscosity (η/cP)	π*	β	α	Refs.
C_4mim	Cl	Good (10 wt% at 85 °C)	–[a]	1.14	0.84	0.44	[22]
Allylmim	Cl	Good (10 wt% at 100 °C)	2,090	1.17	0.83	0.46	[22]
Allylmim	HCO_2	Good (10 wt% at 60 °C)	66	1.08	0.99	0.48	[22]
C_4mim	HCO_2	Good (10 wt% at 35 °C)	–[a]	1.03	1.01	0.56	[14]
C_4mim	CH_3CO_2	Good[b]	–[a]	0.99	1.09	0.55	[14]
C_4mim	$C_3H_7CO_2$	Good[b]	–[a]	0.94	1.10	0.56	[14]
C_4mim	t-$C_4H_9CO_2$	Good[b]	–[a]	0.91	1.19	0.54	[14]
C_2mim	(MeO)HPO_2	Good (4 wt% at 30 °C)	107	1.06	1.00	0.52	[17]
C_2mim	(MeO)$MePO_2$	Good (4 wt% at 40 °C)	510	1.04	1.07	0.50	[17]

[a]Not known by the authors
[b]Detailed conditions of the dissolution temperature not reported
π*: dipolarity
α: hydrogen bond acidity
β: hydrogen bond basicity

[21] are also reported to show high hydrogen-bonding basicity. All carboxylic acid salts dissolve cellulose very well and have better dissolution ability than the corresponding chloride salts. The IL [C_4mim][RCO_2] shows stronger hydrogen-bonding basicity than chloride salts and can dissolve cellulose [14]. Ohno and Fukaya [14] established that the solubilization temperature of cellulose in imidazolium carboxylates depends on the length of alkyl side chains of imidazolium cations and increases for larger alkyl side chains [14]. Since viscosity of the ILs generally increases with imidazolium salts that have long alkyl chain lengths, this clearly indicates that decreased viscosity of ILs greatly affects the high dissolution of cellulose in IL; the solubilization temperature of cellulose, in fact, decreases with decreasing viscosity (Table 4.1)[14]. [C_4mim][t-$C_4H_9CO_2$] has the highest β-value (see Table 4.1), and so it might be expected to dissolve cellulose readily. However, the solubilization temperature of this liquid was higher than [C_4mim][HCO_2] and this is probably due to increased viscosity of the [t-$C_4H_9CO_2$] salt.

Ohno and co-workers conducted a very detailed study of this issue and prepared various [C_2mim] salts combined with various sulfonium and phosphonium anions [17]. They found that all ILs have identical π* values, α values and β values. A comparison of sulfonium salts and phosphonium salts showed the latter to have higher β values and higher cellulose solubility. In particular, it was found that [C_2mim][(MeO)HPO_2] dissolves cellulose even at room temperature [17]. Although β-value of [C_2mim][(MeO)HPO_2] is smaller than that of [C_2mim][(MeO)$MePO_2$], the former liquid dissolved cellulose better at lower temperature than the latter. This can be explained by the decreased viscosity of the former liquid [17].

4.3 Improving the Affinity of Ionic Liquids for Cellulose

Strong hydrogen-bonding basicity (β-value in KAT values) is now recognized as the most important property of ILs with high cellulose dissolution. Viscosity of ILs is the second key factor causing cellulose dissolution at low temperature conditions. However, a rational design of ILs with a dissolution property of cellulose has not yet been established. We discuss in this chapter how to accomplish the design of ILs with high cellulose dissolution from the standpoint of nature.

Remsing et al. [23] reported based on their ^{13}C and $^{35/37}Cl$ NMR studies that there was a stoichiometric interaction between the chloride anion and the cellulose hydroxyl groups, and this might be the key driving force of cellulose dissolution in this IL (Fig. 4.3) [23].

Inspired by their result, we hypothesized that enhanced interaction of a certain anion or cation of ILs between hydroxyl groups in the cellulose molecule might be the key factor causing cellulose dissolution and we might be able to obtain a hint on how to design such anion or cation from nature. Focusing on the structure of hydrolyzing enzyme of cellulose (cellulase), we found that amino acid ILs were strongly capable of dissolving cellulose: N,N-diethyl,N-methyl,N-(2-methoxy) ethylammonium alanate ($[N_{221(ME)}][Ala]$) worked as an excellent solvent for cellulose dissolution among ILs whose anion part was natural amino acid [24].

Hydrolysis of solid cellulose is achieved by cellulases such as endoglucanase (EGs) and cellobiohydrolases (CBHs) [25, 26]. The former can hydrolyze internal β-1,4-glycoside bonds in a cellulose polymer in the amorphous regions within the cellulose micro-fibril, and the latter can act on the free ends of cellulose polymer chains. Both types of cellulases have cellulose-binding modules that facilitate their

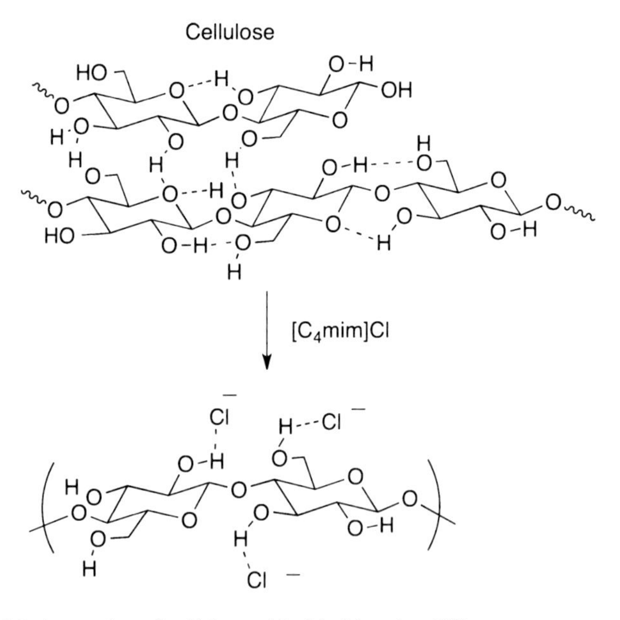

Fig. 4.3 Possible interaction of cellulose with chloride anion [23]

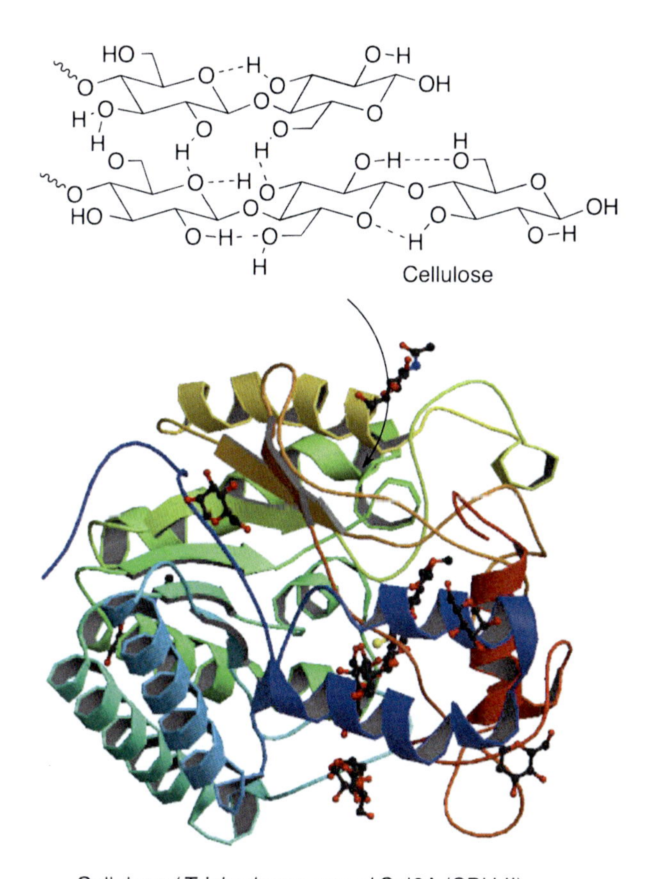

Fig. 4.4 Working hypothesis of the first incorporation of cellulose into cellulase in Ref. [26] with kind permission from Elsevier

adsorption onto crystalline cellulose, bringing the catalytic domains physically close to their site of action (Fig. 4.4) [25, 26]. We looked at what the protein sequences of several cellulases were causing particularly in the area of substrate-binding cleft, and recognized that glucosyl-binding sites of cellulases were frequently formed by the exposed surface of aromatic side-chains of protein residues [25, 26]. Three of the four binding sites making up the enclosed cellulose-binding tunnel reportedly contain the tryptophan (W), asparagine (N), and isoleucine (I) residue side-chains for *Trichoderma reesei* Cel6A (CBH II) [25, 26]. Therefore, it was expected that ILs made from amino acids might have an affinity toward a certain part of cellulose.

Ohno and co-workers prepared ILs that contained amino acid moieties as anion parts [21, 27]. Since the hydrogen-bonding basicity of amino acid salts was reported to be high [27], amino acid ionic liquids are expected to display good cellulose dissolving ability.

Based on these results, the dissolving property of 1-butyl-3-methylimidazolium tryptophan ([C_4mim][Trp]) against cellulose was tested using microcrystalline cellulose (Avicel®) as a model compound. However, the cellulose did not dissolve at all in this IL. Further evaluation of tryptophan salts with ammonium, phosphonium, or pyrridinium cation, revealed that choice of cation was also a key point in designing an IL with cellulose dissolution capability: N-(2-methoxyethyl),N,N-diethyl,N-methylammonium tryptophan ($[N_{221(ME)}]$[Trp]) dissolved cellulose (5 wt% vs. IL) at 100 °C [24]. Encouraged by the results, we prepared $[N_{221(ME)}]$ salts with natural amino acids and carefully evaluated their cellulose dissolution properties against the model cellulose (Avicel) (Table 4.2). Among 20 types of amino acid salts, we found that $[N_{221(ME)}]$[Ala] worked best to dissolve cellulose with 12 wt% versus solvent: the second solvent most effective was lysine salt ($[N_{221(ME)}]$[Lys]) (11 wt%) and the third was ornitin salt ($[N_{221(ME)}]$[Orn]). Threonine ($[N_{221(ME)}]$[Thr]) and isoleucine ($[N_{221(ME)}]$[Ile]) salts also showed similar solubility against the cellulose (7 wt%) [24]. We expected that amino acid might have affinity with a certain part of cellulose and cause its dissolution in the amino acid ionic liquid. The results reached our expectations though the details were slightly different; one of these we had anticipated, however, because there was no alanine residue near the entrance part of the cellulases [25, 26].

Many amino acid ILs dissolved cellulose, except for glutamic acid salt and all amino acids had high β-values [27]. Hence, we fixed the anionic part to alanin, and the cationic portion was re-evaluated (Fig. 4.5). It was thus confirmed that cellulose solubility was strongly dependent on the cationic part. High cellulose solubility was recorded for N,N-bis(2-methoxyethyl),N-ethyl-N-ethylammonium ($[N_{22(ME)2}]$), N,N,N-tris(2-methoxyethyl),N-ethylammonium ($[N_{2(ME)3}]$), N-(2-thiomethoxyethyl), N,N-diethyl,N-methylammonium ($[N_{221(MTE)}]$), N-methyl-N-ethoxyethylpyrrolidium ($[P_{1(ME)}]$) salts, in a range of 12 to 11 wt%. On the contrary, no dissolution of cellulose took place in $[P_{444ME}]$[Ala] or $[Py_{ME}]$[Ala] salt. Interestingly, the presence of the methoxyethyl group on the ammonium cationic part strongly modified cellulose solubility: better dissolution was obtained for $[N_{221ME}]$[Ala], while poor solubility was obtained for N-butyl-N,N-diethyl,N-methylammonium alanine($[N_{4221}]$[Ala]). However, both the methoxyethoxymethyl substituted salt ($[N_{221(MEM)}]$) and N,N,N,N-tetra(methoxyethyl)ammonium ($[N_{(ME)4}]$) alanine showed poor cellulose solubility (Fig. 4.5). These results clearly indicate that cellulose solubility is determined not only by the physical characteristics of the solvent shown as KAT values, but also by the affinity of a certain group interaction of ionic liquids with cellulose might be an important factor of cellulose dissolution in the ILs.

We next investigated cellulose solubility against various ILs which have $[N_{221(ME)}]$ as a cationic part. Although 10 wt% of cellulose dissolves in $[N_{221(ME)}]$Cl, this salt requires a higher temperature (over 120 °C) and a longer mixing time over $[N_{221(ME)}]$[Ala]. Slight decomposition of IL was observed under the conditions used [24]. No dissolution of cellulose was observed when $[N_{221(ME)}]$Br, hexafluorophosphate (PF_6), N,N-bis(trifluoromethyl)sulfonylamide (NTf_2), 2,2,3,3,4,4,5,5-octafluoropentyl sulfate (C5H8), or 2-aminoethylsulfonate (taurine), was used as solvent. $[N_{221(ME)}]$ salts with hydrogen oxide also showed poor dissolution properties [24]. Poor solubility was also recorded for N,N-dimethylalaine or

Table 4.2 $[N_{221(ME)}]$ salts with amino acids that show high cellulose dissolution at 100 °C

Amino acid	Cellulose solubility at 100 °C in wt%
Alanine	12
Lysine	11
Ornitine	8
Threonine	7
Isoleucine	7
Tryptophan, methionine, tyrosine, asparagine, leucine, phenylalanine, valine	5

$[N_{221(ME)}]$: *N*,*N*-diethyl-2-methoxy-*N*-methylethanaminium

Fig. 4.5 Optimization of cations of ILs for dissolution of cellulose

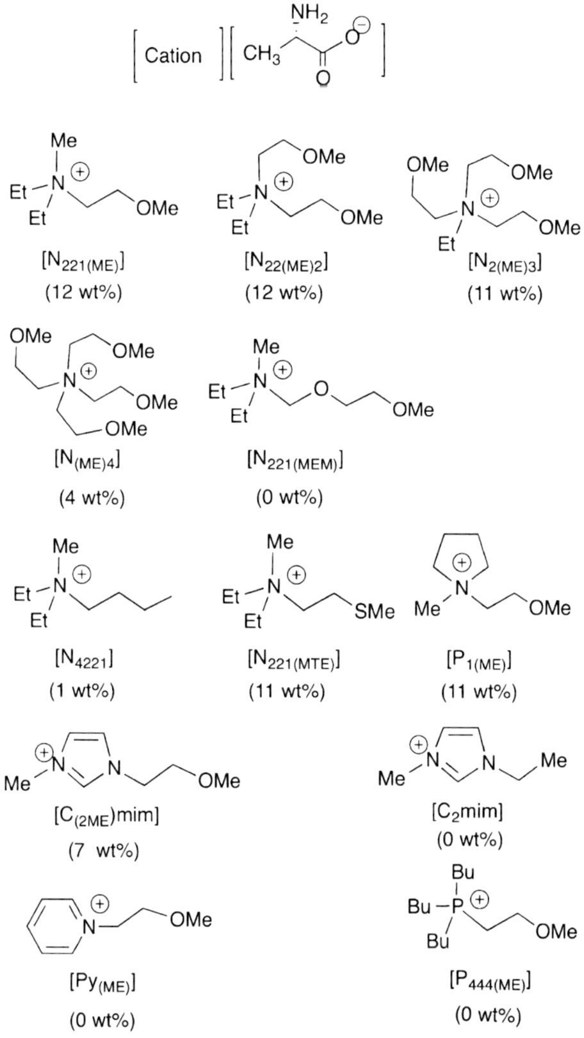

[Cation][$CH_3CH(NH_2)COO^-$]

$[N_{221(ME)}]$ (12 wt%)
$[N_{22(ME)2}]$ (12 wt%)
$[N_{2(ME)3}]$ (11 wt%)
$[N_{(ME)4}]$ (4 wt%)
$[N_{221(MEM)}]$ (0 wt%)
$[N_{4221}]$ (1 wt%)
$[N_{221(MTE)}]$ (11 wt%)
$[P_{1(ME)}]$ (11 wt%)
$[C_{(2ME)}mim]$ (7 wt%)
$[C_2mim]$ (0 wt%)
$[Py_{(ME)}]$ (0 wt%)
$[P_{444(ME)}]$ (0 wt%)

Fig. 4.6 Optimization of anions of ILs for dissolution of cellulose

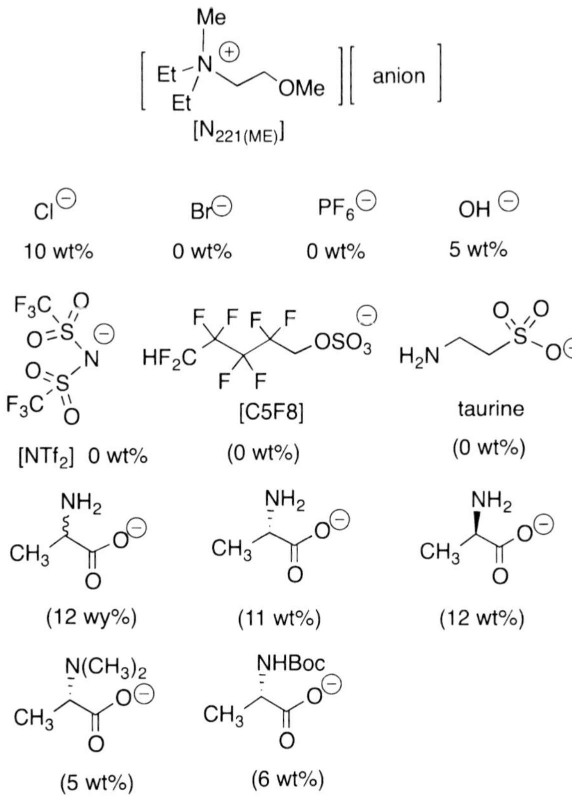

N-Boc-alanine salts compared to the alanine salt (Fig. 4.6). The presence of amino group might play an important role in dissolving cellulose [24]. Therefore, we anticipate that the amino group of [Ala] may interact with a certain part of cellulose and contribute to breaking its hydrogen bond network. Liquid ammonia reportedly changes the crystalline phase of naturally occurring cellulose and dissolves it by allowing the ammonia molecules to penetrate the cellulose fibril [28, 29]. From these results, we assume that the amino group interposed the hydrogen bonding between the cellulose and caused dissolution of cellulose in amino acid ILs. However, since the cellulose solubility is also modified by the cationic part of the IL, cation might play an important cooperative role in the mechanism for cellulose dissolution, although its origin is still unclear.

Addition of an anti-solvent like water or ethanol to the cellulose/IL solution causes precipitation of the dissolved cellulose and the structure of the regenerated cellulose changes to a disordered form. Pretreatment of cellulose increases the surface area accessible to water and cellulases are believed to improve the hydrolysis rate [30, 31]. Therefore, many researchers have attempted to hydrolyze regenerated cellulose in order to improve the hydrolysis rate by a cellulase; the dissolution of microcrystalline cellulose with [C_4mim]Cl and the rapid precipitation with water induced an increase of the amorphous region in the regenerated cellulose and enhanced the initial enzymatic hydrolysis rate [30, 31].

It was reported that cellulose regenerated from ionic liquid solution showed Type II crystalline form [31]. In fact, we confirmed that the regenerated cellulose from $[N_{221(ME)}][Ala]$ solution had only Type II form [24]. The X-ray diffraction patterns of the microcrystalline cellulose film (Avicel®) and that of the regenerated one were compared: the regenerated cellulose exhibited the typical diffraction patterns of Type II cellulose at $2\theta = 20.16°$ and $21.76°$ [32]. The results indicate that the transformation from Type I to Type II occurred after the dissolution and regeneration in $[N_{221(ME)}][Ala]$ [24].

Interestingly, we further found that crystal form of the regenerated cellulose was dependent on the dissolution solvent [24, 33]: 7 wt% of cellulose was dissolved in $[N_{221(ME)}][(MeO)(H)PO_2]$ and the regenerated cellulose was a mixture of Type I and II.

Mizuno et al. [33] recently reported that the disordered chain region was increased in the order of $[N_{221(ME)}][Ala] < [C_2mim][OAc] < [C_2mim][(EtO)_2PO_2] < [C_2mim]Cl$ [33]: regenerated cellulose treated with $[C_2mim]Cl$ contained larger amorphous regions than the others. On the contrary, that of $[N_{221(ME)}][Ala]$ had a larger amount of cellulose II crystalline structure and less susceptibility to enzymatic degradation than others; this suggests that the enzymatic hydrolysis rate of regenerated cellulose should increase by the same order. On the other hand, the order of degree of polymerization of the cellulose was $[N_{221(ME)}][Ala] > [C_2mim][OAc] > [C_2mim][(EtO)_2PO_2] > [C_2mim]Cl$; treatment with $[N_{221(ME)}][Ala]$ is therefore much more suitable to use in preparing regenerated cellulose fiber.

4.4 Cellulose Dissolution in a Mixed Solvent of ILs with Molecular Solvent

Various types of cellulose fibers are used as essential materials for our modern life. The first step in manufacturing cellulose fiber is the dissolution of cellulose in an appropriate solvent, and numerous solvent systems for such dissolution have been developed [1]. As mentioned before, ILs are now well acknowledged as cellulose dissolution agents. Although a great deal of such research has been carried out to develop a pretreatment method of cellulose for bioethanol production, ILs are now seen as attractive solvents for cellulose fiber production [1, 2]. We developed a novel amino acid ionic liquid, N-(2-methoxyethyl),N,N-diethyl,N-methylammonium alanine ($[N_{221(ME)}][Ala]$), and demonstrated that it dissolved cellulose very well [24].

Rinaldi [34] reported that addition of $[C_4mim]Cl$ or $[C_4mim][OAc]$ to DMI, DMF, sulfolane, or DMSO caused effective dissolution of cellulose [34]. Some traditional solvents were composed of a highly polar molecular solvent and an appropriate salt material, such as DMAc/LiCl, DMI/LiCl, or DMSO/Bu_4NF as noted earlier [7, 8]. Inspired by these results, we investigated an appropriate combination of polar solvent, such as DMSO or DMF, with an ionic liquid as a cellulose dissolution solvent. The dissolving property of a mixed solvent of DMSO and [C_4mim]Cl (1:1 (w/w)) against cellulose was first investigated using microcrystalline cellulose as a model compound. However, no dissolution of cellulose took place in the solvent. We next prepared various types of mixed solvent of DMSO with hydrophobic ILs (1:1 (w/w)): [C_4mim][NTf_2], [C_4mim][PF_6], [C_4mim][C5F8] [35], [$N_{221(ME)}$][NTf_2], [$N_{221(ME)}$][PF_6], [$N_{221(ME)}$][C5F8][35], [$P_{444(ME)}$][NTf_2], [$P_{444(ME)}$][PF_6], [$P_{444(ME)}$][C5F8][35], [C_4Py][NTf_2], [C_4Py][PF_6], and [C_4Py][C5F8][35]. These mixed solvents did not dissolve cellulose at all even at 100 °C. On the other hand, it was found that by switching the IL to a hydrophilic liquid like [C_4mim][OAc] or [$N_{221(ME)}$][OAc], the corresponding mixed solvent (DMSO : IL = 1:1 (w/w)) did slightly dissolve cellulose (5 and 7 wt% vs. solvent, respectively) at 100 °C. Since [N_{221ME}][Ala] showed the best dissolution among amino acid ILs [24], we prepared a 1:1 mixed solvent of DMSO and [N_{221ME}][Ala] and found that the resulting solution dissolved cellulose very well (11 wt%) after just 10 min of stirring at room temperature (25 °C); a total of 22 wt% of cellulose was dissolved in this solvent at 100 °C. Furthermore, 23 wt% of cellulose dissolved even at room temperature with 6 h stirring [36].

The solubility depended significantly on the ratio of the IL to DMSO solvent ratio as shown in Fig. 4.7: cellulose did not dissolve in pure DMSO at all, and the highest solubility was recorded for ca. a 1:1 mixture of DMSO and [$N_{221(ME)}$][Ala] (IL molar ratio (χ_{IL}) was 0.25) which coincidentally was the same ratio as our initial testing solvent. The resulting solution coagulated in water or methanol to obtain a transparent regenerated cellulose in quantitative yield and XRD analysis confirmed that the cellulose regenerated from this solution was only Type II form [36].

Rinaldi also reported that β-value of the mixed solvent of DMI/[C_2mim][OAc] increased when the ratio of the IL was increased and reached the highest value at the χ_{IL} = ca, 0.1 for the DMI/[C_2mim][OAc] solvent system, where χ_{IL} indicates the molar ratio of the IL in the solvent system [34]. From around $\chi_{IL} = 0.10$, the values were identical to those of the neat IL [34]. Since χ_{IL} of 1:1 (w/w) mixture of DMSO/[$N_{221(ME)}$][Ala] was calculated as 0.25, the β-value of the solvent might be the same as [$N_{221(ME)}$][Ala]. [$N_{221(ME)}$][Ala] has a high β-value (1.041) [36], which is almost the same as that reported for [C_2mim][Ala] (1.036) [27]. High hydrogen bond basicity of the mixed solvent of DMSO/[$N_{221(ME)}$][Ala] might contribute to breaking the inter or intramolecular hydrogen bonds of cellulose and causing its dissolution in the solvent as proposed by Ohno et al. [14, 18].

It was also reported that instantaneous dissolution of the cellulose (10 wt%) took place when [C_2mim][OAc] was added to DMI at a ratio of over χ_{IL} 0.4 at 100 °C [34]. We confirmed that a 1:1 mixture of DMSO and [C_4mim][OAc] (χ_{IL} 0.31) caused 5.0 wt% dissolution of cellulose at 100 °C, while no dissolution of cellulose took place at room temperature in this solvent. On the other hand, the mixture of DMSO/[$N_{221(ME)}$][Ala] (χ_{IL} 0.25) dissolved cellulose even at room temperature. These results clearly indicated that there was a clear contrast in the dissolution

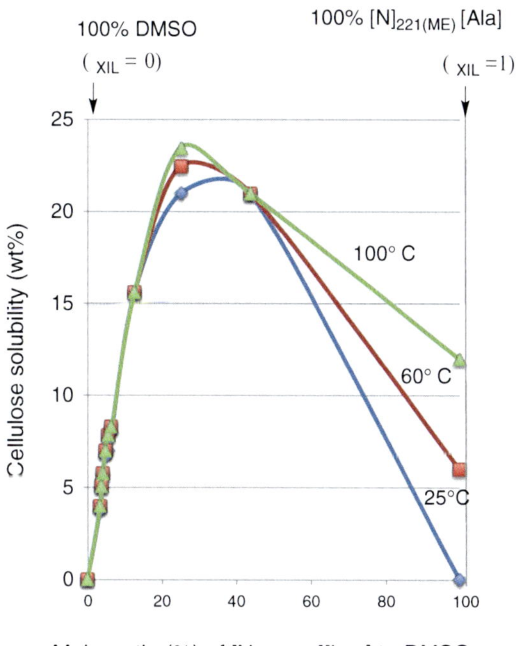

Fig. 4.7 Change in cellulose solubility for a mixed solvent composed of $[N_{221(ME)}]$[Ala] and DMSO at different temperature conditions with 2 h stirring

property between $[N_{221(ME)}]$[OAc] and $[N_{221(ME)}]$[Ala]. Amino acid-based IL is obviously so effective as a co-solvent or salt that it dissolves more cellulose in DMSO than conventional acetate-based ILs [36]. We anticipate that the hydrogen bond acceptor property of [Ala] and [OAc] might be different in the mixed solvent and reflect a different solubility. As mentioned previously, that free amino group of alanine was essential to realize high cellulose dissolution. Therefore, it was supposed that the amino group of [Ala] may interact with a certain part of cellulose and contribute to breaking its hydrogen bond network. However, since the cellulose solubility is also modified by the cationic part of the IL, cation might play an important co-operative role in the mechanism for cellulose dissolution. Further investigation of the scope and limitation of our ionic liquid technology will make it even more beneficial in cellulose science.

4.5 Conclusion

Development of an efficient means to dissolve cellulose in a simple solvent has been a long-standing goal in cellulose chemistry. ILs are now acknowledged as cellulose dissolution agents and are also seen as the most attractive solvents for cellulose fiber production. Strong hydrogen-bonding basicity (β-value in KAT values) is now recognized as the most important property of ILs with high cellulose dissolution. Viscosity of ILs is the second key factor causing cellulose dissolution at low temperature conditions. However, cellulose solubility was not determined only by the physical characteristics of the solvent shown as KAT values, but that affinity of a certain component of ionic liquids with cellulose was an important factor of cellulose dissolution in the ILs. We hypothesized that we might be able to obtain a hint on how to design such anion or cation from nature. Focusing on the structure of hydrolyzing enzyme of cellulose (cellulase), we found that amino acid ILs were strongly capable of dissolving cellulose: N,N-diethyl,N-methyl,N-(2-methoxy) ethylammonium alanate ([$N_{221(ME)}$][Ala]) worked as an excellent solvent for cellulose dissolution among ILs whose anion part was natural amino acid. It should be emphasized that amino acid IL, [$N_{221(ME)}$][Ala], is a halogen free and safe solvent, consisting of non-toxic ammonium cation and natural amino acid. Furthermore, the present results seem to provide an important hint to cellulose chemists to consider the mechanism of how an enzyme interacts with the cellulose surface.

References

1. Libert T. Chapter 1: Cellulose solvents-remarkable history, bright future. In: Liebert T, Heinz TJ, Edgar KJ, editors. Cellulose solvent: for analysis, shaping and chemical modification, ACS symposium series, vol. 1033. Washington, DC: American Chemical Society; 2010. p. 3–54.
2. Hendriksson M, Berglund LA, Isaksson P, et al. Cellulose nanopaper structures of high toughness. Biomacromolecules. 2008;9:1579–85.
3. Pinkert A, Marsh KN, Pang S, et al. Ionic liquids and their interaction with cellulose. Chem Rev. 2009;109:6712–28.
4. Klemm D, Heublein B, Fink H-P, et al. Cellulose: fascinating biopolymer and sustainable raw material. Angew Chem. 2005;44:3358–93.
5. Heinze T, Koschella A. Solvents applied in the field of cellulose chemistry: a mini review. Polimeros. 2005;15(2):84–90.
6. Rosenau T, Potthast A, Sixta H, et al. The chemistry of side reactions and byproduct formation in the system NMMO/cellulose (Lyocell process). Prog Polym Sci. 2001;26:1763–837.
7. (a) McCormick CL, Lichatowich DK. Homogeneous solution reactions of cellulose, chitin, and other polysaccharides to produce controlled-activity pesticide systems. Polym Sci Part B Polym Lett Ed. 1979;17:479–84; (b) Heinz T, Schwikal K, Barthel S. Ionic liquids as reaction medium in cellulose functionalization. Macromol Biosci. 2005;5:520–25.
8. Hermanutz F, Gähr F, Uerdingen E, et al. New developments in dissolving and processing of cellulose in ionic liquids. Macromol Symp. 2008;262:23–7.
9. Abe H, Fukaya Y, Ohno H. Fast and facile dissolution of cellulose with terabutylphosphonium hydroxide containing 40 wt% water. Chem Commun. 2012;48:11808–10.
10. Fischer S, Leipner H, Thüemmler K, et al. Inorganic molten salts as solvents for cellulose. Cellulose. 2003;10:227–36.
11. IL review: (a) Hallett JP, Welton T. Room temperature ionic liquids: solvents for synthesis and catalysis 2. Chem Rev. 2011;111:3508–76; (b) Plaquevent J-C, Levillain J, Guillen F, et al. Ionic liquids: new targets and media for α-amino acid and peptide chemistry. Chem Rev. 2008;108:5035–60.
12. Reviews for enzymatic reactions in IL: (a) van Rantwijk F, Sheldon RA. Biocatalysis in ionic liquids. Chem Rev. 2007;107:2757–85; (b) Itoh T. Chapter 1. In: Matsuda T, editor. Future directions in biocatalysis. Amsterdam: Elsevier Bioscience; 2007. p 3–20; (c) Lozano

P. Enzymes in neoteric solvents: from one-phase to multiphase systems. Green Chem. 2010;12:555–69.
13. Swatloski RP, Spear SK, Holbrey JD, Rogers RD. Dissolution of cellulose with ionic liquids. J Am Chem Soc. 2002;124:4974–5.
14. Ohno H, Fukaya Y. Task specific ionic liquids for cellulose technology. Chem Lett. 2009;38:2–7.
15. Kamlet MJ, Abboud J-L, Taft RW. The solvatochromic comparison method. 6. The.pi.* scale of solvent polarities. J Am Chem Soc. 1977;99:6027–38.
16. Reichardt C. Polarity of ionic liquids determined empirically by means of solvatochromic pyridinium N-phenolate betaine dyes. Green Chem. 2005;7:339–51.
17. Fukaya Y, Hayashi K, Wada M, et al. Cellulose dissolution with polar ionic liquids under mild conditions: required factors for anions. Green Chem. 2008;10:44–6.
18. Abe M, Fukaya Y, Ohno H. Extraction of polysaccharides from bran with phosphonate or phosphinate-derived ionic liquids under short mixing time and low temperature. Green Chem. 2010;12:1274–80.
19. Bonhôte P, Dias A-P, Papageorgiou N, et al. Hydrophobic, highly conductive ambient-temperature molten salts. Inorg Chem. 1996;35:1168–78.
20. Lau RM, Sorgedrager MJ, Carrea G, et al. Dissolution of *Candida antarctica* lipase B in ionic liquids: effects on structure and activity. Green Chem. 2004;6:483–7.
21. Fukumoto K, Yoshizawa M, Ohno H. Room temperature ionic liquids from 20 natural amino acids. J Am Chem Soc. 2005;127:2398–9.
22. Fukaya Y, Sugimoto A, Ohno H. Super solubility of polysaccharides in low viscosity, polar, and halogen-free 1,3-dialkylimidazolium formats. Biomacromolecules. 2006;7:3295–7.
23. Remsing RC, Swatloski RP, Rogers RD, et al. Mechanism of cellulose dissolution in the ionic liquid 1-*n*-butyl-3-methylimidazolium chloride: a ^{13}C and $^{35/37}$Cl NMR relaxation study on model systems. Chem Commun. 2006;2006:1271–3.
24. Ohira K, Abe Y, Suzuki K, et al. Design of cellulose dissolving ionic liquids inspired by nature. ChemSusChem. 2012;5:388–91.
25. Notenboom V, Boraston AB, Kilburn DG, et al. Crystal structures of the family 9 - carbohydrate-binding module from Thermotoga maritima xylanase 10A in native and ligand-bound forms. Biochemistry. 2001;40:6248–56.
26. Zou JY, Kleywegt GJ, Stahlberg J, et al. Crystallographic evidence for substrate ring distortion and protein conformational changes during catalysis in cellobiohydrolase Ce16A from trichoderma reesei. Structure. 1999;7:1035–45.
27. Ohno H, Fukumoto K. Amino acid ionic liquids. Acc Chem Res. 2007;40:1122–9.
28. Gollapalli LE, Dale BE, Rivers DM. Predicting digestibility of ammonia fiber explosion (AFEX)-treated rice straw. Appl Biochem Biotechnol. 2002;98(100):23–35.
29. Bellesia G, Chundawat SPS, Langan P, et al. Probing the early events associated with liquid ammonia pretreatment of native crystalline cellulose. J Phys Chem B. 2011;115:9782–8.
30. Dadi AP, Varanasi S, Schall CA. Enhancement of cellulose saccharification kinetics using an ionic liquid pretreatment step. Biotechnol Bioeng. 2006;95:904–10.
31. Zhao H, Baker GA, Cowins JV. Fast enzymatic saccharification of switchgrass after pretreatment with ionic liquids. Biotechnol Prog. 2011;26:127–33.
32. Kobayashi S, Kashiwa K, Kawasaki T, et al. Novel method for polysaccharide synthesis using an enzyme: the first in vitro synthesis of cellulose via a nonbiosynthetic path utilizing cellulase as catalyst. J Am Chem Soc. 1991;113:3079.
33. Mizuno M, Kachi S, Togawa E, et al. Structure of regenerated cellulose treated with ionic liquids and comparison of their enzymatic digestibility by purified cellulase components. Aust J Chem. 2012;65:1491–6.
34. Rinaldi R. Instantaneous dissolution of cellulose in organic electrolyte solutions. Chem Commun. 2011;47:511–13.

35. Tsukada Y, Iwamoto K, Itoh T, et al. Preparation of novel hydrophobic fluorine-substituted-alkyl sulfate ionic liquids and application as an efficient reaction medium for lipase-catalyzed reaction. Tetrahedron Lett. 2006;47:1801–4.
36. Ohira K, Yoshida K, Itoh T, et al. Amino acid ionic liquid as an efficient cosolvent of dimethyl sulfoxide to realize cellulose dissolution at room temperature. Chem Lett. 2012;41:987–9.

Chapter 5
Ionic Liquids as Solvents for Homogeneous Derivatization of Cellulose: Challenges and Opportunities

Thomas Heinze and Martin Gericke

Abstract The chapter provides a comprehensive overview of the chemical derivatization of cellulose in ionic liquids (ILs). Different types of chemical reactions, including esterification, etherification, and grafting reactions, that have been performed in these novel type of polysaccharide solvents are discussed separately regarding efficiencies and unique characteristics. With respect to the use of ILs in technical scale, specific limitations and open questions are discussed such as the chemical reactivity of certain ILs, their high viscosity and hydrophilicity, and the need to develop efficient recycling strategies. Finally, an outlook on the development of task-specific ILs and IL/co-solvent systems as reaction media for cellulose is presented.

Keywords Cellulose • Ionic liquids • Homogeneous synthesis • Polysaccharide derivatives • Side reactions • Co-solvents • Task-specific solvents

5.1 Introduction

In recent years, ILs have received enormous interest in different areas of polysaccharide research. They are intensively studied in different areas for processing of cellulose and cellulosic biomass:

1. Cellulose is the most abundant bioresource worldwide and ILs can find use in the extraction of cellulose from lignocellulosic biomass and/or the selective separation from other plant components, such as hemicelluloses and lignin [1–3].

T. Heinze (✉) • M. Gericke
Institute of Organic Chemistry and Macromolecular Chemistry,
Centre of Excellence for Polysaccharide Research, Friedrich Schiller University of Jena,
Humboldtstraße 10, D-07743 Jena, Germany
e-mail: thomas.heinze@uni-jenna.de

Z. Fang et al. (eds.), *Production of Biofuels and Chemicals with Ionic Liquids*,
Biofuels and Biorefineries 1, DOI 10.1007/978-94-007-7711-8_5,

2. ILs can also be used for the conversion of biomass into monosaccharides or platform chemicals. They may act either as reaction medium for cellulose hydrolysis or as efficient pretreatment agents to improve saccharification of the polysaccharide [2–4].
3. With respect to the environmental and safety concerns of the viscose and NMMO process, ILs are studied as alternative solvents for shaping of cellulose into fibers, sponges, beads, and other cellulosic objects [5–7].
4. Several ILs could be exploited as efficient homogeneous reaction media for the chemical modification of cellulose [8].

Regarding the complexity of all these topics, the present chapter is only devoted to the latter issue. It should be noted at this point that the ability to dissolve cellulose is not an inherent property of ILs but merely limited to a small fraction of this group with specific structural features. If not explicitly stated otherwise, however, the term 'IL' used in the chapter refers to ones that act as cellulose solvents.

5.2 Ionic Liquids as Reaction Media for Cellulose

Homogeneous chemical modification of cellulose provides several advantages over heterogeneous reactions such as: increased reactivity, uniform product composition, and efficient control over the overall degree of substitution (DS) as well as the distribution of functional groups within the anhydroglucose unit (AGU) and along the polymer chain. A variety of specific polysaccharide solvents that can be used not only for dissolution but also for the derivatization of cellulose have been reported in scientific literature [9]. They could be applied for preparing a broad variety of polysaccharide derivatives with potential applications from multi-kiloton food- and construction material industry to highly engineered materials for medical and biotechnological use. Nevertheless, none of the many cellulose solvents that could be utilized in lab-scale synthesis was found to be suitable for commercially attractive synthesis of cellulose derivatives up to now. Production of cellulose derivatives in technical scales is performed exclusively under heterogeneous conditions. In this context, ILs received a lot of interest because these versatile novel solvents might overcome the limitations of classical cellulose solvents; such as low dissolution power, inefficient solvent recycling, and incompatibility with derivatization reagents. In particular the broad structural diversity of ILs and the possibilities to create task-specific solvents by subtle manipulation of the molecular structure bear huge potential. Moreover, ILs have been employed with high efficiency in low-molecular chemistry as solvents for a vast number of advanced organic reactions that are still waiting to be transferred to cellulose derivatization [10]. Many cellulose derivatives could be prepared already by using ILs as reaction media for cellulose (Table 5.1). A comprehensive overview of the chemical derivatization of cellulose in ILs is provided in the following passages. Most of the synthesis described focused mainly on three particular imidazolium based ILs (Fig. 5.1) or slightly modified analogues. However, some reports on novel IL based reaction media were included as well.

Table 5.1 Overview of cellulose derivatives prepared in ionic liquids

	Cellulose derivative		Reaction conditions		
Entry	Type	DS range[a]	IL[b]	Comments	Refs.
	Cellulose esters				
1	Acetate	0.9–2.8	AMIMCl		[11, 12]
		1.9–3.0	BMIMCl, EMIMCl, BDMIMCl, ADMIMBr		[13, 14]
		0.7–3.0	BMIMCl	Bacterial cellulose	[15]
		2.0–2.7	ABMIMCl		[16]
		1.5–2.8	ABMIMCl	Microwave used	[17]
2	Propionate	0.5–2.9	AMIMCl	Catalyst used	[18]
		1.5–2.3	ABMIMCl	Microwave used	[17]
3	Butyrate	0.5–2.8	AMIMCl	Catalyst used	[18]
		2.4–2.8	ABMIMCl	Microwave used	[17]
4	Pentanoate	2.9	ABMIMCl	Microwave used	[17]
5	Hexanoate	2.7–2.9	ABMIMCl	Microwave used	[17]
6	Laurate	0.3–1.5	BMIMCl	Phase separation	[14]
7	Stearate	2.2–2.6	BMIMCl		[19]
8	Benzoate	1.0–3.0	AMIMCl		[20]
9	Fuorate	0.5–3.0	BMIMCl	CDI activation	[21]
10	Oxy-carboxylic acid ester	0.1–3.0	AMIMCl, BMIMCl, EMIMCl	CDI activation, bacterial cellulose	[22]
11	2-halo-carboxylate	0.6–1.0	AMIMCl	Co-solvent used	[23, 24]
		0.7	AMIMCl	Co-solvent used	[25]
		0.3–1.9	BMIMCl		[26]
12	Succinate	0.2–2.3	BMIMCl	Co-solvent used, catalyst used	[27–29]
13	Phthalate	0.1–2.5	BMIMCl	Catalyst used	[30, 31]
14	Glutarate	0.3–1.2	BMIMCl	Ultrasound used	[32]
15	Sulfate	0.1–1.5	AMIMCl, BMIMCl, EMIMCl	Co-solvent used	[33]
		1.3–1.7	BMIMCl	Co-solvent used	[34]
16	Sulfonate (tosylate)	0.1–1.1	BMIMCl, AMIMCl	Co-solvent used	[35]
		0.8	AMIMCl		[36]

(continued)

Table 5.1 (continued)

	Cellulose derivative		Reaction conditions		
Entry	Type	DS range[a]	IL[b]	Comments	Refs.
	Cellulose ethers				
17	Carboxymethyl	0.5	BMIMCl	Heterogeneous	[13]
18	Hydroxyalkyl	0.1–2.2	BMIMCl, BDMIMCl, BDTAC, EMIMAc	Co-solvent used	[37, 38]
19	Triphenylmethyl (trityl)	0.8, 1.8	AMIMCl	Co-solvent used	[39]
		0.8–1.4	BMIMCl	Co-solvent used	[40]
20	Trimethylsilyl	0.4–2.9	BMIMCl, EMIMAc	Co-solvent used	[41]
		0.2–3.0	BMIMCl, BMIMAc, BMIMBz, BMIMPr, EMIMAc, EMIMDEP	Heterogeneous	[42]
	Miscellaneous derivatives				
21	Phenyl carbamate	0.5–3.0	BMIMCl		[14]
		0.5–3.0	BMIMCl	Bacterial cellulose	[15]
22	*graft*-poly (L-lactide)	0.8–1.0 (1.5–1.7)[c]	AMIMCl	Catalyst used	[43]
		0.7–2.7 (1.4–4.5)[c]	AMIMCl	Catalyst used	[44]
		0.5–2.0 (1.7–2.4)[c]	BMIMCl	Catalyst used	[45]
23	Mixed acetate/ *graft*-poly (L-lactide)	0.4–2.5; 0.2–1.9 (3.5–9.3)[c]	AMIMCl	Catalyst used	[46]
24	*graft*-poly (ε-caprolactone)	0.1–2.4 (2.3–3.1)[c]	BMIMCl	Catalyst used	[47]
25	*graft*-poly (*N*-iospropyl-acrylamide)	n.a.	BMIMCl		[48]
26	*graft*-poly(acrylic acid)	n.a.	BMIMCl	Cross linking	[49]

[a]DS: degree of substitution

[b]Ionic liquids: cations: ADMIM$^+$: 1-allyl-2,3-dimethylimidazolium, AMIM$^+$: 1-allyl-3-methylimidazolium, BDMIM$^+$: 1-butyl-2,3-dimethylimidazolium, BMIM$^+$: 1-butyl-3-methylimidazolium, EMIM$^+$: 1-ethyl-3-methylimidazolium, anions: Ac$^-$: acetate, Bz$^-$: benzoate, Cl$^-$: chloride, DEP$^-$:diethylposphate, Pr$^-$: propionate

[c]Values in braces represent degree of polymerization of the grafted chain

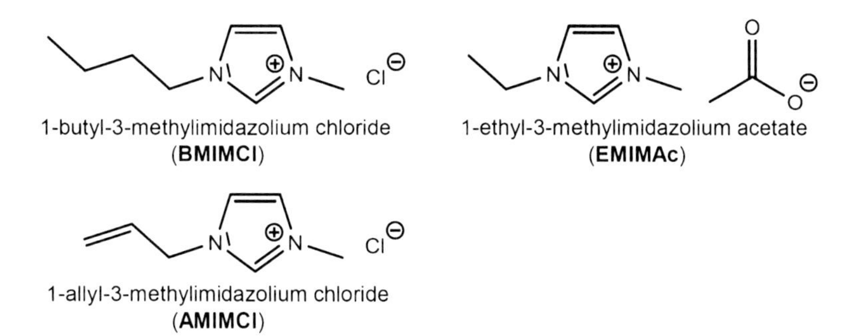

Fig. 5.1 Molecular structures of ionic liquids most frequently applied as homogeneous reaction media for derivatization of cellulose

5.2.1 Esterification

5.2.1.1 Cellulose Esters of Short C_2 to C_6 Carboxylic Acids

Cellulose esters of short chain carboxylic acids, in particular acetates, acetate-propionates, and acetate-butyrates, are of huge commercial importance and found in many applications in the form of fibers, films, coatings, and additives [50, 51]. Nowadays, these derivatives are exclusively prepared by heterogeneous processes. However, homogeneous esterification in ILs is considered as a commercially attractive alternative. Various patents have been published, e.g., by Eastman Chemical Company, that describe the preparation of cellulose esters and mixed esters in imidazolium- and ammonium based ILs, including the synthesis of the solvents and their recycling subsequent to the esterification [52–55].

Conversions of cellulose with carboxylic acid chlorides and anhydrides proved to be very efficient when performed homogeneously in an IL. The reactions are usually performed at elevated temperature ($\geq$80 °C), thus the adverse effect of high viscosity of cellulose/IL solutions is less pronounced. The esterification of cellulose proceeds completely homogeneous even up to a complete derivatization of all hydroxyl groups. Using relatively small amounts of acetic anhydride or chloride (3–5 mol equivalents, Table 5.2), highly functionalized cellulose acetates with DS up to 3 can be obtained within short reaction times (0.5–8 h) [12–14]. Using BMIMCl as homogeneous reaction medium, acetylated derivatives could even be obtained from bacterial cellulose that is usually difficult to dissolve and to chemically modify in other cellulose solvents due to its high degrees of crystallinity and polymerization [15].

In analogue to the homogeneous acetylation in ILs, homologues cellulose esters of higher carboxylic acids, from propionates up to hexanoates, have been prepared in various ILs using the corresponding anhydrides [16–18, 56]. The DS values of cellulose esters were found to decrease successively when increasing the number of carbon atoms in the acyl moiety from 2 to 4 but to increase again with further prolongation of the alkyl chain up to 6 carbon atoms [17]. Cellulose pentanoates

Table 5.2 Degrees of substitution (DS) and solubility of cellulose acetates prepared by homogeneous acetylation in various ionic liquids (IL) and under different reaction conditions

Reaction conditions						Product			
				Reagent			Solubility[c]		
IL[a]	Cellulose type[b]	Temp. [°C]	Time [h]	Type[d]	Ratio[e]	DS	DMSO	$CHCl_3$	Refs.
AMIMCl	DIP	100	3	Anhydride	3:1	1.99	+	–	[11]
AMIMCl	DIP	100	3	Anhydride	4:1	2.09	+	–	[11]
AMIMCl	DIP	100	3	Anhydride	5:1	2.30	+	+	[11]
AMIMCl	CH	100	1	Anhydride	5:1	2.16	+	–	[12]
AMIMCl	CH	100	4	Anhydride	5:1	2.49	+	+	[12]
AMIMCl	CH	100	8	Anhydride	5:1	2.63	+	+	[12]
BMIMCl	MC	80	2	Anhydride	3:1	1.87	+	–	[14]
BMIMCl	MC	80	2	Anhydride	3:1[f]	2.56	+	–	[14]
BMIMCl	MC	80	2	Anhydride	5:1	2.72	+	–	[14]
BMIMCl	MC	80	2	Anhydride	5:1[f]	2.94	+	+	[14]
BMIMCl	MC	80	2	Anhydride	10:1[f]	3.0	+	+	[14]
BMIMCl	MC	80	2	Chloride	3:1	2.81	+	–	[14]
BMIMCl	MC	80	0.25	Chloride	5:1	2.93	+	+	[14]
BMIMCl	MC	80	0.5	Chloride	5:1	3.0	+	+	[14]
BMIMCl	MC	80	2	Chloride	5:1	3.0	+	+	[14]
BMIMCl	MC	80	2	Chloride	5:1[f]	2.93	+	–	[14]
BMIMCl	BC	80	2	Anhydride	1:1	0.69	+	n.a.	[15]
BMIMCl	BC	80	2	Anhydride	2:1	1.66	+	n.a.	[15]
BMIMCl	BC	80	2	Anhydride	3:1	2.25	+	n.a.	[15]
BMIMCl	BC	80	2	Anhydride	5:1	2.50	+	n.a.	[15]
BMIMCl	BC	80	2	Anhydride	10:1	3.0	+	n.a.	[15]

[a]AMIMCl: 1-allyl-3-methylimidazolium chloride, BMIMCl: 1-butyl-3-methylimidazolium chloride
[b]BC: bacterial cellulose (DP: 6,493), CH: cellulose from corn husk (DP: 530), DIP: dissolving pulp (DP: ≈ 650), MC: microcrystalline cellulose (DP: 286)
[c]+: soluble, –: insoluble, n.a.: no information available
[d]Acetic acid derivative used
[e]Molar ration of acetylation reagent to anhydroglucose units
[f]Additionally, 2.5 mol equivalent pyridine

and hexanoates slightly exceeded the DS values of cellulose acetates, prepared under identical reaction conditions. In contrast, the corresponding propionates and butyrate had slightly lower DS values. Cooperative interaction of the long chain anhydrides with the partially substituted, i.e., lipophilic, cellulose chain have been postulated to explain this unexpected finding.

The efficiency of the esterification of cellulose in ILs can be increased by adding pyridine (stoichiometric amounts) or 4-dimethylaminoaminopyridine (DMAP; catalytic amounts) [14, 18]. Moreover, microwave assisted esterification can yield products with an increased DS in comparison to products prepared under conventional heating [17, 56]. Within a microwave field, ILs rapidly heat due to their ionic nature [57]. Thus, the irradiation must be controlled by monitoring power input, pulse length and -interval, and maximum temperature. Efficient mixing is also

Table 5.3 Overview about mixed cellulose esters with different degrees of substitution (DS) prepared in ionic liquids (ILs)

Substituent 1		Substituent 2				
Type	DS range	Type	DS range	Overall DS range	IL[a]	Refs.
Acetate	1.50	Propionate	1.30	2.80	ABIMCl	[17]
Acetate	0.30–0.66	Propionate	0.93–2.46	1.44–2.20	AMIMCl	[58]
Acetate	1.40–2.50	Butyrate	0.40–0.90	2.20–2.90	ABIMCl	[16, 17]
Acetate	0.19–1.16	Butyrate	0.86–2.07	1.05–2.41	AMIMCl	[58]
Acetate	1.40	Pentanoate	1.10	2.50	ABIMCl	[17]
Acetate	1.40	Hexanoate	1.10	2.50	ABIMCl	[17]

[a]ABIMCl: 1-allyl-3-(1-butyl)imidazolium chloride, AMIMCl: 1-allyl-3-methylimidazolium chloride

crucial, especially in case of highly viscous cellulose/IL solutions, in order to avoid local 'hot-spots' of extremely high temperature. Otherwise, degradation of cellulose and carbonization might occur. In addition to microwave assisted cellulose derivatization in ILs, esterification with the aid of ultrasound irradiation has been reported [32].

Parallel conversion of cellulose, dissolved in an IL, with two different carboxylic acid anhydrides yields mixed cellulose esters (Table 5.3) [16, 17, 58]. The product properties (e.g., hydrophobic/hydrophilic character) can be tailored by variation of ester moieties, their partial DS values, and the overall amount of substituents, attached to the cellulose backbone. As already pointed out, the reactivity of carboxylic acid anhydrides is dependent on the length of the alkyl chain [17]. In addition to reaction temperature, time, and amount of acylation reagent, the sequence of adding the two different anhydrides (simultaneous vs. step-wise) is consequently of huge importance. Considering the huge commercial importance of mixed cellulose esters, in particular acetate/propionates, acetate/butyrates, and propionate/butyrates, further comprehensive studies are required in order to evaluate the individual effect of reaction parameters on the product composition, including the distribution of ester moieties within the AGU and along the polymer chain. Moreover, choice of the IL and its recycling subsequent to the reaction (see Sect. 5.3.3) are going to be important issues.

5.2.1.2 Other Organic Esters

In principle, cellulose, dissolved in an IL, can easily be converted also with other types of carboxylic acids, or their corresponding acid derivatives. Cellulose succinates and phthalates, i.e., dicarboxylic acid derivatives could be obtained by homogeneous derivatization in AMIMICl or BMIMCl with the aid of different acylation catalysts, e.g., DMAP, *N*-bromosuccinimide, or iodine [27–31]. Thereby, DMSO has partly been utilized as molecular co-solvent in order to guarantee homogeneous reaction conditions. The homogeneous preparation of cellulose glutarate by acylation in BMIMCl under ultrasound irradiation also has been reported as well [32]. ILs, partly in combination with DMF as co-solvent, have been utilized as reaction media for the preparation of 2-bromo and 2-chloro carboxylic acid ester of cellulose, which could be used as macro initiators for the grafting of poly(styrene) and different types of poly(methacrylate) chains onto the polysaccharide backbone [23–26].

Using AMIMCl as homogeneous reaction medium, a series of cellulose benzoates with high DS values in the range of 1.0–3.0 has been prepared that carried different moieties at the aromatic ring [20]. Also mixed cellulose derivatives, carrying benzoate groups (preferentially at the primary hydroxyl group) and 4-nitrobenzoate moieties (preferentially at the secondary hydroxyl groups) have been prepared in ILs by step-wise conversion with the corresponding acid chlorides. Highly substitute cellulose benzoates, as well as cellulose phenyl carbamates (see Sect. 5.2.3), exhibit high chiral resolving properties and could be utilized as stationary phase in liquid chromatography for separation of enantiomers [59]. Synthesis of these materials is usually performed under heterogeneous starting, e.g., by conversion of cellulose suspended in pyridine or other swelling agents [60]. However, homogeneous derivatization in ILs can facilitate preparation of tailor-made chromatography materials.

In general, cellulose esters are well soluble in ILs even at high DS values meaning that completely homogeneous esterification is feasible. However, in case of fatty acid ester with long, non-polar alkyl chains, the cellulose derivatives become increasingly hydrophobic upon advancing substitution, which renders the products insoluble in the reaction mixture. Cellulose laurates can be obtained in ILs but the derivatives rapidly precipitated upon derivatization of the polysaccharide [14]. Phase separation can be expected also for the higher homologue cellulose stearate although it has not been mentioned explicitly [19].

Compounds containing carboxyl groups are readily available but seldom used directly for esterification with polysaccharides due to their low reactivity. In case conversion into a more reactive acid chloride or anhydride is not feasible, activation of carboxyl groups with *N,N′*-carbonyldiimidazole (CDI) is a very versatile approach for obtaining cellulose esters under mild reaction conditions [61, 62]. Some cellulose derivatives, namely fuorates and water soluble oxy-carboxylic acid ester of cellulose with a broad range of DS from 0.1 to 3.0, have already been obtained in ILs by this procedure [21, 22]. As a first step of the homogeneous derivatization reaction, an imidazolide is formed *in situ* in the IL reaction medium, which reacts as active species with the polysaccharide backbone upon liberation of CO_2 and imidazole as side products (Fig. 5.2). A carbodiimide has also been tested as coupling agent for homogeneous esterification of cellulose with stearic acid in an IL but only small DS values up to 0.16 could be achieved [36]. Another possibility for the activation of carboxyl groups for subsequent esterification with cellulose is the conversion with sulfonic acid chlorides (Fig. 5.2) [63, 64]. Moreover, reactive intermediates are generated by treating carboxylic acids with iminium chlorides, e.g., prepared *in situ* from DMF and oxalyl chloride (Fig. 5.2) [65]. The by-products liberated upon activation and subsequent esterification are either gaseous (CO, CO_2, HCl) or may act as co-solvent (DMF). The last two approaches have been exploited for the

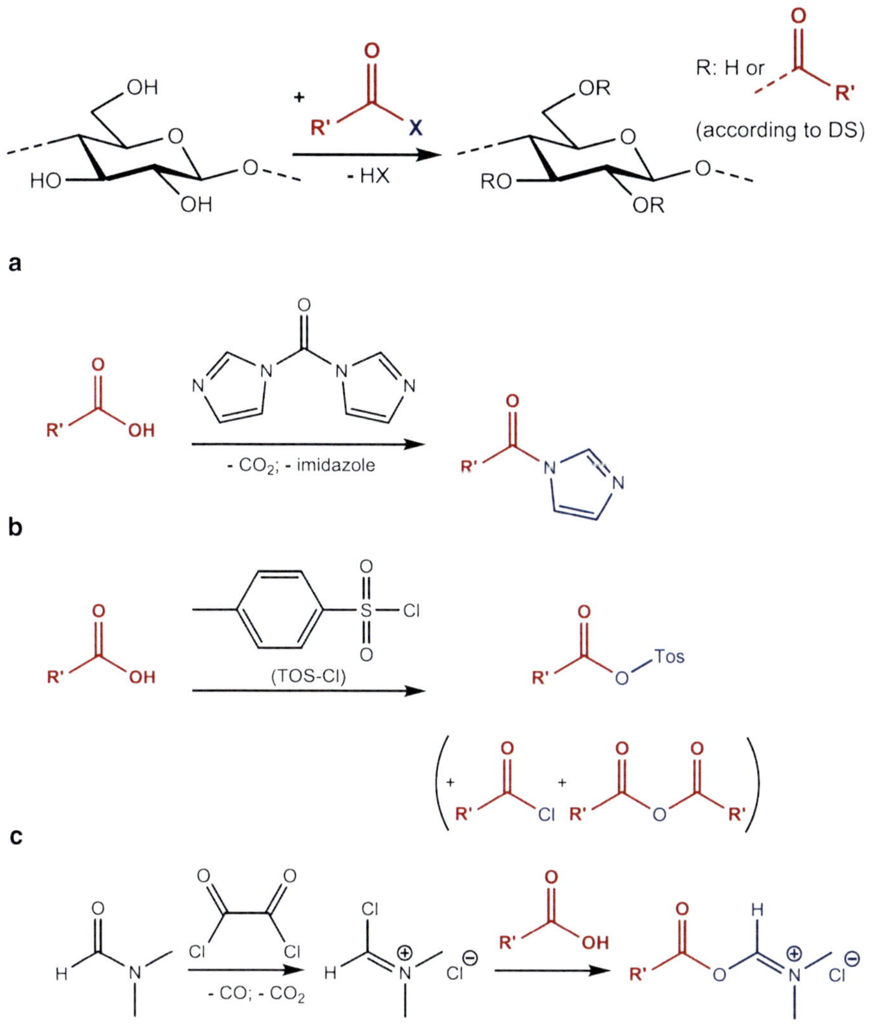

Fig. 5.2 Selected methods for activation of carboxylic acids and for the esterification with cellulose [65]. (**a**) Carbonyldiimidazole activation, (**b**) sulfonic acid chloride activation, (**c**) iminium chloride activation

preparation of various polysaccharide esters but up to now not been adapted for the homogeneous derivatization of cellulose in ILs.

5.2.1.3 Sulfuric and Sulfonic Acid Esters

Cellulose sulfates (CSs), the sulfuric acid half ester of cellulose in their sodium form, are useful compounds for certain biomedical applications and for the formation of polyelectrolyte complex superstructures [66]. However, the homogeneous synthesis of these derivatives in ILs proofed to be more challenging compared to the preparation of organic cellulose esters [33]. The same holds true for toluene sulfonic acid esters of cellulose, usually referred to as tosyl celluloses (TOSCs), that are versatile intermediates for the preparation of polysaccharide derivatives via nucleophilic displacement reaction [35]. Both cellulose derivatives are usually prepared at or below 25 °C to prevent side reactions. Sulfation at elevated temperature often induces severe polymer degradation whereas tosylation might yield chlorinated or cross-linked products. At these low temperatures, the exceptional high viscosity of cellulose/IL solutions (see Sect. 5.3.1) prevents efficient mixing during the reactions, which results in non-uniform product mixtures. Interestingly, EMIMAc, a cellulose dissolving IL with rather low viscosity, cannot be applied as reaction medium. The acetate anion, present in high concentrations, forms mixed anhydrides with the tosylation and sulfation reagents and acetatylation instead of formation of the desired cellulose derivatives occurs (see Sect. 5.3.2) [67]. Certain co-solvents, such as DMSO, DMF, pyridine, or *N,N*-′-dimethylimidazolidinone (DMI) can be added to cellulose/IL solutions to diminish viscosity (see Sect. 5.3.1) [33, 68, 69]. Thus, completely homogeneous sulfation and tosylation of cellulose could be achieved at 25 °C with little polymer degradation using BMIMCl/co-solvent mixtures as reaction media [33, 35]. CSs prepared by this procedure exhibit good water solubility even at low DS values of 0.2–0.3. The derivatization reaction shows higher regioselectivity compared to other heterogeneous or quasi-homogeneous methods for sulfation of cellulose; at $DS < 1$ the vast majority of sulfate groups are located at position 6. Moreover, the DS and consequently the product properties can be tailored by adjusting the molar ratio AGU to sulfation reagent (Fig. 5.3). In a comparable approach, TOSC with $DS_{tosyl} \leq 1.14$ have been prepared by homogeneous derivatization of cellulose in BMIMCl/pyridine and BMIMCl/DMI mixtures [35]. The amount of deoxy-chloro moieties introduced as well as the extent of polymer degradation can be controlled by adjusting the reaction time from 4 to 8 h.

5.2.2 *Etherification*

The first description of the use of ILs as reaction media for etherification of cellulose is a patent related to the carboxymethylation of the polysaccharide [70]. Carboxymethyl cellulose as well as alkyl- and hydroxyalkyl ethers of cellulose are of huge commercial interest but reports on etherification of cellulose in ILs are rare compared to the vast number of publications related to the esterifications in these media. ILs applied for cellulose dissolution are rather hydrophilic and homogeneous etherification of cellulose is more difficult to realize because the reagents/bases applied and/or the derivatives formed are not completely soluble in the reaction mixtures.

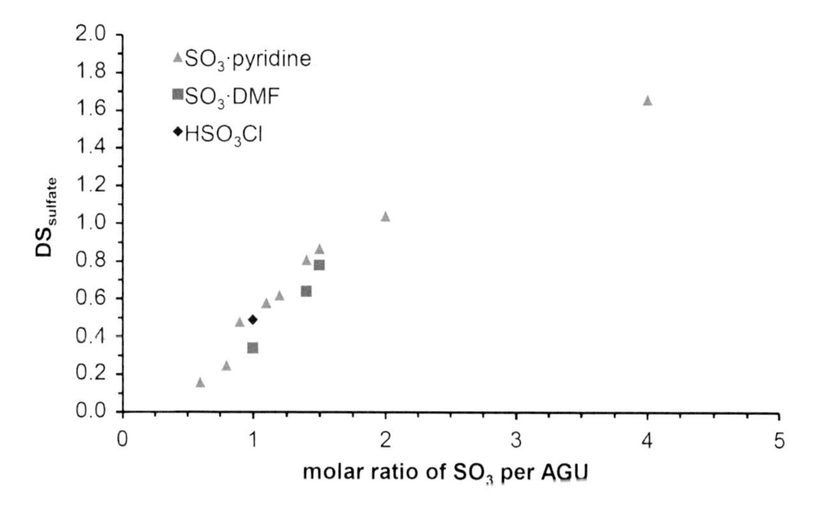

Fig. 5.3 Dependence of degree of substitution (*DS*) on the molar ratio sulfating agent per anhydroglucose unit (*AGU*), obtained by homogeneous derivatization of cellulose in 1-butyl-3-methylimidazolium chloride with *N*,*N*-dimethylformamide as co-solvent

Solid NaOH has been utilized as base for carboxymethylation of cellulose in ILs, i.e., this derivatization occurs heterogeneously despite the fact that cellulose is dissolved in the ILs prior to the reaction. Moreover, the polysaccharide solution forms a gel-like system shortly after the addition of the reagent [13]. As a consequence, DS values obtained are rather low (<0.5). Cellulose dissolving ILs are also immiscible with hexamethyldisilazane (HMDS), a rather hydrophobic reagent utilized for the conversion of cellulose into the corresponding silyl ethers. A biphasic silylation procedure has been reported, in which cellulose is dissolved in an IL and treated with HMDS, dissolved in IL-immiscible toluene [42]. The reaction involves transition of the reagent into the polar IL phase and of the cellulose derivative, which becomes increasingly hydrophobic upon silylation, into the non-polar toluene phase. These phase transitions influence the course of the derivatization reaction, e.g., in terms of overall DS, distribution of silyl groups along the polymer chain, and product homogeneity. These reactions are difficult to control because they strongly depend on multiple reaction conditions, such as temperature, stirring speed, viscosity, and liquid/liquid interface area. Completely homogeneous preparation of trimethylsilyl cellulose (TMSC) with a broad range of DS values from 0.4 to 2.9 has been achieved in EMIMAc/chloroform mixtures [41]. The co-solvent efficiently solubilizes HMDS as well as TMSC formed, which otherwise precipitate from the reaction mixtures at DS > 2.

Hydroxyethyl- and hydroxypropyl celluloses, which are among the most important cellulose derivatives applied commercially, e.g., as additives in paint, cement, and household products, could be prepared in ILs by conversion of cellulose with gaseous ethylene- or propylene oxide [37, 38]. The reaction proceeds heterogeneous and the degree of molecular substitution (MS) could be increased by addition of DMF or DMSO. The increased reactivity is most likely a result of the improved solubilization of the hydroxyalkylation reagents. EMIMAc was found to be the most suitable solvent for this etherification because the acetate anion can catalyze ring opening of the oxiranes, which results in increased MS. By addition of catalytic amounts of sodium/magnesium acetate, cellulose hydroxyalkyl ethers could also be obtained in imidazolium chloride based ILs. In addition, low-melting quaternary ammonium chlorides and formates could be utilized [71].

Triphenylmethyl (trityl) substituents have been exploited in polysaccharide research as protecting group for the primary hydroxyl group that is more accessible for bulky moieties [72, 73]. Ethers with DS values around 1 possessing a pronounced 6-*O*-functionalization, could be obtained by homogeneous conversion of cellulose dissolved in BMIMCl with trityl chloride [40]. By using the more reactive *p*-methoxytrityl chloride, a maximum DS of 1.8 could be achieved in AMIMCl but complete protection of position 6 and 2 was not possible.

5.2.3 Other Reaction Types

Phenyl carbamates of celluloses from different sources, including bacterial cellulose, have been prepared in BMIMCl as solvent by homogeneous conversion with the corresponding isocyanate [14, 15]. No byproducts are formed during this reaction and the DS could be tuned over the whole range, up to a maximum of 3.0, by adjusting reaction time and the amount of derivatization reagent. 'Carbanilation' proceeds with only minor chain degradation and yields organosoluble derivatives. Consequently it is an important tool for determination of the molecular weight of cellulose by size exclusion chromatography [74]. Moreover, cellulose carbamates can be utilized as chiral stationary phase for the chromatographic separation of enantiomers [59].

ILs have been exploited successfully for the homogeneous preparation of several cellulose-*graft*-copolymers. Grafting of poly-L-lactide and poly(ε-caprolactone) chains onto the cellulose backbone has been achieved by ring opening polymerization with DMAP or stannous octanoate as catalyst [43–45, 47]. Depending on the amount of monomer used, different DS values in the range of 0.1–2.7 could be achieved; the average number of repeating units was rather low ($\approx$1–5). Thermoplastic mixed derivatives containing acetyl moieties and grafted poly (L-lactide) chains have been obtained as well in a combined 'one-pot' reaction [46]. Cellulose-*graft*-poly(acrylamide) and -poly(accrylate) have also been prepared in ILs, by radical polymerization, induced by γ-ray irradiation or persulfate initiation [48, 49]. In the latter case, the reaction was performed under microwave heating and a certain amount of bivalent cross linker was present, i.e., a heterogeneous polymer gel was obtained. Also the grafting of copolymers onto cellulose via ATRP has been studied in ILs but no advantages over common molecular solvents in terms of the conversion rate could be observed [26]. Thus, it is important to decide if ILs can be beneficial in this process. At DS values >0.6, the macro initiators required in the first step dissolves, e.g., in DMF, dioxane, or butanone, i.e., ILs or other cellulose solvents are not required [23]. However, if low grafting densities are desired, ILs might prove to be useful as reaction media for homogeneous grafting of polymers onto cellulose via ATRP.

5.3 Challenges and Opportunities

ILs have proved to be efficient solvents for the dissolution of high amounts of cellulose in rather short time. As has been summarized in Sect. 5.2, numerous chemical derivatization reactions have been realized successfully in these novel homogeneous reaction media. However, with increasing gain of knowledge it became apparent that ILs also possess some specific disadvantages that can hamper their use as cellulose solvents in technical scales. Some of these issues are technological aspects that are currently studied and can be expected to be solved in the near future; e.g., IL cost, purity, and recycling availability. Others are intrinsic restrictions related to the physical and chemical properties of ILs, i.e., limitation that cannot be avoided but only be attenuated.

5.3.1 *IL Viscosity and Hydrophobicity*

In comparison to common molecular solvents, the viscosity of ILs, in particular those that can dissolve cellulose, is very high. As an example, at 25 °C the viscosities of DMSO (2 mPa s) and EMIMAc (200 mPa s) differ by two orders of magnitudes. The viscosity further increases when cellulose is dissolved in the ILs. For cellulose solutions in EMIMAc and BMIMCl, transition from a diluted to a semi-diluted regime, in which polymer coils start to entangle, occurred at a polymer content of 1–2 % [75, 76]. Above these concentrations, viscosity is proportional to c^n with n being about 4.0 at 0–40 °C and about 2.5 at 60–100 °C (Fig. 5.4 left). Comparable values for the exponent have been reported for cellulose solutions in DMA/LiCl (n = 3–4) and NMMO (n = 4.6) [77, 78].

With decreasing temperature, the viscosity of cellulose/IL solutions strongly increases (Fig. 5.4 right), which is expected for a polymer solution. However, the increase does not follow an Arrhenius-like behavior (Eq. 5.1) that is usually observed for polymer solutions, including cellulose solutions in NMMO or NaOH/water [79, 80]. As indicated by the strong temperature dependence of the exponent n, the deviation is especially pronounced below 40 °C. The temperature behavior of cellulose/IL solutions can be fitted best by a Vogel Fulcher Tamman

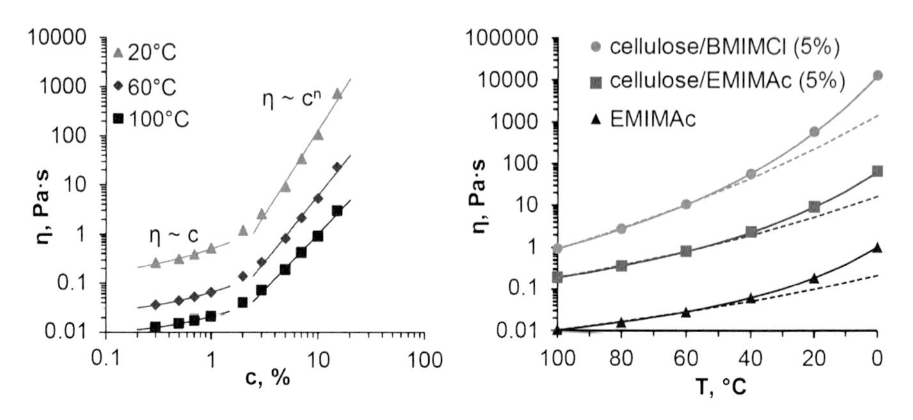

Fig. 5.4 Viscosity of cellulose solutions in 1-ethyl-3-methylimidazolium acetate (EMIMAc, *left* and *right*) and 1-butyl-3-methylimidazolium chloride (BMIMCl, *right*) at different cellulose concentrations (*left*) and temperatures (*right*). *Left*: *straight lines* represent fit according to a linear ($c \leq 1$ %) or exponential equation ($c > 2$ %). *Right*: *dotted lines* represent extrapolation from T = 60–100 °C according to an Arrhenius equation, *straight lines* represent fitting according to a Vogel Fulcher Tamman equation (Adapted with permission from [75], Copyright 2009, American Chemical Society, and from [8], Copyright 2012, MDPI AG)

equation (Eq. 5.2) [75, 76]. Since, many pure ILs show the same curved slopes, it can be concluded that the tremendous viscosity increase is caused by specific interaction within the ILs itself and not with the dissolved polysaccharide [81, 82].

$$\eta = \mathrm{A} \cdot \mathrm{e}^{\frac{\mathrm{E_A}}{\mathrm{R} \cdot T}} \tag{5.1}$$

$$\eta = \mathrm{B} \cdot \sqrt{T} \cdot \mathrm{e}^{\frac{\mathrm{k}}{T - T_0}} \tag{5.2}$$

The high viscosity of cellulose/IL solutions significantly affects the mass transfer in all steps involved in the processing of the polymer. For efficient homogeneous derivatization of cellulose, reactions are usually performed in solutions having a polysaccharide concentration in the range of 2–10 %. Above 80 °C, viscosity of these systems is sufficiently low to guarantee efficient stirring and even distribution of the reactants in the reaction mixture. Thus, homogeneous derivatization reactions performed in ILs at elevated temperatures (e.g., acylation of cellulose) usually yield uniform products with well reproducible properties. In contrast, at temperatures below 40 °C interference of reaction kinetics and mass transport phenomena might occur as result of the significant viscosity increase, in particular if the reagents applied are solids and/or highly reactive (Fig. 5.5a, b).

It has been demonstrated that sulfation and tosylation of cellulose in imidazolium chlorid based ILs, performed at or below 25 °C to prevent undesired side reactions, only yields product mixtures composed of highly functional polysaccharide derivative and unmodified cellulose [33, 35]. In both cases, a rather long period of time is required for a complete mixing and even distribution of the

Fig. 5.5 Cellulose solutions in ionic liquids with different agents, depicting the high viscosity (**a**, **b**) and hydrophilicity (**d**, **e**) of the systems as well as the effect of co-solvents (**c**, **f**)

reagents within the highly viscous system. Although the cellulose/IL solution and the derivatization reagent might eventually form a homogeneous mixture if given sufficient time, they remain as two separated phases due to kinetic hindrance, i.e., the systems is not in a state of thermodynamic equilibrium. Thus, these derivatization reactions can be considered as heterogeneous; cellulose molecules in direct vicinity to the derivatization reagent are converted to derivatives with a high DS whereas other fractions of the polysaccharide remain unmodified. However, efficient homogeneous derivatization of cellulose in ILs could be achieved at room temperature by using co-solvents to diminish the high viscosity (see Fig. 5.5c and Sect. 5.3.4.2).

The chemical modification of cellulose, dissolved in an IL, is a completely homogeneous process only if all reaction partners, including reagents, bases, co-solvents, and intermediate- and final product, remain dissolved in one phase over the whole reaction time. However, cellulose dissolving ILs are known to be rather hydrophilic compounds, which are consequently not miscible with various hydrophobic compounds frequently involved in the chemical derivatization of cellulose (Fig. 5.5d, e). Depending on whether phase separation occurs at the very beginning or later during the reaction, the corresponding reactions proceed either completely or partly heterogeneous. It needs to be considered that this change in the reaction course might affect the reactivity of the derivatization reagent, the regioselectivity, or the distribution of substituents along the polymer chains in comparison to a completely homogeneous conversion of cellulose.

The esterification of cellulose with lauryl chloride in BMIMCl starts homogeneous but results in precipitation of the hydrophobic derivative formed upon increasing substitution [14]. As result of the decreased reactivity in the heterogeneous system formed, no products with DS > 1.5 could be obtained. Due to the hydrophilic nature of common cellulose dissolving ILs, homogeneous preparation of trimethylsilyl cellulose (TMSC) is not feasible [41, 42]. The hydrophobic reagent applied for derivatization is not miscible with the polysaccharide solutions and the silylated products formed precipitate from the reaction mixture at DS > 2. A heterogeneous procedure for silyation of cellulose has been described using toluene, which is immiscible with cellulose/IL solution, to solubilize the hydrophobic reagent [42]. The reaction involves transition of (a) the reagent into the hydrophilic IL phase and (b) the TMSC with a certain DS into the hydrophobic toluene phase, where the derivative is further silylated in a homogeneous reaction. Thus, this particular procedure is most suitable for obtaining highly functionalized TMSC. In contrast, complete homogeneous preparation of TMSC in the DS range of 0.4–2.9 has been achieved in ILs by using chloroform as a hydrophobic but IL-miscible co-solvent [41].

5.3.2 Chemical Reactivity of Ionic Liquids and Their Impurities

ILs are considered as non-derivatizing cellulose solvents, i.e., the dissolution of cellulose is not due to chemical conversion of the polysaccharide [13]. Nevertheless, ILs are not necessarily chemically inert. Both cation and anion can participate in the course of chemical derivatization reactions of cellulose or react with the dissolved polysaccharide. In addition, the effect of typical IL impurities needs to be considered.

The proton at position C-2 of 1,3-dialkylimidazolium based ILs is rather acidic; the p*K*a-values are estimated to be about 21–24, and can be abstracted with bases to yield *N*-heterocyclic singlet carbenes [83, 84]. These reactive species act as nucleophilic intermediates in the catalytic cycles of many organic reaction, which is the reason for surprisingly high yields and/or unexpected products frequently observed for reactions performed in ILs [85, 86]. The presence of carbenes and their influence on the derivatization of cellulose in ILs needs to be considered, in particular when bases are utilized. In addition, it has been reported for low-molecular weight cellulose mimics and later on for cellulose as well that the carbene species attach to the reducing end-group in its open-chain aldehyde form (Fig. 5.6) [87, 88]. Although the conversion is accelerated in the presence of bases, it also occurs upon dissolution of cellulose in pure imidazolium acetates. To a

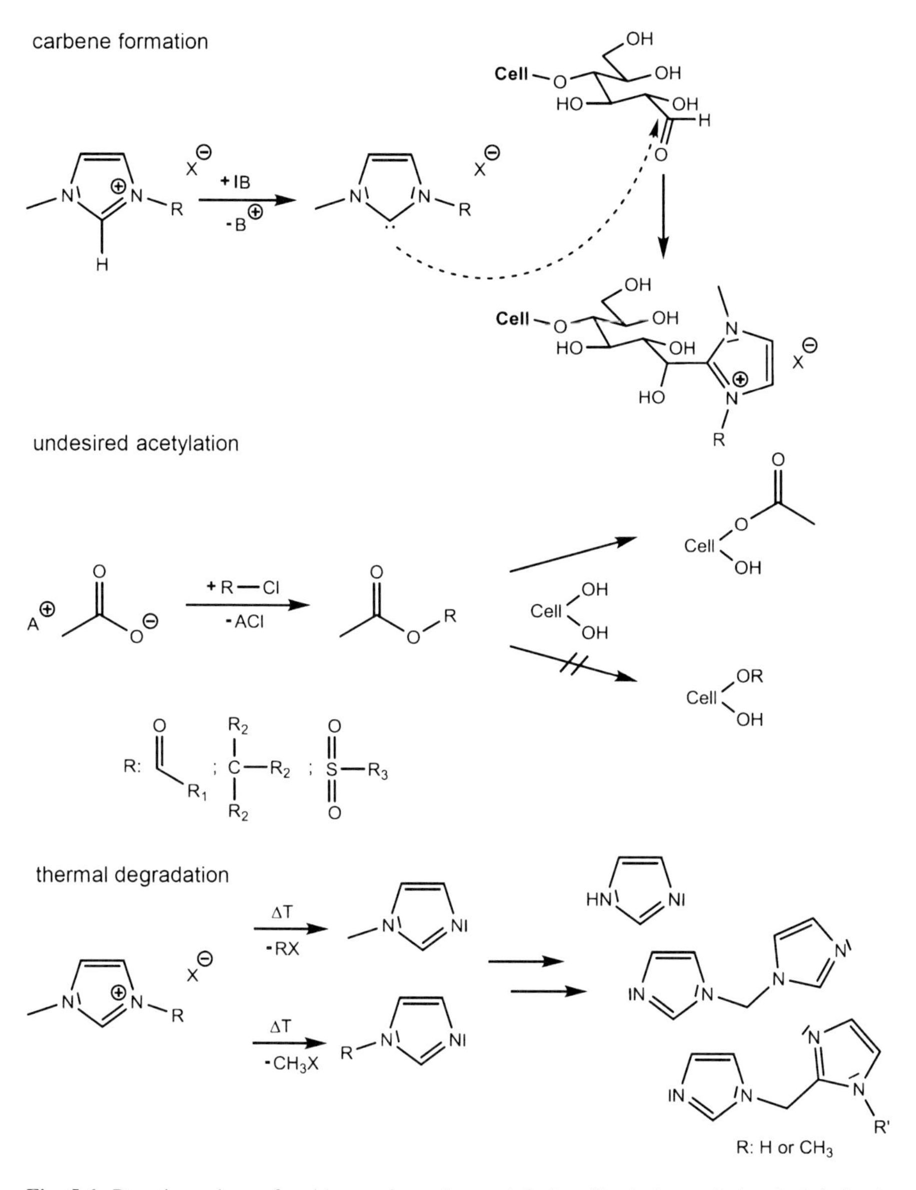

Fig. 5.6 Reaction scheme for side reactions observed during dissolution and chemical derivatization of cellulose in typical imidazolium based ionic liquids (A^+X^-) [67, 87, 88, 91]

certain extent, these particular ILs undergo self-deprotonation as result of the relatively high basicity of their anion [89]. The equilibrium is further shifted by subsequent reaction of the carbenes formed with the cellulose end-group. To avoid the effect of reactive carbene intermediates, imidazolium based ILs that were methylated at C-2 have been proposed as cellulose solvents [14]. However, the methyl group can also be deprotonated to a certain extend [90].

In addition to the IL's cation, the anion can undergo specific side reactions. The conversion of cellulose, dissolved in an imidazolium chloride IL, with furoyl, tosyl, and trityl chloride as well as with SO_3 complexes yields the expected cellulose derivatives 8, 16, 19, and 15 (see Table 5.1) [33, 35]. In contrast, only acetylated products could be obtained when the same derivatization reactions were carried out in an EMIMAc [33, 67]. This unexpected finding has been attributed to the formation of mixed anhydrides of the anion, which is present in high concentrations and not surrounded by a solvent cage, with the reagents applied (Fig. 5.6). These intermediates subsequently react with cellulose and transfer the acetyl group. If well understood, the chemical reactivity of ILs is not necessarily a drawback of this class of cellulose solvents. As an example, the acetate anion acts as a catalyst for the ring opening of oxiranes, which could be exploited for the efficient hydroxyalkylation of cellulose in ILs [37]. Moreover, ILs can become valuable for performing derivatization reactions in which other homogeneous cellulose reaction media, such as DMA/LiCl and DMSO/TBAF, show specific side reactions and cannot be employed.

ILs often contain certain impurities, derived from the synthesis, such as unreacted educts, side products, inorganic salts, and organic acids [92]. These compounds can affect the dissolution and chemical derivatization of cellulose in ILs. *N*-Methylimidazole is a starting material for the synthesis of imidazolium salts, the most frequently applied type of ILs in cellulose research, and one of their major impurities. This heterocyclic base acts as catalyst, e.g., for the silyation of cellulose. Thus, highly silylated products could be obtained using common reagent grade ILs (90–95 % purity) that contain traces of *N*-methylimidazole (0.1–0.5 wt%), whereas no significant derivatization could be achieved when ILs of high purity (>99 %) were applied as reaction media [41].

Even hydrophobic ILs can adsorb rather high amounts of moisture from humid air atmosphere [93]. Thus, the ubiquitous presence of water should not be neglected when using ILs for processing of cellulose. However, handling under protecting gas and strictly anhydrous conditions is not necessary for most applications. Water can directly influence chemical derivatization reactions. Most reagents applied for chemical modification of cellulose are prone to hydrolysis, which leads to an apparent decrease in reaction efficiency. Water also promotes the chain degradation of cellulose especially at high temperatures or if the ILs applied contain acidic impurities [94, 95].

In addition to the influence of water on chemical reactions, water affects the solubility and state of dissolution of cellulose in ILs. In large excess, the protic non-solvent acts as 'precipitation agent', i.e., cellulose is regenerated from ILs upon pouring the solution into five to ten times the volume of water. It has been reported that cellulose solutions in some ILs tolerate rather high amounts of water of up to 20 wt% [96]. However, even traces of water can alter the state of dissolution of cellulose in ILs before any 'macroscopic changes' can be detected. It has been demonstrated that the intrinsic viscosity of cellulose/EMIMAc solutions, which is directly correlated with the size and conformation of the dissolved polysaccharide chains, first increases with increasing water content up to a maximum of 10 wt% and then decreases again until finally reaching the solubility limit [97]. Based on this finding it was concluded that a 'micro-gel', i.e., agglomerates of polymer coils, is formed upon the addition of water to cellulose/IL solutions. This phenomenon has significant influence on the rheological flow behavior of these solutions and might also affect the chemical derivatization of cellulose as well as the processing into cellulosic fibers by spinning processes.

Pure ILs are commonly regarded as highly thermostable with a broad liquid range; some individual representatives of this class withstand temperatures up to 400 °C [10]. Under practical lab-conditions, however, decomposition may already occur at much lower temperatures, in particular in the presence of impurities [98]. For common cellulose/IL solutions, onset temperatures (T_{on}) for the chemical decomposition and liberation of gaseous compounds around 180–220 °C have been observed by means of differential scanning- and reaction calorimetry [99, 100]. The values changed only slightly upon the addition of additives such as silver or charcoal. In contrast, cellulose solutions in NMMO, employed for the production of cellulosic fibers on a technical scale, are significantly less stable ($T_{on} \approx 130$–160 °C) [101, 102]. Moreover, stabilizers are required in order to prevent autocatalytic thermal runaway reactions. Usually, dissolution, shaping, and chemical derivatization of cellulose in ILs is performed below 130 °C, i.e., cellulose/IL solutions are safe to handle at typical processing temperatures. However, it has been noted that the thermostability is reduced significantly when using recycled ILs. This is an indication that already below T_{on}, degradation products are formed. The thermal decomposition of imidazolium-based ILs, which proceeds by a dealkylation mechanisms inverse to the synthesis, yields 1-alkylimidazoles (Fig. 5.6) [103, 104]. These primary products can further decompose, e.g., into imidazole, and/or condensate with other fragments formed [91]. These compounds are highly basic and can significantly affect chemical derivatization of cellulose in ILs. Although they might initially be formed only in small amounts, these heterocyclic degradation products cannot be removed simply by evaporation. Thus, they might accumulate during multiple recycling sequences.

5.3.3 *Recycling of Ionic Liquids*

Recycling of ILs is one, if not the major technological issue that needs to be solved in order to utilize this novel class of solvents in commercialized procedures in general and for the processing of cellulose in particular. ILs can be considered as rather expensive compounds but reutilization of the solvent for multiple processing cycles decreases the impact of IL prize on the overall process costs. Moreover, preventing pollution with potentially hazardous organic compounds is a matter of general concern when developing environmentally benign processes. ILs are often considered as recyclable mostly because of their very low vapor pressure.

Nevertheless, it needs to be emphasized that efficient recycling strategies for ILs used in (1) shaping, (2) biorefinery, or (3) chemical derivatization reactions are still missing.

The major component that needs to be removed after processing of cellulose is the non-solvent (usually water or an alcohol) used to regenerate cellulose (shaping) or the polysaccharide derivative (chemical derivatization). Additional impurities derive from the chemical derivatization reaction, e.g., residual reagents, side products, and co-solvents, but also from thermal decomposition of the IL and cellulose (see Sect. 5.3.2). Volatile compounds can easily be removed by evaporation under reduced pressure yielding crude ILs that, in some cases, could be utilized directly for another dissolution and chemical derivatization cycle. As an example; acylation of cellulose, dissolved in an imidazolium chloride based IL, yields hydrochloric or carboxylic acid when no additional base is added. Both compounds could be removed from the IL together with the volatile precipitation agent and excess acylation reagent [12, 13]. The NMR spectra of the recycled ILs showed no residual impurities and when reused as homogeneous reaction media, results were comparable to derivatization in the initial IL.

IL recycling becomes more complex with increasing number of potential impurities. In particular, the removal of non-volatile compounds proved to be more challenging. If a base is applied during the derivatization, the corresponding protonated acid is usually not removable by evaporation due to its higher boiling point [35, 40]. For comparison: the boiling points of pyridine, which has been applied for acylation, tosylation, and tritylation of cellulose in ILs, and pyridinium hydrochloride are 115 and 223 °C. Recycling can be achieved by neutralization of the IL in an aqueous solution, evaporation of water and the deprotonated base, and subsequent extraction of the crude IL with chloroform, which results in precipitation of inorganic salts that can be removed by filtration [33]. Finally, treatment of the recycled IL with an anion exchanger might be required in case anionic species are generated that cannot be removed by other means because they are too similar to the original IL's anion, e.g., carboxylates differing in their alkyl chain, or because their protonated form is not volatile, e.g., tosylate. NMR spectroscopy can be used to follow the individual recycling steps (Fig. 5.7).

So far, most strategies for the recycling of ILs after cellulose processing focused entirely on evaporation to remove the non-solvent, used for regeneration of the polysaccharide, and as well as side product formed upon derivatization. Up to now, it has not been studied whether this rather energy consuming approach is suitable from an economic and ecologic point of view. Thus, alternative approaches for recovery and purification of ILs are constantly studied not only in the field of polysaccharide research. Upon addition of 'water structuring salts', e.g., phosphates, carbonates, and citrates, or of certain organic compounds such as carbohydrates, amino acids, and surfactants, to an aqueous IL solution, separation into an IL-rich/water-deficient and an IL-deficient/water-rich phase occurs [105, 106]. This 'salting out' phenomenon has been exploited for recovery of AMIMCl, used as reaction medium for acylation of cellulose, in 85 % yield [16]. ^{1}H-NMR spectroscopy was used to confirm the purity of the recycled IL but no information on

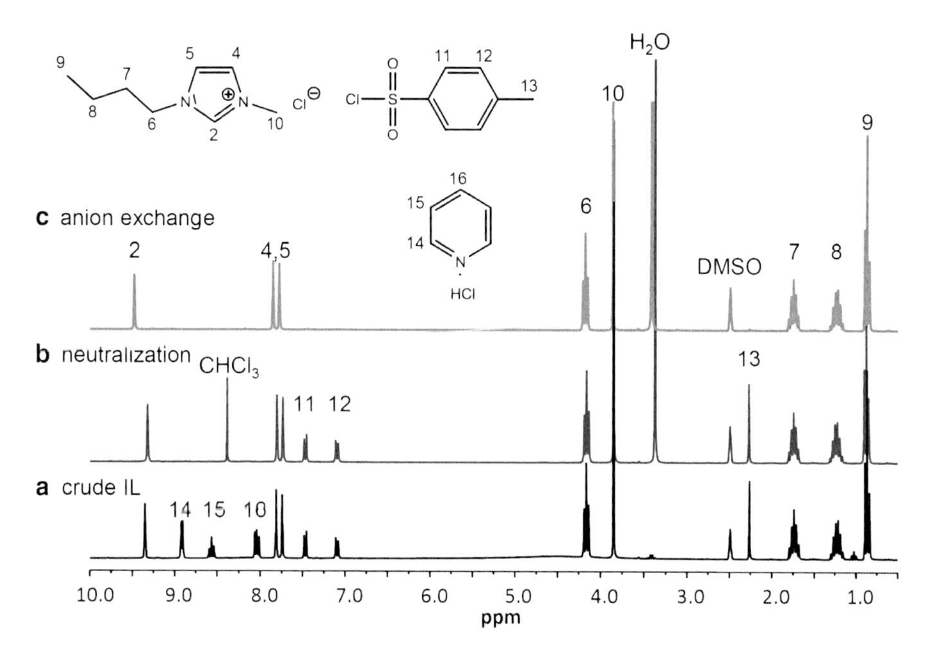

Fig. 5.7 ^{1}H-NMR spectra of ionic liquid (IL) samples, originally used as reaction medium for tosylation of cellulose, from different recycling steps; (**a**) crude IL after evaporation of volatile compounds, (**b**) IL after neutralization and removal of residual base, (**c**) purified IL after removal of ionic impurities by anion exchange (Adapted with permission from [35], Copyright 2012, Elsevier Ltd.)

inorganic impurities, in particular the residual amount of Na_2HPO_4 was provided, which was used to induce phase separation.

Alternative strategies, which are frequently discussed for recycling of ILs in general, are pervaporation, reverse osmosis, and nanofiltration [107, 108]. Only the latter has been studied already for recovery of ILs that were used as cellulose solvent but not as homogeneous reaction medium. A straight forward approach of increasing importance is the development of novel cellulose dissolving ILs that facilitate efficient recycling, e.g., by extraction or induced phase separation (see Sect. 5.3.4). Frequently, utilization of 'distillable ILs' such as guanidinium carboxylates has been proposed [109]. At high temperatures and low pressure, these ILs decompose into volatile compounds that reconstituted upon cooling. This process yields ILs of high purity but is also very energy consuming. Moreover, most of the impurities need to be removed in advance because they would evaporate prior to the IL.

When developing procedures for the chemical derivatization of cellulose in ILs, it is important to take recycling aspects into consideration as well. To give an example; homogeneous esterification of cellulose with a carboxylic acid chloride (very reactive reagent) in EMIMAc (low viscous room temperature liquid IL) will yield cellulose ester in good yield and high DS. However, purification of the IL by evaporation of volatile compounds is not feasible. Hydrochloric acid, formed as side product, will induce protonation of the less acidic carboxylate anion that is removed under reduced pressure. Thus, partial anion exchange from EMIMAc to EMIMCl will occur. A very complex mixtures of the recycled ILs can also be expected when mixed cellulose esters are prepared [58].

5.3.4 Task-Specific Reaction Media

One of the unique trademarks of ILs, which results from the possibility to combine a huge number of potential anions and cations, is their broad diversity in terms of structural features as well as physical and chemical properties. Based on the number of IL cations and anions, which have been reported in literature for the dissolution of the polysaccharide, about 300 ILs can be generated that are likely to act as solvents for cellulose [8]. Nevertheless, the vast majority of studies related to the use of ILs for processing of cellulose or cellulosic biomass focused on only three of those, namely BMIMCl, AMIMCl, or EMIMAc (see Sect. 5.3.1). Although most of the derivatization reactions described above have been carried out more or less successfully in these ILs, they possess certain 'intrinsic restrictions'. With respect to the increasing interest to use novel solvents for cellulose processing in commercial scales, it can be concluded that the development of task-specific IL based reaction media for the homogeneous cellulose chemistry bears huge potential and are going to be a focus of future research in this area.

'Task-specific' is not an absolute term but related to a particular 'task', i.e., homogeneous synthesis of different cellulose derivatives, and a "specific" feature that is enabled or prevented by a particular IL. In general, task-specific reaction media for the chemical modification of cellulose should be characterized by:

1. decreased viscosity
2. tunable hydrophobicity/hydrophobicity
3. efficient solvent recycling
4. the presence of specific chemical functionalities that prevent or catalyze specific derivatization reactions and
5. improved biocompatibility and/or biodegradability. The last two issues have scarcely been considered up to now.

First of all, even those ILs, used thus far for processing of cellulose, can become task-specific reaction media if they are applied consciously according to their specific advantages and disadvantages. This approach is limited to a certain extend and requires fundamental knowledge on the specific properties of ILs, e.g., viscosity and specific side reactions of anion and cation, with respect to the restrictions imposed by the particular derivatization reaction in which they are used as cellulose solvent and the subsequent recycling process. As an example, EMIMAc can be the reaction medium of choice for derivatization of cellulose at low temperatures. It is liquid at room temperature and shows lower viscosity then imidazolium chloride based ILs, which is beneficial in order to guarantee efficient mixing and mass distribution. However, it also needs to be considered that the acetate anion might act as catalyst or react with the derivatization reagents yielding unexpected reaction products (see Sect. 5.3.2) [37, 67]. Novel cellulose solvents can be obtained by systematic synthesis of ILs and low-melting organic salts with tailored properties based on the knowledge gained with common ILs, on one hand (Sect. 5.3.4.1). On the other, molecular co-solvents can be used to alter the properties cellulose/IL solutions towards specific applications (Sect. 5.3.4.2).

5.3.4.1 Novel ILs and Organic Salts for Derivatization of Cellulose

It should be pointed out that it is unlikely to find one particular IL that is the most suited cellulose solvent for all kinds of chemical derivatization reactions. Instead, the reaction media should be adapted to the specific requirements of a particular reaction. The synthesis of novel ILs for the use in cellulose chemistry involves the design of new cations and anions as well as the combination of both to an IL. Thereby, two major limitations need to be considered. It is self-evident that the new ILs (or low melting salts) must be able to dissolve reasonable amounts of cellulose (at least 5–10 %) without serve degradation of the polysaccharide. Moreover, dissolution should be possible at moderate temperatures (<130 °C), which implies that the polysaccharide solvents needs to be liquid in that area. Strictly speaking, the term IL describes compounds with a melting point below 100 °C. However, it has been demonstrated that also several organic salts with slightly higher melting points in the range of 100–140 °C, henceforward designated as low-melting salts, can either directly or in combination with certain co-solvents be useful solvents for derivatization of cellulose.

The dissolution mechanism of cellulose in ILs is still not understood yet and is matter of controversial discussions; as is the question why cellulose is actually insoluble in most molecular solvents (see [8, 110] and references therein). However, it has been demonstrated that in particular the anion has a very important role. It was found by means of solvatochromic experiments that cellulose is only soluble in ILs with a strong hydrogen acceptor capacity ($\beta > 0.8$) [111, 112]. This parameter is mainly determined by the nature of the anion. Thus variation and structure design of the IL's anion is restricted to a certain extent if the ability to dissolve cellulose needs to be retained. Most cellulose dissolving ILs applied so far for chemical derivatization of the polysaccharide contained either chloride or acetate as relatively strong hydrogen bond acceptor ions. Although chloride base ILs proofed to be efficient solvents for cellulose, their use is limited by the high melting points and viscosities. ILs bearing acetate anions are usually less viscous and liquid at room temperature but show some specific side reactions (see Sect. 5.3.2). Following the initial reports on this class of cellulose solvents, several task-specific ILs with alternative anions and beneficial properties have been reported that could be used for dissolution of cellulose. However, up to now only the minority has been studied as homogeneous reaction media.

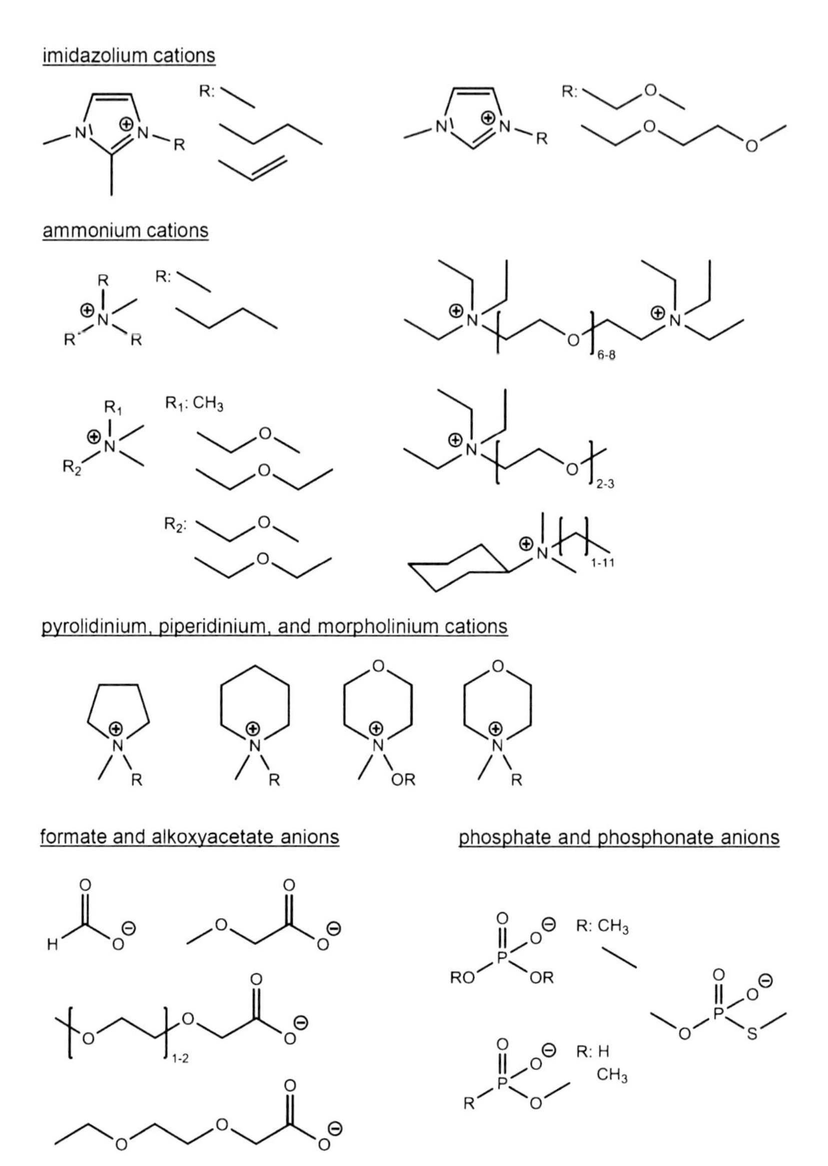

Fig. 5.8 Molecular structures of task-specific ionic liquids reported for dissolution and processing of cellulose, arranged by type of cation and anion

In addition to acetate, other carboxylates have been proposed as anionic species in cellulose dissolving ILs (Fig. 5.8). In particular imidazolium formates have been reported to possess lower viscosities compared to the corresponding chloride and acetate analogues [113]. With respect to the specific side reactions of the acetate anion, these low viscous ILs might be useful reaction media, e.g., for derivatization reactions that do not tolerate high temperatures. Several low melting ammonium formates have been prepared that could dissolve some polysaccharides including cellulose [71]. These low melting salts could successfully be applied also for the preparation of carboxymethyl cellulose with a rather high DS of about 1.6 and a non-statistic functionalization pattern.

Imidazolium- and ammonium salts with alkylphosphonate or dialkylphopshate as anions possess high hydrogen bond acceptor capacities and could be utilized as cellulose solvents [114, 115]. These ILs are liquid at room temperatures and their viscosities are in the range of 100–500 mPa s, which is comparable to the corresponding acetates. It has been proposed that substitution of oxygen by sulfur will result in a decreased viscosity due to the reduced symmetry of the anion [116]. First results on the use of phosphate ILs as homogeneous reaction media indicate that choice of the cation is crucial for the derivatization reaction. Whereas tetraalkylammonium dialkylphosphates could be used as solvents for the homogeneous preparation of cellulose esters and mixed ester, the corresponding imidazolium salts appear to be less suitable for the derivatization of cellulose because gelation of the reaction mixture occurs shortly after the addition of derivatization reagents [35, 54].

Tailoring the nature of the IL cation, e.g., by choosing among different general types of cations or by modifying length, degree of branching, and flexibility of side chains, offers access to a very broad structural diversity (Fig. 5.8). Nevertheless, up to now cellulose research has been focused mainly on 1-alkyl-3-methylimidazolium based ILs, bearing mostly ethyl, butyl, or allyl in the side chain. The cation has a significant effect on the ILs physical properties, e.g., melting point and viscosity, as well as its chemical reactivity. If and to what extent it is also directly involved in the cellulose dissolution mechanism is matter of current scientific discussion [8]. However, it appears that the choice of potential cations is less restricted compared to the limitation in terms of the possible anions.

Regarding commonly applied 1,3-dialkylimidazolium IL, task-specific solvents with interesting properties can be generated be substituting also the other positions of the aromatic ring [117]. The possibility to suppress deprotonation and carbene formation by methylation at position 2 has already been described (see Sect. 5.3.2). Another possibility is to increase flexibility of the side chains by introducing oxy-alkyl groups that also facilitate additional hydrogen bond interactions [56]. Replacing the butyl group in BMIMCl by a 2-methoxyethyl group seems to improve the solubility of cellulose and also resulted in an increased reactivity for acetylation reactions performed in this IL. Further studies are required to fully understand the reasons for these improvements, in particular because the opposite effect was observed when changing from a heptyl- to a 2-(2-ethoxyethoxy)ethyl substituted IL.

Taking into account the specific side reactions of dialkylimidazolium ILs, quaternary ammonium salts can be expected to become more important as solvents for dissolution and chemical modification of cellulose. A certain degree of cation asymmetry is required to obtain ILs with low melting points [118]. Despite that limitation, quaternary ammonium salts exhibit a much broader structural diversity than imidazolium ILs because in principle four side chains can be tailored.

Moreover, they can easily be prepared by complete alkylation of primary, secondary, and tertiary amines, precursors that are inexpensive and readily available.

Several pyrolidinium, piperidinium, and morpholinium ILs with different anions have been reported for the use as cellulose solvents [115, 119]. However, comprehensive comparison of their dissolution power as well as chemical and physical properties in comparison to ILs derived from aromatic or acyclic amines are scarcely found in literature. Especially morpholinium salts are of considerable scientific interest due to their relationship with NMMO. Low melting triethyl- and tributylmethylammonium formates could be obtained from the commercially available methyl carbonates by conversion with formic acid [71]. In the presence of a small excess of acid, these salts could be used as solvents for the homogeneous carboxymethylation of cellulose. Cellulose dissolving ammonium ILs with rather low viscosities (30–220 mPa s at 25 °C) have been obtained by introducing a cyclohexyl side chain to decrease symmetry of the cation [120]. The ILs carrying either acetate or alkoxyacetates could dissolve 1–9 % cellulose, which might be sufficient for specific applications. The solubility could be increased by adding DMSO.

Ether functionalized IL cations are expected to be better biodegradable, less toxic, and less viscous than their aliphatic analogues [121]. Several quaternary di- and trimethylammonium acetates bearing one or two alkoxy side chains have been prepared that showed melting points of 30–40 °C. Compared to imidazolium acetate, the viscosities of these ILs were slightly higher but they could dissolve similar amounts of cellulose and might consequently be interesting alternative reaction media for the derivatization of the polysaccharide [122]. Mono- and bicationic ammonium ILs with longer alkoxy groups and acetate as counter ion could be derived from poly(ethylene glycols) (PEG) and PEG monomethylethers [123]. According to the same approach, imidazolium and piperidinium acetates could be prepared as well. Depending on the number of C_2H_4O-units, some of these PEG functionalized ILs could dissolve approximately 8–12 % cellulose.

5.3.4.2 IL/Co-solvent Mixtures as Cellulose Solvents and Reaction Media

The systematic preparation of ILs and low-melting organic salts is a viable approach for obtaining tailored task-specific reaction media but it is only applicable within certain limitations. As an example, the viscosities of cellulose dissolving ILs are very high, compared to molecular solvents but also other ‘low-viscous’ ILs. It is not likely that a reduction of two or three orders of magnitudes, which seems to be required for certain derivatization reactions, can be achieved by modification of the molecular structure of ILs. One main reason for the high viscosity is the strong coulomb interaction between anions and cations, which is an ‘intrinsic feature’ of ILs and salts in general. Another factor is the formation of internal hydrogen bonds. Reduction in viscosity can be achieved by decreasing the strength of hydrogen

Table 5.4 Hydrogen bond acceptor ability (β), viscosity at 25 °C (η), and ability to dissolve cellulose of 1-butyl-3-methylimidazolium salts with different anions

Anion[a]	β	η, in Pa·s	Dissolves cellulose	Refs.
Ac^-	1.20	140	Yes	[111, 124]
Cl^-	0.83	142[b]	Yes	[111, 125]
Fo^-[c]	0.99	66	Yes	[113]
DMP^-[d]	1.00	265	Yes	[114]
$MeSO_4^-$	0.67	188	No	[111, 124]
$N(CN)_2^-$	0.64	29	No	[111, 124]
TfO^-	0.46	83	No	[124, 126]
BF_4^-	0.38	104	No	[126, 127]
$N(TfO)_2^-$	0.24	51	No	[124, 126]

[a]Ac^-: acetate, BF_4^-: tetrafluoroborate, Cl^-: chloride, DMP^-: diemethylphosphate, Fo^-: formate, $MeSO_4^-$: methylsulfate, $N(CN)_2^-$: dicyanamide, $N(TfO)_2^-$: bis(trifluoromethylsulfonyl)-imide, TfO^-: trifluoromethanesulfonate
[b]Recorded at 80 °C
[c]1-allyl-3-methylimidazolium salt
[d]1 ethyl-3-methylimidazolium salt

bonding between the two ionic species. Reducing the hydrogen bond donor ability (β) of an IL, which is mainly determined by the nature of the anion, can result in a significant decrease in viscosity (Table 5.4). Imidazolium salts with more hydrophobic cations, e.g., dicyanamide or bis(trifluoromethylsulfonyl)imide, are less viscous compared to the corresponding chlorides, carboxylates, and phosphates that are rather strong hydrogen bond acceptors. However, these low viscous ILs also lack the ability to dissolve cellulose. It has been demonstrated that cellulose is only soluble in ILs with rather high β values above >0.8, i.e., those that exhibit strong hydrogen bonding between anions and cations and consequently possess higher viscosities.

The addition of co-solvents is an efficient way to diminish 'intrinsic limitations' of ILs that are predetermined in a narrow frame by the molecular structure, such as viscosities, polarities, densities, and melting points [128]. It has been demonstrated that the viscosity of cellulose/IL solutions can be reduced drastically by addition of DMSO or DMF (Fig. 5.9) [33]. The decrease proceeds nearly exponentially with the weight fraction of dipolar aprotic co-solvent in the mixtures; at a typical mixing ration of 1 g co-solvent per g cellulose/IL solution, the viscosity is two order of magnitudes lower compared to an undiluted cellulose/IL solution of the same polymer content. Pyridine has been utilized as well to reduce the viscosity of cellulose/IL solutions [69]. For dilute and concentrated cellulose solutions in BMIMCl and AMIMCl it was demonstrated that the conformation of the dissolved polysaccharide is not impaired upon the addition of DMSO as a co-solvent [68]. The viscosity of the IL/DMSO mixtures as a function of the mole fraction of DMSO (x_{DMSO}) could be predicted precisely by the following simple equation; with k being a constant (0.12 for BMIMCl and 0.15 for AMIMCl):

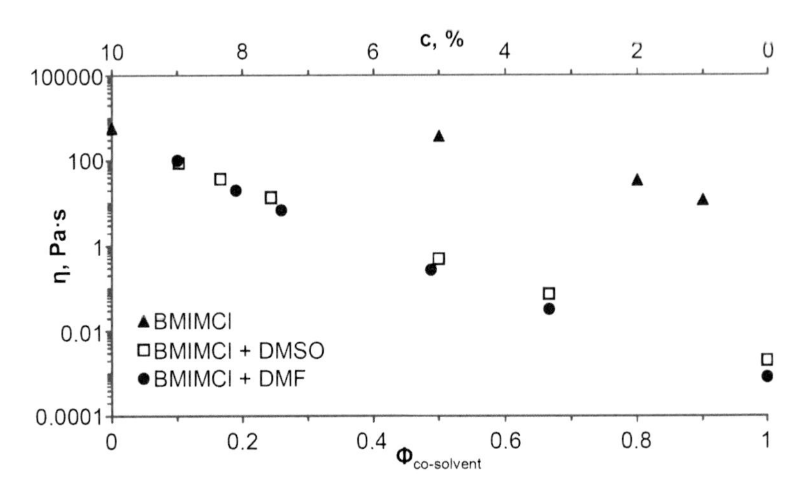

Fig. 5.9 Viscosity of cellulose dissolved in BMIMCl/co-solvent mixtures at 25 °C in comparison to undiluted cellulose/BMIMCl solutions, depending on the mass fraction (Φ) of co-solvent (*bottom scale*) and cellulose content (*top scale*) (Reprinted with permission from Gericke et al. [33], Copyright 2009, WILEY-VCH Verlag GmbH & Co. KGaA, Weinheim)

$$\ln \frac{\eta}{\eta_{IL}} = \frac{x_{\mathrm{DMSO}}}{k} \tag{5.3}$$

Co-solvents can also be applied to tailor hydrophobicity of ILs along with their hydrogen bond donor/acceptor properties [129–131]. Thus, miscibility of hydrophilic cellulose/IL solutions with hydrophobic derivatization reagents could be improved by adding nonpolar co-solvents such as chloroform [41].

The addition of a solvent to cellulose/IL solutions can result in complete miscibility but also phase separation might occur. Moreover, it might induce precipitation of cellulose. Semi-quantitative predictions on these phenomena can be gained based on the solvatochromic parameters of potential co-solvents (Table 5.5) [132]. Only compounds with a high normalized empirical polarity ($E_T^N > 0.3$) were found to be miscible with the relatively hydrophilic ILs used as cellulose solvents. Precipitation of cellulose is induced by solvents with a strong hydrogen bond donor acidity ($\alpha < 0.5$). In contrast, compounds that are strong hydrogen bond donor ($\beta > 0.4$) are very efficient co-solvents for cellulose/IL solutions; the amount of co-solvent that is tolerated before precipitation increases with increasing β-value. Solvatochromic measurements have been employed to characterize IL/co-solvent mixtures of different compositions that were employed for direct dissolution of cellulose [133, 134]. It could be demonstrated that the molar fraction of ILs in the mixtures can be reduced drastically (<0.3) if suitable dipolar aprotic co-solvents such as DMSO and amides are applied. In these mixtures, the high β-value of the IL, which seems to be important for the dissolution of cellulose, is almost unaffected even in case of an excess of co-solvent [133].

Table 5.5 Miscibility of cellulose/ionic liquid (IL) solutions with different co-solvents, adapted from [132]

Solvent	Solvent parameters[a]			IL[b]	Miscibility[c]	Max. equivalent of co-solvent[d]
	E_T^N	α	β			
Protic solvents[e]	>0.5	>0.8	0.5–0.8	AMIMCl	+/p	<0.2
				BMIMCl	+/p	<0.2
				EMIMAc	+/p	<0.2
Acetonitrile	0.46	0.19	0.40	AMIMCl	+/p	0.6
				BMIMCl	+/p	0.8
				EMIMAc	+/p	2
DMSO	0.44	0.0	0.76	AMIMCl	+	>10
				BMIMCl	+	>10
				EMIMAc	+	>10
DMF	0.39	0.0	0.69	AMIMCl	+/p	4
				BMIMCl	+/p	9
				EMIMAc	+/p	5
Dichloromethane	0.31	0.13	0.10	AMIMCl	+/g	1
				BMIMCl	+/g	1
				EMIMAc	+/p	3
Pyridine	0.30	0.0	0.64	AMIMCl	+/p	0.8
				BMIMCl	+/p	3
				EMIMAc	+/p	2
Chloroform	0.26	0.20	0.10	AMIMCl	+/g	1
				BMIMCl	+/g	1
				EMIMAc	+/g	1
Non-polar solvents[f]	≤0.23	0.0	≤0.55	AMIMCl	–	<0.2
				BMIMCl	–	<0.2
				EMIMAc	–	<0.2
DMF/methanol (7:3)	0.61	0.48	0.70	AMIMCl	+/p	0.8
				BMIMCl	+/p	1
				EMIMAc	+/p	1
DMF/methanol (5:5)	0.68	0.64	0.69	AMIMCl	+/p	0.6
				BMIMCl	+/p	0.6
				EMIMAc	+/–	0.8

[a]Solvent parameters: E_T^N: normalized solvent polarity, α: hydrogen bond donor ability, β: hydrogen bond acceptor ability

[b]AMIMCl: 1-allyl-3-methylimidazolium chloride, BMIMCl: 1-butyl-3-methylimidazolium chloride, EMIMAc: 1-ethyl-3-methylimidazolium acetate

[c]+: miscible, –: immiscible, +/p transition from miscible to precipitation, +/g transition from miscible to gelation, +/– transition from miscible to immiscible

[d]Amount of co-solvent (g per g cellulose/IL solutions) that can be added without permanent precipitation/gelation/immiscibility

[e]Water, methanol, ethanol, and 2-propanol

[f]Ethyl acetate, tetrahydrofuran, dioxane, diethyl ether, toluene, and hexane

In contrast, alcohols that are used for precipitation significantly decreased the hydrogen bond acceptor basicity. For IL/water mixtures it has been demonstrated that cellulose dissolves if the 'net basicity' ($0.35 < \beta - \alpha < 0.90$) of the solvent falls into a specific range [134].

IL/co-solvent mixtures can be exploited for the homogeneous preparation of highly engineered cellulose derivatives. As described above, CS and TOSC can be prepared at low temperatures (≤25 °C) in a completely homogeneous reaction in ILs but utilization of co-solvents is indispensable due to the high viscosity of the undiluted cellulose solutions [33, 35]. Homogeneous preparation of hydrophobic cellulose silyl ethers was realized in IL/chloroform mixtures [41]. Another example for the beneficial use of co-solvents is the hydroxyalkylation of cellulose in EMIMAc with gaseous oxiranes is a heterogeneous process that yields higher DS values if DMSO is added to decrease viscosity and increase solubility of the etherification reagents [37]. In fact, many of the derivatization reactions reported in literature have been carried out not in pure ILs but with the aid of co-solvents (see Tables 5.1 and 5.2) although it has not been mentioned explicitly in each case. Several patented procedures that are of interest for the preparation of commercially attractive cellulose derivatives were performed in IL/co-solvent mixtures [38, 54, 135–137].

In addition to physical and dissolution properties, co-solvents can also alter the chemical behavior of IL based reaction media by either facilitating or preventing specific derivatization reactions. As an alternative to commonly applied imidazolium chlorides, 1-allyl-3-methylimidazolium fluoride has been prepared as a novel solvent for cellulose [138]. Although the IL could be used for dissolution and acetylation of cellulose, the DS values of the products obtained were low because the fluoride anion induced cleavage of the ester bond. In the presence of DMSO, however, highly substituted cellulose esters could be obtained, presumably because of strong solvation of the anion by hydrogen bonding with the co-solvent. Pyridine, on the other hand, is an attractive co-solvent for derivatization reactions that require bases, e.g., tosylation, and tritylation, because it reduces viscosity of cellulose/IL solutions and likewise promotes the chemical conversion [35, 40]. Thus, no additional bases are required because the co-solvent is applied in a slight excess. An interesting approach is to exploit the co-solvent as derivatization reagent. A series of cationic cellulose esters has been prepared in mixtures of BMIMCl with different lactames, e.g., *N*-methyl-2-pyrolidine and ε-caprolactam, as co-solvents [135, 139]. The cyclic amides were activated with tosyl chloride and subsequently reacted with the polysaccharide backbone under ring opining.

Combination of the above described aspects, i.e., synthesis of novel ILs or low melting salts and the use of co-solvents to improve dissolution and processing of cellulose, is a logical consequence with huge potential for a broad range of applications. Interestingly this approach has already been described in the early 1930s in a number of patents [140, 141]. *N*-Alkylpyridinium salts have been utilized therein as cellulose solvents and pyridine was utilized to decrease their melting points of 120–130 °C and achieve dissolution at ambient temperature. Mixtures of these pyridinium salts and co-solvents have also been applied for shaping and chemical derivatization of cellulose [142, 143]. Although, these publications rose little interest during their time, they were rediscovered within the frame of the increasing interest for novel polysaccharide solvents. These reports are frequently considered as the first attempts to use ILs for dissolution of cellulose which is strictly speaking incorrect regarding the fact that the high melting points of the salts applied lie above 100 °C.

It is self-evident that co-solvents represent additional components that need to be considered in the recycling process. Volatile co-solvents, e.g., pyridine and chloroform, can be recovered by evaporation. Subsequently, the crude IL can be purified using suitable techniques. In contrast, co-solvents with relatively low vapor pressures, in particular dipolar aprotic ones like DMF, DMSO, and DMI, will remain in the crude IL after evaporating the non-solvent used for precipitation of the cellulose derivative. After removal of impurities that are harmful for dissolution or chemical modification of cellulose, the polysaccharide can be dissolved directly in the recycled IL/co-solvent mixture.

5.4 Conclusion

Compared to other cellulose solvents that are useful for cellulose derivatization yet restricted to small lab-scale synthesis, ILs bear huge potential for the homogeneous preparation of highly engineered cellulose derivatives on a commercially attractive scale. They can rapidly dissolve large amounts of cellulose and can be utilized as reaction media for numerous derivatization reactions. Nevertheless, it has been demonstrated in this chapter that several very specific issues that are related to the unique physical and chemical properties of ILs need to be considered. Regarding the limitations of currently applied ILs, it can be expected that a 'next generation' of task-specific ILs and related reaction media will be advanced in the near future, e.g., by including alternative types of cations and anions as well as IL/co-solvent systems. Another open question is IL recycling. In this context, the thermal behavior and chemical reactivity of ILs during dissolution and chemical derivatization of cellulose must be considered even more. If the specific side reaction of ILs are recognized and fully understood, they can be avoided or even exploited for preparing novel types of cellulose derivatives. As concluding remark it should be pointed out that research on the use of ILs for processing of cellulose will benefit a lot from interdisciplinary contributions from areas such as general organic, physical, and theoretical chemistry, chemical- and process engineering, biochemistry, toxicology, material testing, and other related fields.

References

1. Brandt A, Grasvik J, Hallett JP, Welton T. Deconstruction of lignocellulosic biomass with ionic liquids. Green Chem. 2013;15(3):550–83.
2. Stark A. Ionic liquids in the biorefinery: a critical assessment of their potential. Energy Environ Sci. 2011;4(1):19–32.

3. Sun N, Rodriguez H, Rahman M, Rogers RD. Where are ionic liquid strategies most suited in the pursuit of chemicals and energy from lignocellulosic biomass? Chem Commun. 2011;47 (5):1405–21.
4. Mora-Pale M, Meli L, Doherty TV, Linhardt RJ, Dordick JS. Room temperature ionic liquids as emerging solvents for the pretreatment of lignocellulosic biomass. Biotechnol Bioeng. 2011;108(6):1229–45.
5. Wendler F, Kosan B, Krieg M, Meister F. Possibilities for the physical modification of cellulose shapes using ionic liquids. Macromol Symp. 2009;280(1):112–22.
6. Sescousse R, Gavillon R, Budtova T. Aerocellulose from cellulose–ionic liquid solutions: preparation, properties and comparison with cellulose–NaOH and cellulose–NMMO routes. Carbohydr Polym. 2011;83(4):1766–74.
7. Gericke M, Trygg J, Fardim P. Functional cellulose beads: preparation, characterization, and applications. Chem Rev. 2013;113(7):4812–36.
8. Gericke M, Fardim P, Heinze T. Ionic liquids – promising but challenging solvents for homogeneous derivatization of cellulose. Molecules. 2012;17(6):7458–502.
9. Liebert T, Heinze T, Edgar KJ, editors. Cellulose solvents: for analysis, shaping and chemical modification, ACS symposium series. Washington, DC: American Chemical Society; 2010.
10. Wasserscheid P, Welton T, editors. Ionic liquids in synthesis. 2nd ed. Weinheim: Wiley-VCH; 2007.
11. Wu J, Zhang J, Zhang H, He J, Ren Q, Guo M. Homogeneous acetylation of cellulose in a new ionic liquid. Biomacromolecules. 2004;5(2):266–8.
12. Cao Y, Wu J, Meng T, Zhang J, He JS, Li HQ, Zhang Y. Acetone-soluble cellulose acetates prepared by one-step homogeneous acetylation of cornhusk cellulose in an ionic liquid 1-allyl-3-methylimidazolium chloride (AmimCl). Carbohydr Polym. 2007;69(4):665–72.
13. Heinze T, Schwikal K, Barthel S. Ionic liquids as reaction medium in cellulose functionalization. Macromol Biosci. 2005;5(6):520–5.
14. Barthel S, Heinze T. Acylation and carbanilation of cellulose in ionic liquids. Green Chem. 2006;8(3):301–6.
15. Schlufter K, Schmauder HP, Dorn S, Heinze T. Efficient homogeneous chemical modification of bacterial cellulose in the ionic liquid 1-N-butyl-3-methylimidazolium chloride. Macromol Rapid Commun. 2006;27(19):1670–6.
16. Fidale LC, Possidonio S, El Seoud OA. Application of 1-allyl-3-(1-butyl)imidazolium chloride in the synthesis of cellulose esters: properties of the ionic liquid, and comparison with other solvents. Macromol Biosci. 2009;9(8):813–21.
17. Possidonio S, Fidale LC, El Seoud OA. Microwave-assisted derivatization of cellulose in an ionic liquid: an efficient, expedient synthesis of simple and mixed carboxylic esters. J Polym Sci A Polym Chem. 2010;48(1):134–43.
18. Luan Y, Zhang J, Zhan M, Wu J, Zhang J, He J. Highly efficient propionylation and butyralation of cellulose in an ionic liquid catalyzed by 4-dimethylminopyridine. Carbohydr Polym. 2013;92(1):307–11.
19. Huang K, Xia J, Li M, Lian J, Yang X, Lin G. Homogeneous synthesis of cellulose stearates with different degrees of substitution in ionic liquid 1-butyl-3-methylimidazolium chloride. Carbohydr Polym. 2011;83(4):1631–5.
20. Zhang J, Wu J, Cao Y, Sang S, Zhang J, He J. Synthesis of cellulose benzoates under homogeneous conditions in an ionic liquid. Cellulose. 2009;16(2):299–308.
21. Köhler S, Heinze T. Efficient synthesis of cellulose furoates in 1-*N*-butyl-3-methylimidazolium chloride. Cellulose. 2007;14(5):489–95.
22. Dorn S, Pfeifer A, Schlufter K, Heinze T. Synthesis of water-soluble cellulose esters applying carboxylic acid imidazolides. Polym Bull. 2010;64(9):845–54.
23. Meng T, Gao X, Zhang J, Yuan J, Zhang Y, He J. Graft copolymers prepared by atom transfer radical polymerization (ATRP) from cellulose. Polymer. 2009;50(2):447–54.
24. Sui X, Yuan J, Zhou M, Zhang J, Yang H, Yuan W, Wei Y, Pan C. Synthesis of cellulose-graft-poly(N, N-dimethylamino-2-ethyl methacrylate) copolymers via homogeneous ATRP and their aggregates in aqueous media. Biomacromolecules. 2008;9(10):2615–20.

25. Xin T-T, Yuan T, Xiao S, He J. Synthesis of cellulose-graft-poly(methyl methacrylate) via homogeneous ATRP. Bioresources. 2011;6(3):2941–53.
26. Chun-xiang L, Huai-yu Z, Ming-hua L, Shi-yu F, Jia-jun Z. Preparation of cellulose graft poly (methyl methacrylate) copolymers by atom transfer radical polymerization in an ionic liquid. Carbohydr Polym. 2009;78(3):432–8.
27. Li WY, Jin AX, Liu CF, Sun RC, Zhang AP, Kennedy JF. Homogeneous modification of cellulose with succinic anhydride in ionic liquid using 4-dimethylaminopyridine as a catalyst. Carbohydr Polym. 2009;78(3):389–95.
28. Liu C-F, Zhang A-P, Li W-Y, Yue F-X, Sun R-C. Homogeneous modification of cellulose in ionic liquid with succinic anhydride using N-bromosuccinimide as a catalyst. J Agric Food Chem. 2009;57(5):1814–20.
29. Liu CF, Zhang AP, Li WY, Yue FX, Sun RC. Succinoylation of cellulose catalyzed with iodine in ionic liquid. Ind Crop Prod. 2010;31(2):363–9.
30. Liu CF, Sun RC, Zhang AP, Ren JL. Preparation of sugarcane bagasse cellulosic phthalate using an ionic liquid as reaction medium. Carbohydr Polym. 2007;68(1):17–25.
31. Li W, Wu L, Chen D, Liu C, Sun R. DMAP-catalyzed phthalylation of cellulose with phthalic anhydride in [bmim]Cl. Bioresources. 2011;6(3):2375–85.
32. Ma S, Xue X-l, Yu S-j, Wang Z-h. High-intensity ultrasound irradiated modification of sugarcane bagasse cellulose in an ionic liquid. Ind Crop Prod. 2012;35(1):135–9.
33. Gericke M, Liebert T, Heinze T. Interaction of ionic liquids with polysaccharides, 8 – synthesis of cellulose sulfates suitable for polyelectrolyte complex formation. Macromol Biosci. 2009;9(4):343–53.
34. Wang Z-M, Xiao K-J, Li L, Wu J-Y. Molecular weight-dependent anticoagulation activity of sulfated cellulose derivatives. Cellulose. 2010;17(5):953–61.
35. Gericke M, Schaller J, Liebert T, Fardim P, Meister F, Heinze T. Studies on the tosylation of cellulose in mixtures of ionic liquids and a co-solvent. Carbohydr Polym. 2012;89 (2):526–36.
36. Granström M, Kavakka J, King A, Majoinen J, Mäkelä V, Helaja J, Hietala S, Virtanen T, Maunu S-L, Argyropoulos D, Kilpeläinen I. Tosylation and acylation of cellulose in 1-allyl-3-methylimidazolium chloride. Cellulose. 2008;15(3):481–8.
37. Köhler S, Liebert T, Heinze T, Vollmer A, Mischnick P, Mollmann E, Becker W. Interactions of ionic liquids with polysaccharides 9. Hydroxyalkylation of cellulose without additional inorganic bases. Cellulose. 2010;17(2):437–48.
38. Möllmann E, Heinze T, Liebert T, Köhler S. Homogeneous synthesis of cellulose ethers in ionic liquids, US 20090221813 A1; 2009.
39. Granström M, Olszewska A, Mäkelä V, Heikkinen S, Kilpeläinen I. A new protection group strategy for cellulose in an ionic liquid: simultaneous protection of two sites to yield 2,6-di-O-substituted mono-p-methoxytrityl cellulose. Tetrahedron Lett. 2009;50(15):1744–7.
40. Erdmenger T, Haensch C, Hoogenboom R, Schubert US. Homogeneous tritylation of cellulose in 1-butyl-3-methylimidazolium chloride. Macromol Biosci. 2007;7(4):440–5.
41. Köhler S, Liebert T, Heinze T. Interactions of ionic liquids with polysaccharides. VI. Pure cellulose nanoparticles from trimethylsilyl cellulose synthesized in ionic liquids. J Polym Sci Part A Polym Chem. 2008;46(12):4070–80.
42. Mormann W, Wezstein M. Trimethylsilylation of cellulose in ionic liquids. Macromol Biosci. 2009;9(4):369–75.
43. Dong H, Xu Q, Li Y, Mo S, Cai S, Liu L. The synthesis of biodegradable graft copolymer cellulose-graft-poly(l-lactide) and the study of its controlled drug release. Colloids Surf B: Biointerfaces. 2008;66(1):26–33.
44. Yan C, Zhang J, Lv Y, Yu J, Wu J, Zhang J, He J. Thermoplastic cellulose-graft-poly(l-lactide) copolymers homogeneously synthesized in an ionic liquid with 4-dimethylaminopyridine catalyst. Biomacromolecules. 2009;10(8):2013–8.
45. Guo Y, Wang X, Shu X, Shen Z, Sun R-C. Self-assembly and paclitaxel loading capacity of cellulose-graft-poly(lactide) nanomicelles. J Agric Food Chem. 2012;60(15):3900–8.

46. Luan Y, Wu J, Zhan M, Zhang J, Zhang J, He J. "One pot" homogeneous synthesis of thermoplastic cellulose acetate-graft-poly(l-lactide) copolymers from unmodified cellulose. Cellulose. 2013;20(1):327–37.
47. Guo Y, Wang X, Shen Z, Shu X, Sun R. Preparation of cellulose-graft-poly(ε-caprolactone) nanomicelles by homogeneous ROP in ionic liquid. Carbohydr Polym. 2013;92(1):77–83.
48. Hao Y, Peng J, Li J, Zhai M, Wei G. An ionic liquid as reaction media for radiation-induced grafting of thermosensitive poly (*N*-isopropylacrylamide) onto microcrystalline cellulose. Carbohydr Polym. 2009;77(4):779–84.
49. Lin C-X, Zhan H-Y, Liu M-H, Fu S-Y, Huang L-H. Rapid homogeneous preparation of cellulose graft copolymer in BMIMCL under microwave irradiation. J Appl Polym Sci. 2010;118(1):399–404.
50. Eastman Chemical Company. Available online: http://www.eastman.com/Literature_Center/E/E325.pdf. Accessed 1 June 2011.
51. Celanese Corporation Home Page. Available online: http://www.celanese.com/index/productsmarkets_index/products_markets_acetate.html. Accessed 1 June 2011.
52. Buchanan CM, Buchanan NL. Reformation of ionic liquids in cellulose esterification, WO2008100569A1; 2008.
53. Buchanan CM, Buchanan NL, Hembre RT, Lambert JL. Cellulose esters and their production in carboxylated ionic liquids, WO2008100566A1; 2008.
54. Buchanan CM, Buchanan NL, Guzman-Morales E. Cellulose solutions comprising tetraalkylammonium alkylphosphate and products produced therefrom, WO2010120268A1; 2010.
55. Hembre RT, Buchanan NL, Buchanan CM, Lambert JL, Donelson ME, Gorbunova MG, Kuo T, Wang B. Regioselectively substituted cellulose esters produced in a carboxylated ionic liquid process and products produced therefrom, WO2010019244A1; 2010.
56. El Seoud OA, da Silva VC, Possidonio S, Casarano R, Arêas EPG, Gimenes P. Microwave-assisted derivatization of cellulose, 2 – the surprising effect of the structure of ionic liquids on the dissolution and acylation of the biopolymer. Macromol Chem Phys. 2011;212 (23):2541–50.
57. Hoffmann J, Nuchter M, Ondruschka B, Wasserscheid P. Ionic liquids and their heating behaviour during microwave irradiation – a state of the art report and challenge to assessment. Green Chem. 2003;5(3):296–9.
58. Huang K, Wang B, Cao Y, Li H, Wang J, Lin W, Mu C, Liao D. Homogeneous preparation of cellulose acetate propionate (CAP) and cellulose acetate butyrate (CAB) from sugarcane bagasse cellulose in ionic liquid. J Agric Food Chem. 2011;59(10):5376–81.
59. Yashima E. Polysaccharide-based chiral stationary phases for high-performance liquid chromatographic enantioseparation. J Chromatogr A. 2001;906(1–2):105–25.
60. Toga Y, Hioki K, Namikoshi H, Shibata T. Dependence of chiral recognition on the degree of substitution of cellulose benzoate. Cellulose. 2004;11(1):65–71.
61. Hussain MA, Liebert T, Heinze T. Acylation of cellulose with N, N′-carbonyldiimidazole-activated acids in the novel solvent dimethyl sulfoxide/tetrabutylammonium fluoride. Macromol Rapid Commun. 2004;25(9):916–20.
62. Liebert TF, Heinze T. Tailored cellulose esters: synthesis and structure determination. Biomacromolecules. 2005;6(1):333–40.
63. Sealey JE, Samaranayake G, Todd JG, Glasser WG. Novel cellulose derivatives. IV. Preparation and thermal analysis of waxy esters of cellulose. J Polym Sci Part B: Polym Phys. 1996;34(9):1613–20.
64. Heinze T, Schaller J. New water soluble cellulose esters synthesized by an effective acylation procedure. Macromol Chem Phys. 2000;201(12):1214–8.
65. Heinze T, Liebert T, Koschella A. Esterification of polysaccharides. Berlin/Heidelberg: Springer; 2006.
66. Heinze T, Daus S, Gericke M, Liebert T. Semi-synthetic sulfated polysaccharides – promising materials for biomedical applications and supramolecular architecture. In: Tiwari A, editor. Polysaccharides: development, properties and applications, Polymer science and technology. New York: Nova Science Publishers; 2010.

67. Köhler S, Liebert T, Schöbitz M, Schaller J, Meister F, Günther W, Heinze T. Interactions of ionic liquids with polysaccharides 1. Unexpected acetylation of cellulose with 1-ethyl-3-methylimidazolium acetate. Macromol Rapid Commun. 2007;28(24):2311–7.
68. Lv Y, Wu J, Zhang J, Niu Y, Liu C-Y, He J, Zhang J. Rheological properties of cellulose/ionic liquid/dimethylsulfoxide (DMSO) solutions. Polymer. 2012;53(12):2524–31.
69. Vitz J, Yevlampieva NP, Rjumtsev E, Schubert US. Cellulose molecular properties in 1-alkyl-3-methylimidazolium-based ionic liquid mixtures with pyridine. Carbohydr Polym. 2010;82(4):1046–53.
70. Myllymaeki V, Aksela R. Etherification of cellulose in ionic liquid solutions, WO2005054298A1; 2005.
71. Köhler S, Liebert T, Heinze T. Ammonium-based cellulose solvents suitable for homogeneous etherification. Macromol Biosci. 2009;9(9):836–41.
72. Kondo T, Gray DG. The preparation of *O*-methyl- and *O*-ethyl-celluloses having controlled distribution of substituents. Carbohydr Res. 1991;220:173–83.
73. Petzold-Welcke K, Kötteritzsch M, Heinze T. 2,3-*O*-methyl cellulose: studies on synthesis and structure characterization. Cellulose. 2010;17(2):449–57.
74. Saake B, Patt R, Puls J, Philipp B. Molecular weight distribution of cellulose. Papier (Darmstadt). 1991;45:727–35.
75. Gericke M, Schlufter K, Liebert T, Heinze T, Budtova T. Rheological properties of cellulose/ionic liquid solutions: from dilute to concentrated states. Biomacromolecules. 2009;10(5):1188–94.
76. Sescousse R, Le KA, Ries ME, Budtova T. Viscosity of cellulose-imidazolium-based ionic liquid solutions. J Phys Chem B. 2010;114(21):7222–8.
77. Matsumoto T, Tatsumi D, Tamai N, Takaki T. Solution properties of celluloses from different biological origins in LiCl center dot DMAc. Cellulose. 2001;8(4):275–82.
78. Blachot J-F, Brunet N, Navard P, Cavaillé JY. Rheological behavior of cellulose/monohydrate of n-methylmorpholine n-oxide solutions Part 1: Liquid state. Rheol Acta. 1998;37(2):107–14.
79. Roy C, Budtova T, Navard P. Rheological properties and gelation of aqueous cellulose-NaOH solutions. Biomacromolecules. 2003;4(2):259–64.
80. Gavillon R, Budtova T. Kinetics of cellulose regeneration from cellulose-NaOH-water gels and comparison with cellulose-*N*-methylmorpholine-*N*-oxide-water solutions. Biomacromolecules. 2007;8(2):424–32.
81. Froba AP, Kremer H, Leipertz A. Density, refractive index, interfacial tension, and viscosity of ionic liquids [EMIM][$EtSO_4$], [EMIM][NTf_2], [EMIM][$N(CN)_2$], and [OMA][NTf_2] in dependence on temperature at atmospheric pressure. J Phys Chem B. 2008;112(39):12420–30.
82. Okoturo OO, VanderNoot TJ. Temperature dependence of viscosity for room temperature ionic liquids. J Electroanal Chem. 2004;568(1–2):167–81.
83. Alder RW, Allen PR, Williams SJ. Stable carbenes as strong bases. J Chem Soc Chem Commun. 1995;12:1267–8.
84. Amyes TL, Diver ST, Richard JP, Rivas FM, Toth K. Formation and stability of N-heterocyclic carbenes in water: the carbon acid pKa of imidazolium cations in aqueous solution. J Am Chem Soc. 2004;126(13):4366–74.
85. Canal JP, Ramnial T, Dickie DA, Clyburne JAC. From the reactivity of *N*-heterocyclic carbenes to new chemistry in ionic liquids. Chem Commun. 2006;17:1809–18.
86. Enders D, Niemeier O, Henseler A. Organocatalysis by *N*-heterocyclic carbenes. Chem Rev. 2007;107(12):5606–55.
87. Liebert T, Heinze T. Interaction of ionic liquids with polysaccharides. 5. Solvents and reaction media for the modification of cellulose. Bioresources. 2008;3(2):576–601.
88. Ebner G, Schiehser S, Potthast A, Rosenau T. Side reaction of cellulose with common 1-alkyl-3-methylimidazolium-based ionic liquids. Tetrahedron Lett. 2008;49(51):7322–4.
89. Rodriguez H, Gurau G, Holbrey JD, Rogers RD. Reaction of elemental chalcogens with imidazolium acetates to yield imidazole-2-chalcogenones: direct evidence for ionic liquids as proto-carbenes. Chem Commun. 2011;47(11):3222–4.

90. Handy ST, Okello M. The 2-position of imidazolium ionic liquids: substitution and exchange. J Org Chem. 2005;70(5):1915–8.
91. Liebner F, Patel I, Ebner G, Becker E, Horix M, Potthast A, Rosenau T. Thermal aging of 1-alkyl-3-methylimidazolium ionic liquids and its effect on dissolved cellulose. Holzforschung. 2010;64(2):161–6.
92. Gordon CM, Muldoon MJ, Wagner M, Hilgers C, Davis JH, Wasserscheid P. Synthesis and purification. In: Ionic liquids in synthesis. Weinheim: Wiley-VCH Verlag GmbH & Co. KGaA; 2008. p. 7–55.
93. Anthony JL, Maginn EJ, Brennecke JF. Solution thermodynamics of imidazolium-based ionic liquids and water. J Phys Chem B. 2001;105(44):10942–9.
94. Hsu W-H, Lee Y-Y, Peng W-H, Wu KCW. Cellulosic conversion in ionic liquids (ILs): effects of H_2O/cellulose molar ratios, temperatures, times, and different ILs on the production of monosaccharides and 5-hydroxymethylfurfural (HMF). Catal Today. 2011;174 (1):65–9.
95. Zhang Z, Wang W, Liu X, Wang Q, Li W, Xie H, Zhao ZK. Kinetic study of acid-catalyzed cellulose hydrolysis in 1-butyl-3-methylimidazolium chloride. Bioresour Technol. 2012;112:151–5.
96. Mazza M, Catana D-A, Vaca-Garcia C, Cecutti C. Influence of water on the dissolution of cellulose in selected ionic liquids. Cellulose. 2009;16(2):207–15.
97. Le K, Sescousse R, Budtova T. Influence of water on cellulose-EMIMAc solution properties: a viscometric study. Cellulose. 2012;19(1):45–54.
98. Kosmulski M, Gustafsson J, Rosenholm JB. Thermal stability of low temperature ionic liquids revisited. Thermochim Acta. 2004;412(1–2):47–53.
99. Dorn S, Wendler F, Meister F, Heinze T. Interactions of ionic liquids with polysaccharides-7: thermal stability of cellulose in ionic liquids and *N*-methylmorpholine-*N*-oxide. Macromol Mater Eng. 2008;293(11):907–13.
100. Wendler F, Todi L-N, Meister F. Thermostability of imidazolium ionic liquids as direct solvents for cellulose. Thermochim Acta. 2012;528:76–84.
101. Wendler F, Graneß G, Heinze T. Characterization of autocatalytic reactions in modified cellulose/NMMO solutions by thermal analysis and UV/VIS spectroscopy. Cellulose. 2005;12(4):411–22.
102. Wendler F, Konkin A, Heinze T. Studies on the stabilization of modified lyocell solutions. Macromol Symp. 2008;262(1):72–84.
103. Kroon MC, Buijs W, Peters CJ, Witkamp G-J. Quantum chemical aided prediction of the thermal decomposition mechanisms and temperatures of ionic liquids. Thermochim Acta. 2007;465(1–2):40–7.
104. Chambreau SD, Boatz JA, Vaghjiani GL, Koh C, Kostko O, Golan A, Leone SR. Thermal decomposition mechanism of 1-ethyl-3-methylimidazolium bromide ionic liquid. J Phys Chem A. 2011;116(24):5867–76.
105. Gutowski KE, Broker GA, Willauer HD, Huddleston JG, Swatloski RP, Holbrey JD, Rogers RD. Controlling the aqueous miscibility of ionic liquids: aqueous biphasic systems of water-miscible ionic liquids and water-structuring salts for recycle, metathesis, and separations. J Am Chem Soc. 2003;125(22):6632–3.
106. Ventura SPM, Sousa SG, Serafim LS, Lima ÁS, Freire MG, Coutinho JAP. Ionic liquid based aqueous biphasic systems with controlled pH: the ionic liquid cation effect. J Chem Eng Data. 2011;56(11):4253–60.
107. Hazarika S, Dutta NN, Rao PG. Dissolution of lignocellulose in ionic liquids and its recovery by nanofiltration membrane. Sep Purif Technol. 2012;97:123–9.
108. Haerens K, Van Deuren S, Matthijs E, Van der Bruggen B. Challenges for recycling ionic liquids by using pressure driven membrane processes. Green Chem. 2010;12(12):2182–8.
109. King AWT, Asikkala J, Mutikainen I, Järvi P, Kilpeläinen I. Distillable acid–base conjugate ionic liquids for cellulose dissolution and processing. Angew Chem Int Ed. 2011;50 (28):6301–5.

110. Glasser W, Atalla R, Blackwell J, Malcolm Brown Jr R, Burchard W, French A, Klemm D, Nishiyama Y. About the structure of cellulose: debating the Lindman hypothesis. Cellulose. 2012;19(3):589–98.
111. Brandt A, Hallett JP, Leak DJ, Murphy RJ, Welton T. The effect of the ionic liquid anion in the pretreatment of pine wood chips. Green Chem. 2010;12(4):672–9.
112. Xu AR, Wang JJ, Wang HY. Effects of anionic structure and lithium salts addition on the dissolution of cellulose in 1-butyl-3-methylimidazolium-based ionic liquid solvent systems. Green Chem. 2010;12(2):268–75.
113. Fukaya Y, Sugimoto A, Ohno H. Superior solubility of polysaccharides in low viscosity, polar, and halogen-free 1,3-dialkylimidazolium formates. Biomacromolecules. 2006;7(12):3295–7.
114. Fukaya Y, Hayashi K, Wada M, Ohno H. Cellulose dissolution with polar ionic liquids under mild conditions: required factors for anions. Green Chem. 2008;10(1):44–6.
115. Fukaya Y, Hayashi K, Kim Seung S, Ohno H. Design of polar ionic liquids to solubilize cellulose without heating. In: Cellulose solvents: for analysis, shaping and chemical modification, ACS symposium series, vol. 1033. Washington, DC: American Chemical Society; 2010. p. 55–66.
116. Hummel M, Froschauer C, Laus G, Roder T, Kopacka H, Hauru LKJ, Weber HK, Sixta H, Schottenberger H. Dimethyl phosphorothioate and phosphoroselenoate ionic liquids as solvent media for cellulosic materials. Green Chem. 2011;13(9):2507–17.
117. Bara JE, Shannon MS. Beyond 1,3-difunctionalized imidazolium cations. Nanomater Energy. 2012;1:237–42.
118. Zhou Z-B, Matsumoto H, Tatsumi K. Low-melting, low-viscous, hydrophobic ionic liquids: aliphatic quaternary ammonium salts with perfluoroalkyltrifluoroborates. Chem Eur J. 2005;11(2):752–66.
119. Hummel M, Laus G, Schwärzler A, Bentivoglio G, Rubatscher E, Kopacka H, Wurst K, Kahlenberg V, Gelbrich T, Griesser Ulrich J, Röder T, Weber Hedda K, Schottenberger H, Sixta H. Non-halide ionic liquids for solvation, extraction, and processing of cellulosic materials. In: Cellulose solvents: for analysis, shaping and chemical modification, ACS symposium series, vol. 1033. Washington, DC: American Chemical Society; 2010. p. 229–59.
120. Pernak J, Kordala R, Markiewicz B, Walkiewicz F, Poplawski M, Fabianska A, Jankowski S, Lozynski M. Synthesis and properties of ammonium ionic liquids with cyclohexyl substituent and dissolution of cellulose. RSC Adv. 2012;2(22):8429–38.
121. Tang S, Baker GA, Zhao H. Ether- and alcohol-functionalized task-specific ionic liquids: attractive properties and applications. Chem Soc Rev. 2012;41(10):4030–66.
122. Chen Z, Liu S, Li Z, Zhang Q, Deng Y. Dialkoxy functionalized quaternary ammonium ionic liquids as potential electrolytes and cellulose solvents. New J Chem. 2011;35(8):1596–606.
123. Tang S, Baker GA, Ravula S, Jones JE, Zhao H. PEG-functionalized ionic liquids for cellulose dissolution and saccharification. Green Chem. 2012;14(10):2922–32.
124. McHale G, Hardacre C, Ge R, Doy N, Allen RWK, MacInnes JM, Bown MR, Newton MI. Density – viscosity product of small-volume ionic liquid samples using quartz crystal impedance analysis. Anal Chem. 2008;80(15):5806–11.
125. Fendt S, Padmanabhan S, Blanch HW, Prausnitz JM. Viscosities of acetate or chloride-based ionic liquids and some of their mixtures with water or other common solvents. J Chem Eng Data. 2010;56(1):31–4.
126. Wu YS, Sasaki T, Kazushi K, Seo T, Sakurai K. Interactions between spiropyrans and room-temperature ionic liquids: photochromism and solvatochromism. J Phys Chem B. 2008;112(25):7530–6.
127. Harris KR, Kanakubo M, Woolf LA. Temperature and pressure dependence of the viscosity of the ionic liquid 1-butyl-3-methylimidazolium tetrafluoroborate: viscosity and density relationships in ionic liquids. J Chem Eng Data. 2007;52(6):2425–30.
128. Heintz A. Recent developments in thermodynamics and thermophysics of non-aqueous mixtures containing ionic liquids. A review. J Chem Thermodyn. 2005;37(6):525–35.

129. Mellein BR, Aki SNVK, Ladewski RL, Brennecke JF. Solvatochromic studies of ionic liquid/organic mixtures. J Phys Chem B. 2007;111(1):131–8.
130. Palomar J, Torrecilla JS, Lemus J, Ferro VR, Rodriguez F. A COSMO-RS based guide to analyze/quantify the polarity of ionic liquids and their mixtures with organic cosolvents. Phys Chem Chem Phys. 2010;12(8):1991–2000.
131. Khupse ND, Kumar A. Delineating solute – solvent interactions in binary mixtures of ionic liquids in molecular solvents and preferential solvation approach. J Phys Chem B. 2011;115 (4):711–8.
132. Gericke M, Liebert T, Seoud OAE, Heinze T. Tailored media for homogeneous cellulose chemistry: ionic liquid/co-solvent mixtures. Macromol Mater Eng. 2011;296(6):483–93.
133. Rinaldi R. Instantaneous dissolution of cellulose in organic electrolyte solutions. Chem Commun. 2011;47(1):511–3.
134. Hauru LKJ, Hummel M, King AWT, Kilpeläinen I, Sixta H. Role of solvent parameters in the regeneration of cellulose from ionic liquid solutions. Biomacromolecules. 2012;13 (9):2896–905.
135. Brackhagen M, Heinze T, Dorn S, Koschella A. Method for manufacturing cellulose derivatives containing amino groups in ionic liquids, EP 2072530 A1; 2009.
136. Liebert T, Heinze T, Gericke M. Production water-soluble, low-substituted cellulose sulfates, DE102007035322 B4; 2009.
137. Granström M, Mormann W, Frank P. Method for chlorinating polysaccharides or oligosaccharides, WO2011086082A1; 2011.
138. Casarano R, El Seoud OA. Successful application of an ionic liquid carrying the fluoride counter-ion in biomacromolecular chemistry: microwave-assisted acylation of cellulose in the presence of 1-allyl-3-methylimidazolium fluoride/DMSO mixtures. Macromol Biosci. 2013;13(2):191–202.
139. Zarth C, Koschella A, Pfeifer A, Dorn S, Heinze T. Synthesis and characterization of novel amino cellulose esters. Cellulose. 2011;18(5):1315–25.
140. Gesellschaft für Chemische Industrie in Basel. Verfahren zur Herstellung einer neuen Zelluloselösung und neue Zelluloselösung, CH153446; 1932.
141. Graenacher C. Cellulose solution, US 1,943,176; 1934.
142. Linko Y-Y, Viskari R, Pohjola L, Linko P. Preparation and performance of cellulose bead-entrapped whole cell glucose isomerase. J Solid-Phase Biochem. 1977;2(3):203–12.
143. Husemann VE, Siefert E. *N*-äthyl-pyridinium-chlorid als lösungsmittel und reaktionsmedium für cellulose. Die Makromolekulare Chemie. 1969;128(1):288–91.

Chapter 6
Fractionation of Lignocellulosic Materials with Ionic Liquids

Timo Leskinen, Alistair W.T. King, and Dimitris S. Argyropoulos

Abstract Ionic liquids (ILs) have been recognized as a promising way to fractionate lignocellulosic biomass. During recent years, a number of publications have introduced a variety of technical developments and solvent systems based on several types of ILs to fractionate lignocellulose into individual polymeric components, after full or partial dissolution. In this chapter we briefly review the latest developments and knowledge in this field of study and introduce an alternative fractionation method based on the controlled regeneration of components from 1-allyl-3-methylimidazolium chloride ([amim]Cl). Norway spruce (*Picea abies*) and *Eucalyptus grandis* woods were dissolved in their fibrous state or by utilizing ball milling to improve solubility. The resulting wood solutions were precipitated gradually into fractions by addition of non-solvents, such as acetonitrile and water. Further water extraction of the crude fractions resulted in better separations. By analyzing molecular weight distributions of the fractions, together with their chemical composition, we have obtained fundamental information concerning the mechanisms of wood fractionation with ILs. Fractionation efficiency is found to be highly dependent on the modification of the wood cell wall ultrastructure and the degree of reduction of the molecular weights of the main components, arising from mechanical degradation.

T. Leskinen
Departments of Forest Biomaterials and Chemistry, North Carolina State University, Raleigh, NC 27695-8005, USA

A.W.T. King
Department of Chemistry, University of Helsinki, Helsinki 00014, Finland

D.S. Argyropoulos (✉)
Departments of Forest Biomaterials and Chemistry, North Carolina State University, Raleigh, NC 27695-8005, USA

Center of Excellence for Advanced Materials Research (CEAMR), King Abdulaziz University, P.O. Box 80203, Jeddah 21589, Saudi Arabia
e-mail: dsargyro@ncsu.edu

Z. Fang et al. (eds.), *Production of Biofuels and Chemicals with Ionic Liquids*, Biofuels and Biorefineries 1, DOI 10.1007/978-94-007-7711-8_6,

Isolation of cellulose enriched fractions was archived with Spruce sawdust and ball milled Eucalyptus, evidently following from distinct dissolution mechanisms.

Keywords Wood • Cellulose • Lignin • LCC • Ionic liquids • 1-allyl-3-methylimidazolium chloride • Fractionation • Extraction • Separation • Molecular weight • Pulp • Biofuels

6.1 Introduction

In a relatively short time, the research area of ionic liquid-mediated fractionation and pretreatment of wood has emerged from the interest of a small group of scientists into a noteworthy and diverse field of study. Global interest in lignocellulosic biomass is experiencing a renaissance, not only because of the growing financial potential in lignocellulose-based liquid biofuels, but also because it represents a source of bio-based materials and chemicals. Ionic liquids (ILs) have been recognized to have potential in many applications that can be categorized under the advanced utilization of grassy and woody biomass. The ability to dissolve various biopolymers and a general status as a green alternative to organic solvents makes IL platform technologies attractive to industry. This is mostly in areas pertaining to the manufacture of novel polymeric materials, by derivatization or blending of cellulose. Alternatively, in biomass pretreatments, including structural or compositional alteration of plant cell walls and acid catalysed hydrolysis of plant polysaccharides, for the purposes of biofuel production [1–3]. Aside from the use of ILs as a media for modification, fractionation of lignocellulosic biomass can be integrated into a variety of applications as it can also be used as a method to obtain purified or specified polymeric raw materials, for further use [4]. IL-mediated fractionation is suitable for the general concept of a biorefinery, serving the demand of component separation for subsequent multiple product streams. Ideally this method should be tunable. However, the selectivity of fractionation of native woods using ILs is still poorly developed or understood, from a mechanistic point of view.

Understanding the fundamentals of the separation of polymeric cell wall components has improved after initial publications concerning cellulose and whole wood dissolution into ILs [5–7]. Ideally, there are two ways to fractionate lignocellulose: (1) complete pre-dissolution of biomass followed by selective precipitation of the sought components as purified fractions, by addition of a non-solvent, or (2) selective extraction of components from the biomass. The first efforts to isolate purified fractions using ILs can be roughly categorized under either of the aforementioned approaches [6, 8–10]. However, the complex recalcitrant structure of wood greatly hinders a complete dissolution and efficient fractionation. During the last few years, a variety of new methods have emerged resulting in enhanced fractionation. In addition to introducing our work on wood fractionation, in this effort we will also present a brief overview of the latest technical advances in the IL-based fractionation systems and of our findings related to the mechanisms controlling the dissolution and fractionation of the complex materials of the wood cell wall.

Our work concerning wood fractionation has focused on dissolution of Norway spruce (softwood) and *Eucalyptus grandis* (hardwood) woods, as completely as possible under mild conditions, followed by a stepwise regeneration of wood components, with the addition of non-solvents. From initial screening, non-solvents were chosen in an attempt to enhance the selectivity of component precipitation. In this selection we considered the optical brightness of the precipitated samples, the ease of recovery (defined precipitate vs. emulsions) and the ability to fractionally precipitate the dissolved material. In our overall work we have selected the IL 1-allyl-3-methylimidazolium chloride ([amim]Cl) for the fractionation experiments, which has been demonstrated to have a good dissolution capability for cellulose and wood materials [7, 11–13]. The starting materials were either coarse TMP softwood pulp or sawdust and fine ball-milled powders from soft- or hardwood. In contrast to the usual approach, in component regeneration from IL solutions, we did not use excess of non-solvent causing rapid precipitation, but gradually increased the amount of a single polar non-solvent to control the amount of precipitated material. By this method, only a fraction of the dissolved material was precipitated, while the rest of the material remained in solution. Careful gel-permeation chromatographic analyses of the fractions offered a visualization of the molecular weight and the distribution of species within the dissolved components. Ball-milled wood dissolved completely in the IL, but the coarse sawdust or TMP pulp preparations were not fully soluble on a microscopic level. It has been noticed earlier that the solubility of softwood in [amim]Cl and subsequent phosphitylation of all hydroxyl functionalities is greatly dependent on the preliminary mechanical treatment [14]. Surprisingly, this partial insolubility of sawdust has enabled a more efficient component separation, by selective extraction of components, compared to the soluble fine powder preparations, which separate according to molecular weight distributions. From the coarse material, a cellulose-rich fraction was extracted and the rest of the lignin-hemicellulose matrix could be isolated as an insoluble material. In the case of increased pulverization, the observed better solubility was rationalized by the fragmentation of the matrix formed by lignin-carbohydrate complexes (LCC). An increase in the amount of water extractable lignin from crude fractions of milled wood, after the IL treatment, points to the presence of soluble fragments, originating from an LCC matrix.

6.2 Advances Towards the Efficient Fractionation of Wood with Ionic Liquids

Our understanding of the mechanisms operating during fractionation of wood polymers using IL-based solvent systems is continuously improving. An increasing number of publications have appeared in the recent literature utilizing increasingly sophisticated and target-selective treatment systems. The fundamentals of action of the ILs during these treatments have been explored, and the importance of factors such as treatment conditions, solvent system reactivity, and structural features of plant cell walls have all been discussed to varying degrees and will be reviewed in this chapter. A large number of these citations have focused on the utility of ILs, as pretreatment method prior to biofuel production. Selective separation of the components in these pretreatments may not typically be the ultimate goal of the pretreatment, but frequently delignification takes place. Nevertheless, the publications describing work oriented to increase enzyme activity on wood have offered useful insights into the action of ILs, which can also be utilized for the design of fractionation systems.

6.2.1 New Generations of Ionic Liquids for Fractionation Applications

New types of ILs specifically designed to wood fractionation applications have been introduced to the field, and some of the new generations of ILs have also being designed to be more suitable for processing steps, such as recycling of the IL after the treatment. Promising alternatives have been found to the most commonly used dialkylimidazolium acetates and chlorides.

One successful effort that has been made towards more sustainable systems is the development of 'distillable' ILs capable of dissolving cellulose, by King et al. [15] This group of ILs is based on the acid–base conjugates of 1,1,3,3-tetramethylguanidine (TMG) and common carboxylic acids, such as acetic or propionic acid. The distillation ability arises from the fact that the acid–base equilibrium can be shifted to a sufficient extent, at high temperature, to produce volatile neutral species. Yet, the applicability to wood fractionation has not yet been examined. Anugwom et al. have achieved the selective extraction of hemicelluloses using 'switchable' ILs (SIL), that are not capable of dissolving cellulose or lignin from the wood matrix [16, 17]. SILs can be formed by reacting CO_2, and 1,8-diazabicyclo-[5.4.0]-undec-7-ene (DBU) with alcohols. They can be converted back to neutral solvents by removal of CO_2, under reduced pressure or bubbling with nitrogen. Reversing the IL equilibrium back to volatile components is an interesting potential method for solvent recycling.

In other areas of the IL field, work has been done with the aim of tuning the hydrogen-bonding properties of traditional ILs, to be more selective towards specific wood components. An example of this, by Froschauer et al. [18] demonstrates modified properties of dialkyl phosphate ILs by using sulfur or selenium to replace one of the oxygen atoms in the anion structure. As a result, the hydrogen-bond accepting ability, as described by the Kamlet-Taft parameter β, was reduced. The new type of anions showed a selective dissolution of hemicelluloses out of hemicellulose-rich pulp. In this regard, Kamlet-Taft parameterization of a range of ILs and co-solvents is starting to allow for a better understanding of solvent and wood biopolymer interactions. In work by Hauru et al., the fundamentals of cellulose dissolution and recovery were presented. They have proposed a different approach to interpret the solvent properties of ILs. Rather than simple evaluation of hydrogen bonding accepting properties (β), the donor ability (α) should be considered as well, resulting in an effective or net basicity value (β−α) that better describes the cellulose dissolution capability of ILs [19]. Undoubtedly an in-depth understanding of the hydrogen-bonding properties has a key role in sophisticated design of ILs that can selectively extract various main wood components. For lignin isolation, Pinkert et al. have applied dialkylimidazolium sulfamates [20], which do not possess any ability to dissolve cellulose and thus can be used to extract lignin out of biomass. Of note, for this class of IL's, is an earlier application of acesulfamate as a food additive, which makes them promising from an ecotoxicity point-of-view.

6.2.2 *Defining the Mechanisms of Plant Cell Wall Dissolution in First Generation Ionic Liquids*

While the action of new designer ILs on wood is still under investigation, many details about the mechanisms of wood solubility in commonly utilized dialkylimidazolium ILs have been revealed. In the early publications covering the area of wood dissolution, a complete dissolution is mentioned to take place [7, 8, 12], at least for certain wood species. Since it has been demonstrated that all of the wood components are soluble in ILs in their individual purified forms [5, 21, 22], the naive simplification was made that wood is soluble in such ILs. However, later it was determined that wood dissolution under mild dissolution conditions was partial, rather than complete. In light of our results and other recent publications, it seems clear that mild dissolutions (<100 °C) in chloride-based ILs, such as [amim]Cl, are actually not able to provide the driving force to completely dissolve the wood cell wall, in its native state [14, 23–25].

Even if each of the polymeric components, in their purified form, has good solubility in the selected ILs, this does not ensure efficient solubilization of intact fibers. This is because of the complexity of the native wood. The structure of the plant cell wall is complex and highly orientated with many physically and chemically distinct regions, such as primary and secondary (S1, S2, and S3) cell walls and the presence of middle lamella. The variety of polymeric backbone structures, the presence of multiple functionalities and in particular the covalent/physical interactions between the three main wood components make things considerably more complicated. Nature's design of the cell wall is resistant towards physical stress and controls the diffusion of fluids inside the fiber. Inefficient mass transfer of IL or solvated polymers can greatly hinder the dissolution process, even in ILs having high capability for dissolution [26]. This is true especially in the cases where single components are isolated by an extraction type of mechanism. For ILs unable to dissolve cellulose, delignification has been observed to take place mainly on the outer surface of the fibers, due to low accessibility inside the bulky secondary wall [27]. This is analogous to kraft pulping where initial delignification occurs on the fiber surface in the middle lamella and lumen. The different cell types in corn stover show drastic differences in relative lignin solubility, during IL treatment, which further demonstrates the significance of cell wall ultrastructure or composition [26]. Even the early- and late woods in the same sample piece of Sugi wood showed very different response to swelling in IL [28]. They showed that the empty lumen allows for flexibility that protects the fiber from swelling induced physical defects. There also seem to be fundamental differences in the overall solubility, or at least in the kinetics of the dissolution, between soft- and hardwood species [6, 8, 24]. Apparently the natural design of tracheids and fibers between soft- and hardwood offers different resistance towards dissolution. All these reports demonstrate that there is a need to focus on revealing the degrees of recalcitrance, related to physical (macroscopic cell wall structure and crosslinking) and chemical factors (polymer interactions and covalent bonding) prior to making generalizations about the efficacy of fractionation of a single IL system on a single species.

Evidently the distribution and structure of lignin in cell walls has a crucial effect on wood solubility. This is not surprising as lignin is considered to be a branched polymer formed by random radical driven crosslinking, thus resembling a complex networked structure [29]. A fundamental property of crosslinked polymer structures is their inability to form true solutions. There are several publications about lignin isolation from lignocellulose via extraction-type mechanisms, using various ILs [9, 10, 20, 30, 31]. It is also not surprising that depolymerization of the isolated lignins have been observed [9, 20, 31]. Depolymerization can be a result of several mechanisms, including covalent bond scission via pulverization operations. This can be controlled by the conditions and choice of IL, as discussed in the following section.

The incomplete solubility of cell wall in an unreactive IL does not rule out component fractionation. It seems that dialkylimidazolium chlorides have a much lower reactivity with lignin than other commonly used IL types, such as dialkylimidazolium acetates. A group of publications has shown that this property can be utilized in isolation of cellulose rich fractions [14, 23–25]. In particular hemicelluloses and lignin are not separated completely in such systems. Physical or chemical methods to alter cell wall ultrastructure and polymeric networks, prior to or simultaneously with dissolution, may be crucial for complete fractionation. Support of this hypothesis can be found from studies where lignin structure was altered by (1) excessively heating the mixtures until thermal decomposition reactions start to take place [32], or (2) use of an oxidative catalyst to partially degrade the lignin polymer backbone [33]. Enhanced separation of lignin from polysaccharide was achieved, compared to a previous method [8]. What is not known and not taken into account are the above enumerated considerations and the molecular weight distributions of the resulting pulps.

6.2.3 *Role of Ionic Liquid Reactivity in the Dissolution of Wood*

Alterations to the polymeric structures, arising from reactions where ILs act as catalysts or even a reactive species to form adducts, may have an important role to play for the complete solubility of the composite structures of plant polymers. Recently a number of publications have demonstrated various types of reactivities for ILs, which were at first commonly thought to be relatively inert (with the dialkylimidazolium acetates being a prime example), under the conditions that are typically used in biomass treatments. These results may explain the observed differences between many ILs classes when it comes to mechanisms of cell wall dissolution.

Acid catalyzed reactions seem to be detrimental for both carbohydrates and lignin. Both of these wood components have been shown to partially depolymerize during acidic IL treatments [41]. These include reactions where the presence of the acids in the systems is intentional and where acid has originated from impurities or side reactions, at elevated temperatures. One destructive dissolution IL class are the so-called protic ILs, such as 1-H-3-alkylimidazolium ILs. Cox et al. have found this IL class to readily hydrolyze the β–O–4 ether bonds, in model compounds. Yields were found to be dependent on the anion in the protic IL [34]. The same was also observed with isolated oak wood lignin under conditions above 110 °C [35].

Similar reactions to those observed with protic ILs, degradation has been found to take place even in aprotic ILs, that shouldn't in theory contain significant concentrations of acidic protons. At a temperature of 120 °C, Kubo et al. have found [36] that β–O–4 type model compounds undergo elimination at the α-β-position [36]. In two chloride based ILs, 1-butyl-3-methylimidazolium chloride ([bmim]Cl) and [amim]Cl, enol ether formation, without cleavage of the β–O–4 bond, was the predominant reaction. Part of the original structure was left intact after 72 h incubation. Conversely, in acetate IL 1-ethyl-3-methylimidazolium acetate ([emim][OAc]) all of the original β–O–4 structures were gone after 72 h, forming only a low amount of enol ethers and the majority of materials as other unidentified structures. In agreement with the aforementioned data, George et al. [37] have found that acetate and sulfate ILs dramatically reduce the molecular weight of organosolv lignin, while only minor changes were observed with several chlorides tested under the same conditions. In any case, based on the mechanism suggested by Cox et al., enol or vinyl ethers may hydrolyze further, even in the chloride ILs. In the acidic environment presence of trace moisture levels may lead to depolymerisation [34]. Formation of condensed structures in lignin have also been detected by HSQC-NMR after [emim][OAc] treatment at 155 °C [38].

There is a notable difference in depolymerization between different lignin preparations, such as alkali and organosolv lignins, where the β-O-4 type of linkages seem to be among the most reactive linkages [37]. Based on structural features it may be reasonable to expect lower reactivity of softwood lignin, compared to hardwood lignin, as softwood lignins have a higher abundance of condensed linkages and hardwoods are rich in the more labile β–O–4 ethers [39].

Impurities remaining from the IL synthesis are likely catalysts for certain reactions. According to Li et al. [32], the delignification of bagasse in [emim][OAc], at temperatures reported to be above the lignin glass-transition temperature, does not happen anywhere near the same efficiency when recycled IL is used. Loss of the ILs delignification capability, after recycling [13], may allude to the presence of reactive species in the IL as an impurity, which are consumed during the first treatment step.

Cellulose is known to be labile towards acid hydrolysis in the dissolved state. The IL environment has been demonstrated to be effective for such reactions [40, 41]. The fact that rapid depolymerisation can also take place in technical ILs, even without added acid catalyst, should be surprising. Gazit and Katz have demonstrated that cellulose hydrolysis can happen under relatively mild conditions, in commercial-grade dialkylimidazolium chlorides and acetates [42]. The higher purity grades of ILs were also degrading cellulose even faster than lower grades. The catalyst for the reaction was found to be a trace amount acid, formed during the treatment. This could be scavenged by 1-methylimidazole, which is a very typical impurity in low grade commercial ILs. Such observations about acid formation strongly suggest the use of mild temperatures in IL treatments.

Fine control of temperature and dissolution atmosphere may be necessary to reduce the depolymerization of wood during the treatments. According to Miyafuji et al., the depolymerization of carbohydrates in 1-ethyl-3-methylimidazolium chloride ([emim]Cl) can be mostly prevented using temperatures below 90 °C [43]. On the other hand, the use of mild conditions effects negatively to the degree of wood dissolution. Lignin showed much higher resistance towards degradation and any low molecular weight fragments were observed only at temperatures above 110 °C. Reactions in chloride-based ILs, induced by atmospheric impurities, have been investigated by Nakamura et al. [44]. Oxygen was found to facilitate the dissolution of lignin, in addition to solubilization of carbohydrates.

Depolymerization is not the only reaction type taking place when lignocellulose is treated with ILs. Addition of dialkylimidazolium cations, at the C2 position, to the reducing ends of polysaccharides has been reported from acetate ILs [45–47]. The same effect has been observed with isolated lignin [31]. The reaction follows from self-deprotonation of the imidazolium cation, forming a carbene. The carbene formation and following reactions with electrophiles will lead to conversion of anions to their conjugate acid form [48, 49]. Acid species formed in these types of reactions are suggested to be responsible for the depolymerization of cellulose [42], but basic impurities may also capture the released proton. Alternatively, acetylation of cellulose by the IL anion has been observed to happen to low degree in pure [emim][OAc] at high temperature (150 °C)[47]. The mechanism of formation of such structures is still controversial. Çetinkol et al. [50] have reported deacetylation of hemicelluloses and acetylation of lignin taking place when wood is treated in [emim][OAc]. This suggests a transacetylation mechanism [48, 49], but it is still unproven and alternative mechanisms may prevail.

Reactivity of ILs with wood polymers is certainly an important issue related to green processing of wood in such media. Not only for its effects to recycling and atom economy of the processes, but also to the yield, purity, and overall quality of the resulting materials. The use of mild conditions and possible additives, or co-solvents, in IL systems may help to gain control over unwanted side reactions.

6.2.4 Use Co-solvents in Ionic Liquid Based Dissolution Systems

Research has been performed on the use of organic co-solvents in ILs. This is mainly to alter the properties of the dissolution system, such as viscosity reduction. As protic solvents like water or alcohols tend to prevent cellulose dissolution and are working as efficient non-solvents for dissolved components, the group of polar aprotic organic solvents typically will not decrease the solvation efficiency of ILs towards cellulose, in co-solvent concentrations up to 50 m% [51, 52]. The use of co-solvents may enhance the kinetics of the dissolution process, by accelerated diffusion. This allows the use of lower dissolution temperatures that in turn prevent the unwanted depolymerization reactions. Enhancement of wood dissolution kinetics when using co-solvents, compared to pure IL, can be notable, as demonstrated by Qu et al., who aided dissolution of milled Fir with pyridine and DMAc (as co-solvents), at the low temperature of 30 °C [53]. However, much longer dissolution times were needed than for the typical high temperature dissolution. Co-solvents can also enhance wood dissolution at higher temperatures. An article by Xie et al. has demonstrated that complete dissolution of corn stover can be achieved using NMP: [emim][OAc] solutions at the higher temperature of 140 °C, in under 60 min [54]. This of course is not designed for material production, where molecular weights are maintained, but rather biofuel production where maintaining molecular weights is not critical. The majority of the available co-solvents will eventually turn into non-solvent when a limiting concentration is reached [51] and so it may even be possible to maintain binary solvent mixtures with ILs throughout the process. In fractionation processes aiming at the manufacture of derivatized products, certain co-solvents can also act as catalysts for subsequent modification reactions without need for product isolation, in between the unit processes.

Water in the IL systems may also be termed as a limiting solvent, instead of a non- or co-solvent. It has been used as a way of limiting cellulose solubility in certain ILs, while close to complete delignification with removal of hemicelluloses can still take place. The presence of acidic species have been stated to be essential for delignification in these kinds of systems and they can be added as catalysts [55] or originate from the natural acidity of the IL [56]. Depending on the anion of IL, the aqueous solutions can be relatively acidic [55, 57]. It remains uncertain whether this is related to impurities specific to pure ILs, technical preparations of ILs, as a natural property of IL-water solutions [57], or from reactions leading to acidic products. Zhang et al. have reported that from a neutral pH of the pretreatment solvent down to pH 3.4, all of the wood components are regenerating close to their natural compositions from the aqueous IL-system, without resulting in delignified pulp [55]. This is in slight contradiction with the results from Fu et al., who have used neutral aqueous solutions to basic [emim][OAc] efficiently, without any added acid catalysts [58]. The fibrous structure of wood cells still exists in the solid cellulose enriched fractions afforded by treatments in aqueous-IL solutions [55], resembling traditional chemical pulps. This is due to the inability of ILs to dissolve crystalline cellulose, once high enough water contents are added. Thus, only the amorphous parts of the fiber are accessible to the acidic solvent.

The efficiency of the IL-water solvent system was highly dependent on the type of treated biomass as grass-type feedstocks, such as Miscanthus or Triticale, were found to be highly responsive. Nearly complete delignification and glucose digestibility are observed for grasses, followed by mediocre efficiencies for hardwoods and significantly lower response for softwoods [55, 56].

6.3 Fractionation of Wood by Solvation in [amim]Cl and Fractional Precipitation with a Non-solvent

In this section of the chapter we report data of the fractionation work that was performed in our laboratories. The original focus of our work was to study the mild fractionation and the molecular weight distributions of the resulting precipitated fractions. The reasoning for this was to assess whether it was possible to get technically useful fractions, with suitable molecular weight distributions, by avoiding depolymerisation. Furthermore, the interaction of wood biopolymers and how this affects fractionation, is always a fundamental question that needs answering. [amim]Cl was used as the solvating IL. It has a low tendency to react with lignin, whilst being an efficient solvent for isolated preparations of all wood components. In agreement with earlier results from our laboratory, King et al. [14], it was found that only heavily pulverized starting materials were completely soluble in the IL of choice. The use of coarse materials, such as sawdust, only offered partial solubility. The reasons behind the solubility differences will be discussed in more detail in this chapter. For the regeneration of dissolved components we have applied an alternative method for the non-solvent addition. For this we have used gradual increases in non-solvent volume instead of rapid excessive addition to the IL-solution. As a result, the regeneration event is well controlled and follows the principles of traditional molecular weight distribution-related polymer fractionation. By applying a derivatization procedure developed in our laboratory, Zoia et al. [59], we have been able to obtain soluble lignocellulose derivatives for size-exclusion chromatography (SEC). This has allowed us to characterize the total molecular weight distributions for majority of the precipitated fractions. Combining the molecular weight information with composition analysis (acid soluble lignin analysis and IR-spectroscopy), we have been able to observe two fundamentally different mechanisms that apply during component separation, related to the degree of interaction of the wood biopolymers. These are found to be dependent on the degree of pulverization (from extensive milling) and therefore the solubility of the resulting materials. As mentioned previously solubility of wood is dependent on the degree of pulverization. This therefore influences whether the fractionation is extraction based or solvation (and subsequent selective precipitation) based. An acetonitrile non-solvent was able to regenerate majority of the dissolved materials but additional non-solvents, such as water and methanol, allowed for further component separation. The motives for selecting this non-solvent, and more comprehensive discussion about our data can be found from our earlier publications [23, 60]. It was also found that further purification of isolated crude fractions with water resulted in secondary separations and it was possible to recover more water-soluble materials in their own fraction. A complete flow diagram of the fractionation scheme is presented in Fig. 6.1. Selected starting materials (see Table 6.1) have undergone different mechanical pre-treatment processes. Particle sizes and properties changed accordingly with the preparation method. Nearly intact fibrous structures have remained after TMP pulping, but were notably fragmented during sawdust preparation. Ball-milled materials represent highly pulverized wood that has lost all fibrous characteristics.

As mentioned previously, fractions were analyzed by Klason lignin analysis, ATR-IR and the molecular weight distributions of some fractions were determined by re-dissolution into [amim]Cl, benzoylation and SEC analysis (Zoia et al. [59]). The analysis results are presented in Table 6.2.

The fractionation procedure was performed roughly as follows: Wood samples were heated with [amim]Cl for the specified period and temperature. Crude fraction 1 was precipitated from IL using acetonitrile as non-solvent and was washed with water and dried. Fraction 2 was precipitated from the residual IL solution by addition of further acetonitrile and further washed, using the same procedure as for fraction 1. Fraction 3 was precipitated from IL-solution by water addition, after the acetonitrile had been removed by evaporation. Fraction 4 was prepared from the combined aqueous extracts from fractions 1 and 2. The aqueous extracts were combined, concentrated, precipitated with methanol and dried. Fraction 5 was prepared by concentration of the remaining water solution, from fraction 3, precipitation with methanol and drying.

6.3.1 Fractionation Based on Molecular Weight

It is well known that polymers of different molecular weight have different solubility in solvents. This means that controlling the precipitation of polymers, of high polydispersity, from any solution can be used to separate them into fractions of decreasing molecular weight [62]. The main components in wood have distinctively different average degrees of polymerization (DP). Isolated softwood

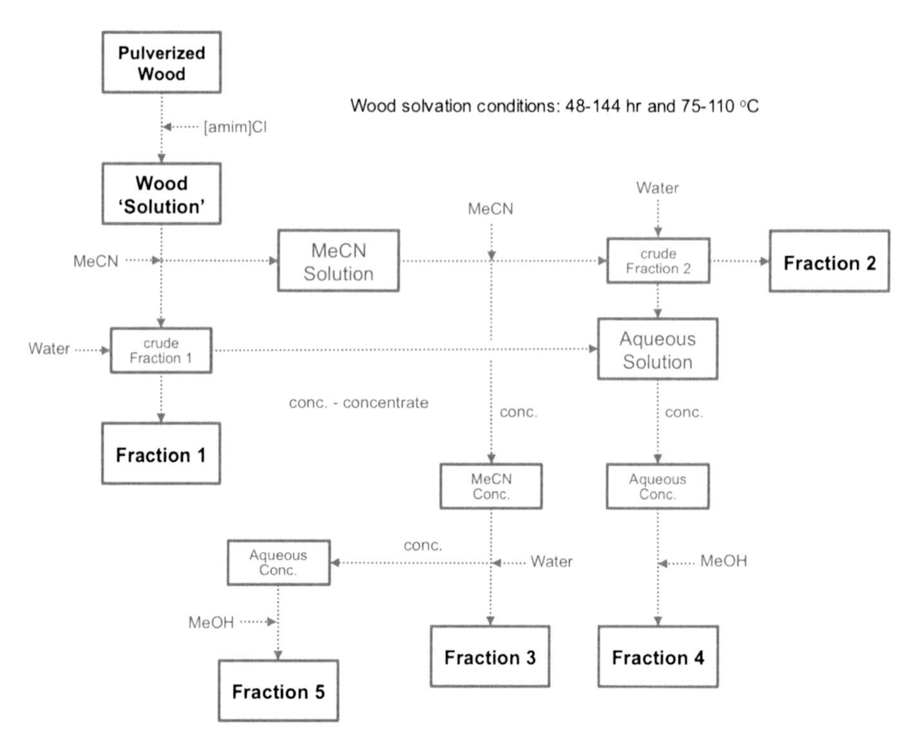

Fig. 6.1 Total fractionation procedure. Fraction 5 was not isolated in every experiment (Reprinted with permission from [23]. Copyright © 2013 American Chemical Society)

Table 6.1 Starting materials, their upper particle diameter limits, and lignin contents

Material	Particle diam. (μm)	Lignin content (%)
Wiley milled Norway spruce TMP pulp	<400	28.7
Norway spruce sawdust	<200	26.6
Ball milled Norway spruce TMP pulp	<75	28.7
Ball milled *Eucalyptus grandis*	<75	25.8[a]

[a]Value from literature [61]

celluloses have been measured to have average molecular weights from 730 kDa [63] up to 1,550 kDa [64]. Hemicelluloses are of typically lower DP than cellulose and isolated hemicelluloses consist of polymers on average from 18 to 80 kDa [65, 66], depending on the isolation method. Lignin preparations that represent as close to native lignin as we can isolate with current methods, have molecular weights between 52 and 98 kDa [67]. Differences of such magnitude, including differences in chemical composition of the polymers, should offer plenty of opportunity for separation of lignin from hemicellulose from cellulose by controlled addition of a nonsolvent into ionic liquid. Articles by Lee at al. and Lateef et al. [10, 68], have shown that mixtures of the purified polymers can be highly selectively precipitated

Table 6.2 All fractions collected in this study including yields, lignin contents and analyses performed. Fraction 5 is not included in the table or in yield calculations

Starting materials and conditions	Y_{total} (%)	Fraction number	$Y_{Fraction}$ (%)	Lignin cont. (%)	Y_{Lignin} (%)	$Y_{Carb.}$ (%)
[amim]Cl 6 % 40 mesh	86	1	57.4	37.7	75	50
		2	25.9	12.1	11	32
TMP spruce 144 h 100 °C		3	0.5	–	–	–
		4	1.8	–	–	–
[amim]Cl 4 % spruce sawdust 120 h 110 °C	93	1	53.2	45.0	90	40
		2	33.2	6.2	8	42
		3	1.3	–	–	–
		4	4.8	–	–	–
[amim]Cl: 10 % 28 days	86	1	45.7	28.4	45	46
		2	21.6	36.5	27	19
Rotary-milled TMP spruce 48 h		3	2.4	–	–	–
75 °C		4	16.3	20.9	12	18
[amim]Cl: 10 % 48 h	59[a]	1	42.1	9.0	15	52
Ball-milled eucalyptus 48 h		2	12.7	–	–	–
75 °C		3	4.2	–	–	–

[a]Fraction 4 not included
Y_{total}: Total yields of precipitated material
$Y_{Fraction}$: Yield of precipitated fraction from starting material
Lignin cont: Lignin content of fraction that includes Klason lignin + acid soluble lignin
Y_{Lignin}: Yield of lignin in fraction from total lignin content in starting material
$Y_{Carb.}$: Yield of carbohydrates in fraction including cellulose and hemicelluloses
–: Value not determined

from IL solutions. In actual fractionation processes, using minimally treated wood, this efficiency is never observed.

If we examine at the molecular weight distributions of the fractions from the highly pulverized, 28 days rotary milled and 48 h ball-milled, samples from spruce and Eucalyptus respectively (Fig. 6.2), we can see that there is a distinct precipitation based on molecular weight. In both cases, the molecular weights of the fractions decrease from fractions 1 to 3. Fraction 4 overlaps with fraction 2 for spruce, due to the fact that the majority of the material in the water-soluble fraction was originally dissolved from crude fraction 2 (see Fig. 6.1). If we look at the lignin contents for the main fractions 1 and 2 for spruce (see Table 6.2 entry for milled TMP), there is very little change in the lignin contents from the native wood. Thus there is clear evidence for precipitation based on molecular weight and very poor separation of lignin from polysaccharide, contrary to previous reports on the separation of mixtures of the purified polymers [10, 68]. It seems evident that LCCs are preventing the separation of the lignin and polysaccharide portion of this fully soluble pulverized wood. Seemingly, disintegration of the LCC matrix during pulverization creates fragments that have similar molecular weights to cellulose, and that have been extensively depolymerized during milling. As a result, a mixture of similar sized LCC polymers precipitate in order of molecular weight.

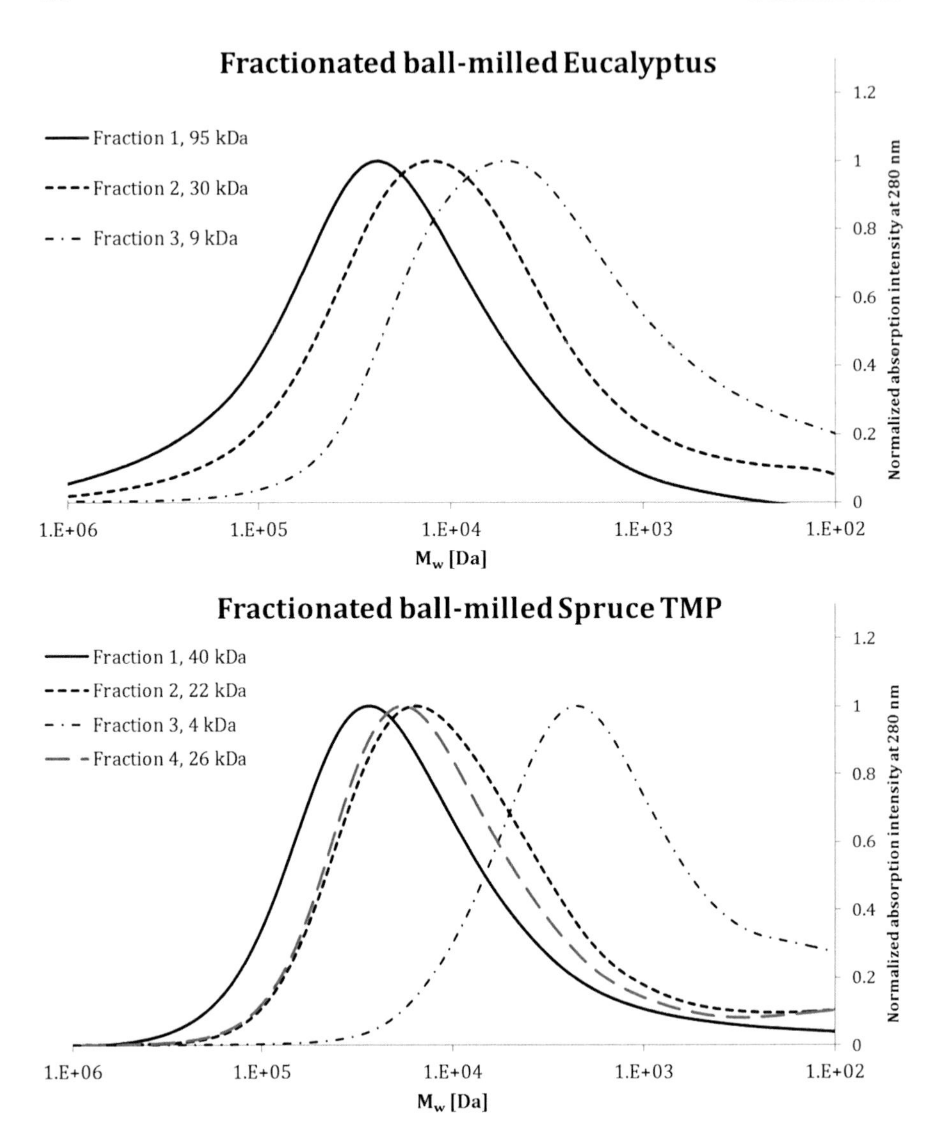

Fig. 6.2 The molecular weight distributions of the isolated fractions, from fully soluble ball-milled wood. The regeneration order was largely controlled by the molecular weight of the components. Fractions from (**a**) milled Eucalyptus (**b**) milled Spruce TMP

6.3.2 Effect of Particle Size on Fractionation Mechanisms

For fully soluble finely pulverized materials, the dissolution is relatively rapid for all components. Dissolution is much faster compared to coarse materials and wood converts to a completely solvated state. Solvation was not complete with the coarse

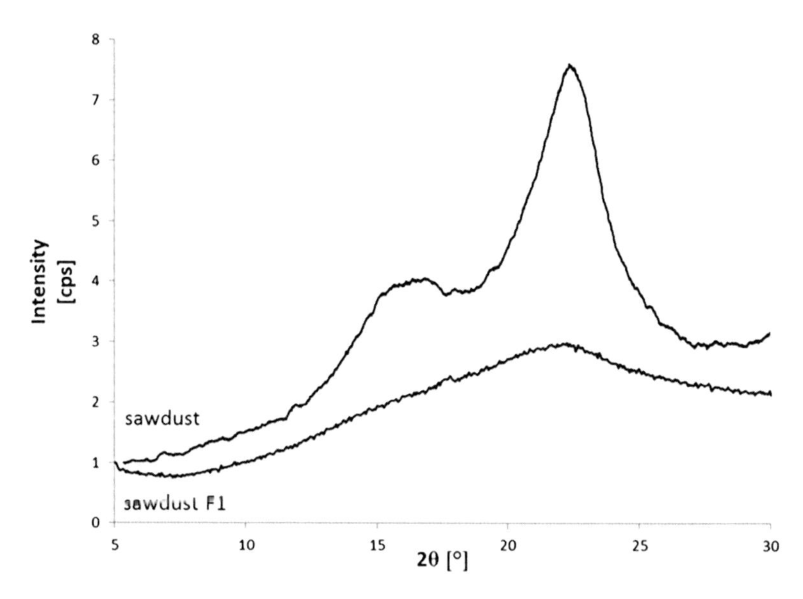

Fig. 6.3 X-ray powder diffractograms of spruce sawdust, starting material (*above*), and fraction 1, recovered from [amim]Cl (Reprinted with permission from [23b]. Copyright © 2013 American Chemical Society)

materials. The following dissolution and fractionation mechanism seemed to differ greatly from the pulverized materials, based on compositional analyses of isolated fractions (Table 6.2) and XRD-analysis of the sawdust fraction 1 and the original sawdust (Fig. 6.3).

For the incomplete dissolution of the coarser materials (sawdust and TMP), fraction 1 was mostly composed of the materials that remained solid (but seemingly swollen) during the whole dissolution/extraction period. This was determined to be mainly lignin and polysaccharide. Fraction 2 was determined to be mainly cellulose, based on Klason lignin and ATR-IR analyses (Fig. 6.4). As fraction 2 was 33 % of the original sawdust fraction, meaning that most of the cellulose is extracted from the partially soluble wood sample leaving an insoluble matrix of lignin and hemicellulose. Further evidence of this is found after XRD analysis of the regenerated fraction 2, in comparison to the starting sawdust. After extraction of cellulose, the amorphous LCC network was remaining. One should ask the questions, 'Why is there an insoluble fraction when the purified polymers are all soluble in the IL?' and 'Why can lignin not be separated efficiently from the polysaccharide, even when the finely pulverized samples are completely soluble?'.

Both questions can be addressed by the explanation of precipitation based on molecular weight. However, most solvents will preferentially precipitate one component over another and this is simply not the complete picture. Both questions can be answered if you take in to consideration that wood is suggested to contain LCCs and it is actually the LCC network that is preventing dissolution. Only material that is not bound into the LCC network is extractable, under mild dissolution conditions.

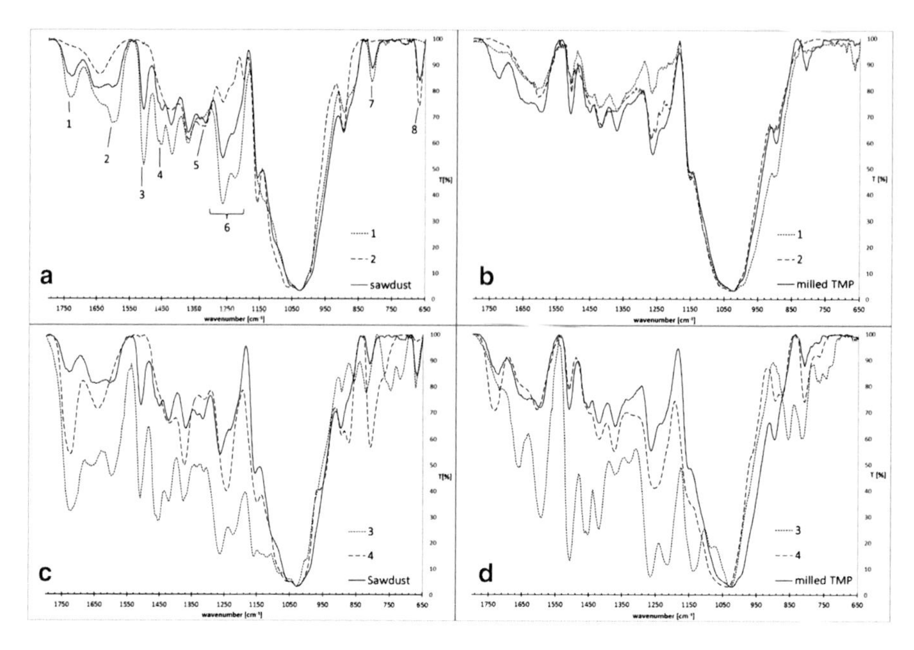

Fig. 6.4 FT-IR spectra of fractions 1–4 precipitated from solutions of sawdust and milled TMP, compared to the starting materials. (**a**) Sawdust fractions 1 and 2, (**b**) Milled TMP fractions 1 and 2, (**c**) Sawdust fractions 3 and 4, (**d**) Milled TMP fractions 3 and 4. *Band assignments*: 1 = Carbonyl groups from hemicelluloses and lignin [70, 71]; 2 = Carboxylic acids from xylan and lignin [70, 71]; 3 = Lignin [72]; 4 = Xylan [71]; 5 = Cellulose [72]; 6 = Carbonyl groups from hemicelluloses and lignin [71]; 7 = Glucomannan [73]; 8 = Cellulose [72]. Reprinted with permission from [23b]. Copyright © 2013 American Chemical Society

This is entirely consistent with a report by Lawoko et al. [69] showing that almost all isolatable LCCs from Norway spruce consist of lignin, which is chemically bonded with hemicelluloses. Whereas, only a minor portion of spruce LCCs have been found to contain lignin-cellulose type LCCs. With this literature confirmation it is no wonder that we can extract cellulose from an insoluble LCC matrix.

As anticipated, it was nearly impossible to derivatize and dissolve any further material from fraction 1 for SEC, due to its insolubility. In case of highly pulverized wood, physical degradation of all polymeric components seem to take place and overall polydispersity decreases. Fractions from ball milled TMP pulp further gave more evidence about the close association with lignin and carbohydrates and how these structures could control the total dissolution of wood. The FT-IR analysis offered some details about the carbohydrate compositions in isolated fractions. Neutral sugar analysis is traditionally used to characterize the carbohydrate moieties in lignocellulose, but in this work FT-IR was used instead as a fast, non-destructive, and semi-quantitative analytical method. IR spectra of the isolated fractions from two spruce materials, ball-milled TMP and sawdust, are presented in Fig. 6.4. When spectra of fraction 1 from the two materials are compared, the significant presence of hemicelluloses can be seen for sawdust, whereas in milled

TMP the carbohydrates seem to be mostly cellulose. The hemicelluloses are more present in the lower molecular weight fraction 2 for the case of milled TMP. Other significant differences can be found in composition and yield of the water-soluble fraction 4. For sawdust it seems that the majority of the hemicelluloses have remained totally water insoluble in fraction 1, for some yet unknown reason. For milled wood around half of the original hemicelluloses were converted to be water soluble and dissolved together with relatively large portions of lignin. No lignin was observed to be water soluble from sawdust crude fractions.

Once again, this observation could be explained by the covalent attachment of hemicelluloses that are released during the milling via the fragmentation of the supporting lignin polymers, that otherwise would prevent them from being extracted during the water washing.

Sawdust and TMP pulp preparations represent structurally quite unaltered wood. Our results suggest that swelling and dissolution of native or relatively intact fibers start from the amorphous and crystalline domains of cellulose. The solvated cellulose polymers then diffuse to the bulk solution (*fraction 2*) leaving behind the lignin-hemicellulose matrix that remains in a rather swollen form and is restricted from complete dissolution (*fraction 1*). Only minor fractions of lignin (*fraction 3*) or hemicelluloses (*fraction 4*) seem to be unbound and transfer to bulk solution. Molecular weight analysis showed that the isolated cellulose-rich fraction 2 had a significantly lower molecular weight than reported for e.g. softwood pulps.

From the dissolution treatment, it is hard to estimate if there is significant depolymerization during the 48 h dissolution period. For the sawdust fractionation, some depolymerization of the carbohydrate components during the dissolution seems evident. Molecular weight of the isolated fraction 2 is low, considering the fact that this fraction was composed mostly of cellulose, as our analysis revealed (see Table 6.2 and Fig. 6.4). In light of a recent study by Gazit and Katz [42], the depolymerization of cellulose during long dissolution periods, even in purified IL, is not surprising. Their results indicate that trace level formation of acidic by-products can take place during dissolution, even below the temperatures that were used in our work. The unfortunate fact is that technical pulps have very defined specifications, in terms of molecular weight distributions, and not only their lignin and hemicellulose contents. This means that controlling acidic and oxidative impurities during an IL-mediated fractionation will be critical in the future to obtain technically useful pulps that fit existing value-chains. In many cases even the present commercial ILs contain these impurities. More must be done to quantify and understand the effects of these impurities.

The single experiment that was performed with Eucalyptus resulted in a lignin poor fraction that was the first to precipitate from the IL-solution (*fraction 1*). As mentioned previously, precipitation was dependent on molecular weight for Eucalyptus, as well as milled spruce. It remains as a topic for further studies how differences in covalent structures between lignin and polysaccharide will affect the selectivity of separation. Other ILs like [emim][OAc] have been reported to fully dissolve hardwood [8]. This may be due to chemical degradation of LCC-matrix during the treatment, as discussed earlier. It remains possible that other ILs could facilitate higher yields of cellulose-enriched materials by gradual precipitation, only if the LCC cleavage and carbohydrate depolymerization remains at a low level during the dissolution.

6.4 Conclusions

Due to the rapidly expanding field of IL mediated wood processing, our knowledge in this area has increased to a new level. Many new technical advances are apparent, including more refined ILs, electrolytes, pre-treatments and processing techniques. However, the application of fundamental knowledge related to the connectivity of wood biopolymers, wood morphology, wood ultrastructure and even the solubility of wood in ILs seems to have been largely neglected. Increasing awareness related to IL reactivity has brought both challenges and possibilities to wood fractionation. Depolymerization during fractionation can result in undesired products. This is most relevant when molecular weight distributions should be maintained, e.g. for the production of cellulosic pulps. However, in some cases degradation may be beneficial, e.g. for dissolving the LCC network or reducing the recalcitrance of wood for biofuel production.

Based on our work, with sawdust and highly pulverized spruce wood, we have demonstrated that wood is not completely soluble in [amim]Cl in its native state. This is confusing as isolated lignin, cellulose and hemcellulose preparations have been dissolved efficiently in several publications. One possible reason for this is the presence of an extended LCC matrix in wood that is simply of too high molecular weight and is too interconnected to dissolve. This property can be utilized to extract cellulose, as it is not covalently bound to the insoluble LCC matrix. Cellulose is extracted and by careful control of non-solvent addition, the insoluble lignin-hemicellulose rich fraction can be first isolated, followed by regeneration of relatively pure cellulose. This cellulose extraction procedure is not yet at a stage that would yield a technically useful pulp, due to apparent depolymerization, in comparison to technical pulps and holo-cellulose. However, the more we learn about the stability of wood and lignocellulose, in technical and pure ILs, the better are our chances of yielding close-to native polymers.

The implications for biofuels production are more straight-forward, in regard to pre-treatment mechanisms. Certain ionic liquids are excellent media for cellulose dissolution and regeneration to a state, which is easier to process. Presence of impurities or intentionally added catalysts, that may depolymerize the biopolymers during this process, are beneficial, provided the IL is somewhere between 99 and 100 % recoverable. This is a function of the high cost of ILs, at present. The method of biopolymer regeneration, to enhance separate lignin from polysaccharide, is therefore quite important. If degradation is significant enough breakage of covalent linkages between lignin and polysaccharide should facilitate this. IL recyclability is a major challenge here due to the buildup of monomers, dimers, oligomers, silicates and other inorganics. Therefore improving IL recyclability will greatly enhance the chances of success.

6.5 Experimental

6.5.1 Materials

Synthesis of the [amim]Cl was performed according to a method adapted from Wu et al. [74]. Allyl chloride (200 mL, 2.51 mol) and 1-methylimidazole (160 mL, 2.01 mol) were added to a flask under nitrogen atmosphere. The mixture was refluxed at 50 °C with stirring under positive pressure of nitrogen for 18 h. The reaction was determined to be complete by ^{1}H NMR. The mixture was transferred under nitrogen atmosphere to a rotary evaporator, attached to a high vacuum pump. The excess of allyl chloride was removed at 50 °C. The cloudy crude product was further purified by heating, at 80 °C for 18 h, with activated charcoal (3.0 g) and water (200 mL). The mixture was then filtered through Celite in agrade-3 sinter. Water was removed at 65 °C by rotary evaporation over 18 h, under high vacuum, to yield [amim]Cl as a pale yellow viscous oil. ^{1}H NMR (300 MHz, CDCl3) δ 3.97 (3H, s,NCH3), 4.86 (2H, d, J = 6.4 Hz, NCH2), 5.33-5.26 (2H, m,C=CH–C), 5.86 (1H, ddt, J = 16.9, 10.3, 6.5 Hz,C=CH2), 7.40 (1H, s, C=CH), 7.65 (1H, s, C=CH), 10.39(1H, s, NCHN).

Unbleached Norway spruce (*Picea abies*) thermomechanical pulp (TMP) was donated from a Swedish mill. Norway spruce sawdust (particle size <0.2 mm by sieving) was prepared with a belt grinder (grade 60), in-house. *Eucalyptus grandis* was supplied by Novozymes, NC, USA. Prior to ball-milling treatments, Norway spruce TMP was first milled in a Wiley mill with a 20 mesh (0.84 mm) sieving screen. After Wiley milling the 20 mesh powder was extracted in a Soxhlet extractor for 48 h with acetone. A portion of this fibrous material was further sieved to pass a coarse 40 mesh (0.40 mm) sieve. The remaining extracted 20 mesh Norway spruce powder was rotary ball milled in a ceramic plated 5.5 L steel jar with 470 ceramic balls (diameter 0.9 cm) and a rotation speed 60 rpm, for 28 day period. After milling, the fine powder was dried in vacuum oven. The average particle size was determined to be less than 200 mesh (75 μm).

Eucalyptus chips were Soxhlet extracted with acetone for 48 h. Remaining tannins were removed by refluxing in 0.075 M NaOH solution (1:50 w/v ratio) for 1 h prior to milling of the dried sample. Milling was performed in a Fritsch Pulverette planetary ball-mill, with a 20 mL tungsten carbide grinding bowl and steel balls, at a rotation speed of 420 rpm for 48 h in total. The total milling time was made up of a repetitive milling cycle of 30 min milling time and 20 min brake, to avoid burning of the sample. All the wood materials were dried in vacuum oven over night at 40 °C prior to their use.

6.5.2 Solvation of Wood with [amim]Cl

Lignocellulosic material (typically *ca.* 1 g) was quickly added to a flask containing dry [amim]Cl (typically *ca.* 20 g) under nitrogen atmosphere. The mixture was homogenized with vortex mixer until an even dispersion was obtained. Dissolution was performed in a temperature controlled oil bath using a three-necked flask under positive pressure of nitrogen. This was equipped with an overhead mechanical stirrer with steel blade. A positive pressure of nitrogen gas was maintained during the whole dissolution period. Solvation conditions and quantities of materials were varied according each experiment performed, ranging from 48 h at 80 °C to 122 h at 110 °C. Rotary milled Norway spruce powder generally dispersed and gave a clear solution in a short period of time. Wiley milled and sawdust materials remained slightly cloudy even after extensive heating at 100 or 110 °C.

6.5.3 Fractionation of Solvated Wood by Non-solvent Addition

Preparation of fractions 1–4 was carried out as follows; Crude fraction 1 was precipitated from IL using acetonitrile as non-solvent. The crude fraction was separated using centrifuge and washed with water, so that filtrate was retained. The solid residue was dried to give fraction 1. For preparation of fraction 2 additional acetonitrile was added into IL-solution followed by similar separation and washing procedure. Fraction 3 was precipitated from IL-solution by water addition, after the acetonitrile had been concentrated by evaporation. Fraction 4 was prepared from the aqueous filtrates retained from the purification of fractions 1 and 2. Filtrates were concentrated down to a volume *ca.* 2 mL, which was followed by addition of MeOH. Formed fluffy precipitate was filtered and purified with MeOH. Fraction 5 was prepared as follows. The aqueous IL-solution left from the precipitation of fraction 3 was concentrated so that nearly all water had removed. The remaining IL-solution was precipitated with MeOH (1:10 v/v). Formed precipitation was filtered in grade-3 sinter, washed with MeOH and the filtrand was dried in a vacuum over for 18 h at 40 °C to give fraction 5.

6.5.4 Analysis of Precipitated Fractions

For molecular weight determination the fractionated samples were derivatized with benzoyl chloride in [amim]Cl solution following the procedure introduced by Zoia et al. [59] and analyzed using HP G1312A pump connected to Waters HR5E and HR1 columns with a Waters 484 UV-absorbance detector calibrated using polystyrene standards. Acid insoluble (Klason) lignin and acid soluble lignin were determined by method modeled from one published by Dence [75]. The acid that was used to hydrolyze the samples was diluted from conc. sulfuric acid corresponding to 72 ± 0.1 %. The lignocellulose samples were dried in a vacuum oven at 40 °C overnight. *ca.* 100 mg of the samples were measured accurately and mixed with sulfuric acid solution (100 mg per 2 mL) using magnetic stirring and vortex mixer. After 2 h hydrolysis at room temperature with occasional manual mixing the samples were diluted with 50 mL of deionized water and transferred into sealable bottles. The bottles were placed into a commercial pressure cooker and heated at elevated pressure for 90 min. The solid residues were filtered with a grade-3 sinter and washed with 40 mL of water. The filtrate was retained for acid soluble lignin determination. The solid residue was further washed with 60 mL of water, so that filtrate was neutral, and after air-drying the sample was placed into a vacuum oven for 20 h. Acid soluble lignin was determined spectrophotometrically from the retained filtrates. The filtrates were first diluted to precisely 100 mL and then the absorbance was measured at 205 nm wavelength in a 1 cm pathlength cuvette. The concentrations were calculated using extinction co-efficient of 110 L/g.cm.

FT-IR spectra were recorded from finely powdered samples that were dried for 20 h at 50 °C in a vacuum oven, using Perkin-Elmer Spectrum One AT-IR spectrometer. Processing was carried out using PE Spectrum One software. The spectra were processed with baseline correction, noise elimination and normalization.

References

1. Sun N, Rodríguez H, Rahman M, Rogers RD. Where are ionic liquid strategies most suited in the pursuit of chemicals and energy from lignocellulosic biomass? Chem Commun. 2011;47:1405–21.
2. Mäki-Arvela P, Anugwom I, Virtanen P, Sjöholm R, Mikkola JP. Dissolution of lignocellulosic materials and its constituents using ionic liquids—a review. Ind Crop Prod. 2010;32:175–201.
3. Mora-Pale M, Meli L, Doherty TV, Linhardt RJ, Dordick JS. Room temperature ionic liquids as emerging solvents for the pretreatment of lignocellulosic biomass. Biotechnol Bioeng. 2011;108:1229–45.
4. Tadesse H, Luque R. Advances on biomass pretreatment using ionic liquids: an overview. Energy Environ Sci. 2011;4:3913–29.
5. Swatloski RP, Spear SK, Holbrey JD, Rogers RD. Dissolution of cellulose with ionic liquids. J Am Chem Soc. 2012;124:4974–5.
6. Fort DA, Remsing RC, Swatloski RP, Moyna P, Moyna G, Rogers RD. Can ionic liquids dissolve wood? Processing and analysis of lignocellulosic materials with 1-n-butyl-3-methylimidazolium chloride. Green Chem. 2007;9:63–9.
7. Kilpeläinen I, Xie H, King A, Granström M, Heikkinen S, Argyropoulos DS. Dissolution of wood in ionic liquids. J Agric Food Chem. 2007;55:9142–8.
8. Sun N, Rahman M, Qin Y, Maxim ML, Rodríguez H, Rogers RD. Complete dissolution and partial delignification of wood in the ionic liquid 1-ethyl-3-methylimidazolium acetate. Green Chem. 2009;11:646–55.
9. Tan SSY, MacFarlane DR, Upfal J, Edye LA, Doherty WOS, Patti AF, Pringle JM, Scott JL. Extraction of lignin from lignocellulose at atmospheric pressure using alkylbenzenesulfonate ionic liquid. Green Chem. 2009;11:339–45.
10. Lee SH, Doherty TV, Linhardt RJ, Dordick JS. Ionic liquid-mediated selective extraction of lignin from wood leading to enhanced enzymatic cellulose hydrolysis. Biotechnol Bioeng. 2009;102:1368–76.

11. Zhang H, Wu J, Zhang J, He J. 1-allyl-3-methylimidazolium chloride room temperature ionic liquid: a new and powerful nonderivatizing solvent for cellulose. Macromolecules. 2005;38:8272–7.
12. Zavrel M, Bross D, Funke M, Buchs J, Spiess AC. High-throughput screening for ionic liquids dissolving (ligno-)cellulose. Bioresour Technol. 2009;100:2580–7.
13. Li B, Asikkala J, Filpponen I, Argyropoulos DS. Factors affecting wood dissolution and regeneration of ionic liquids. Ind Eng Chem Res. 2010;49:2477–84.
14. King A, Zoia L, Filpponen I, Olszewska A, Xie H, Kilpeläinen I, Argyropoulos DS. In situ determination of lignin phenolics and wood solubility in imidazolium chlorides using ^{31}P NMR. J Agric Food Chem. 2009;57:8236–43.
15. King AWT, Asikkala J, Mutikainen I, Järvi P, Kilpeläinen I. Distillable acid–base conjugate ionic liquids for cellulose dissolution and processing. Angew Chem Int Ed. 2011;50:6301–5.
16. Anugwom I, Mäki-Arvela P, Virtanen P, Damlin P, Sjöholm R, Mikkola J. Switchable Ionic liquids (SILs) based on glycerol and acid gases. RSC Adv. 2011;452.
17. Anugwom I, Mäki-Arvela P, Virtanen P, Willför S, Sjöholm R, Mikkola J. Selective extraction of hemicelluloses from spruce using switchable ionic liquids. Carbohydr Polym. 2012;87:2005–11.
18. Froschauer C, Hummel M, Laus G, Schottenberger H, Sixta H, Weber HK, Zuckerstätter G. Dialkyl phosphate-related ionic liquids as selective solvents for xylan. Biomacromolecules. 2012;13:1973.
19. Hauru LKJ, Hummel M, King AWT, Kilpeläinen I, Sixta H. Role of solvent parameters in the regeneration of cellulose from ionic liquid solutions. Biomacromolecules. 2012;13:2896.
20. Pinkert A, Goeke DF, Marsh KN, Pang S. Extracting wood lignin without dissolving or degrading cellulose: investigations on the use of food additive-derived ionic liquids. Green Chem. 2011;13:3124.
21. Pu Y, Jiang N, Ragauskas AJ. Ionic liquid as a green solvent for lignin. J Wood Chem Technol. 2007;27:23–33.
22. Peng X, Ren JL, Sun RC. Homogeneous esterification of xylan-rich hemicelluloses with maleic anhydride in ionic liquid. Biomacromolecules. 2010;11:3519–24.
23. (a) Leskinen T, King AWT, Kilpeläinen I, Argyropoulos DS. Fractionation of lignocellulosic materials with ionic liquids. 1. Effect of mechanical treatment. Ind Eng Chem Res. 2011;50:12349–57; (b) Leskinen T, King AWT, Kilpeläinen I, Argyropoulos DS. Fractionation of lignocellulosic materials using ionic liquids: Part 2. Effect of particle size on the mechanisms of fractionation. Ind Eng Chem Res. 2013;52:3958–66.
24. Wang X, Li H, Cao Y, Tang Q. Cellulose extraction from wood chip in an ionic liquid 1-allyl-3-methylimidazolium chloride (amimCl). Bioresour Technol. 2011;102:7959–65.
25. Casas A, Alonso MV, Oliet M, Santos TM, Rodriguez F. Characterization of cellulose regenerated from solutions of pine and eucalyptus woods in 1-allyl-3-methilimidazolium chloride. Carbohydr Polym. 2013;92:1946–52.
26. Sun L, Li C, Xue Z, Simmons BA, Singh S. Unveiling high-resolution, tissue specific dynamic changes in corn stover during ionic liquid pretreatment. RSC Adv. 2013;3:2017–27.
27. Doherty TV, Mora-Pale M, Foley SE, Linhardt RJ, Dordick JS. Ionic liquid solvent properties as predictors of lignocellulose pretreatment efficacy. Green Chem. 2010;12:1967–75.
28. Miyafuji H, Suzuki N. Morphological changes in sugi (Cryptomeria japonica) wood after treatment with the ionic liquid, 1-ethyl-3-methylimidazolium chloride. J Wood Sci. 2012;58:222–30.
29. Boerjan W, Ralph J, Baucher M. Lignin biosynthesis. Annu Rev Plant Biol. 2003;54:519–46.
30. Fu D, Mazza G, Tamaki Y. Lignin extraction from straw by ionic liquids and enzymatic hydrolysis of the cellulosic residues. J Agric Food Chem. 2010;58:2915–22.
31. Kim J, Shin E, Eom I, Won K, Kim YH, Choi D, Choi I, Choi JW. Structural features of lignin macromolecules extracted with ionic liquid from poplar wood. Bioresour Technol. 2011;102:9020–5.

32. Li W, Sun N, Stoner B, Jiang X, Lu X, Rogers RD. Rapid dissolution of lignocellulosic biomass in ionic liquids using temperatures above the glass transition of lignin. Green Chem. 2011;13:2038–47.
33. Sun N, Jiang X, Maxim ML, Metlen A, Rogers RD. Use of polyoxometalate catalysts in ionic liquids to enhance the dissolution and delignification of woody biomass. ChemSusChem. 2011;4:65–73.
34. Cox BJ, Jia S, Zhang ZC, Ekerdt JG. Catalytic degradation of lignin model compounds in acidic imidazolium based ionic liquids: Hammett acidity and anion effects. Polym Degrad Stab. 2011;96:426–31.
35. Cox BJ, Ekerdt JG. Depolymerization of oak wood lignin under mild conditions using the acidic ionic liquid 1-H-3-methylimidazolium chloride as both solvent and catalyst. Bioresour Technol. 2012;118:584–8.
36. Kubo S, Hashida K, Yamada T, Hishiyama S, Magara K, Kishino M, Ohno H, Hosoya S. A characteristic reaction of lignin in ionic liquids; glycelol type enol-ether as the primary decomposition product of β-O-4 model compound. J Wood Chem Technol. 2008;28:84–96.
37. George A, Tran K, Morgan TJ, Benke PI, Berrueco C, Lorente E, Wu BC, Keasling JD, Simmons BA, Holmes BM. The effect of ionic liquid cation and anion combinations on the macromolecular structure of lignins. Green Chem. 2011;13:3375–85.
38. Torr KM, Love KT, Çetinkol ÖP, Donaldson LA, George A, Holmes BM, Simmons BA. The impact of ionic liquid pretreatment on the chemistry and enzymatic digestibility of Pinus radiata compression wood. Green Chem. 2012;14:778–87.
39. Argyropoulos D, Lignin MS, editors. Biotechnology in the pulp and paper industry, vol. 57. Berlin/Heidelberg: Springer; 1997. p. 127–58.
40. Rinaldi R, Meine N, Vomstein J, Palkovits R, Schüth F. Which controls the depolymerization of cellulose in ionic liquids: the solid acid catalyst or cellulose? ChemSusChem. 2010;3:266–76.
41. Li B, Filpponen I, Argyropoulos DS. Acidolysis of wood in ionic liquids. Ind Eng Chem Res. 2010;49:3126–36.
42. Gazit OM, Katz A. Dialkylimidazolium ionic liquids hydrolyze cellulose under mild conditions. ChemSusChem. 2012;5:1542–8.
43. Miyafuji H, Miyata K, Saka S, Ueda F, Mori M. Reaction behavior of wood in an ionic liquid, 1-ethyl-3-methylimidazolium chloride. J Wood Sci. 2009;55:215–9.
44. Nakamura A, Miyafuji H, Saka S. Influence of reaction atmosphere on the liquefaction and depolymerization of wood in an ionic liquid, 1-ethyl-3-methylimidazolium chloride. J Wood Sci. 2010;56:256–61.
45. Liebert T, Heinze T. Interaction of ionic liquids with polysaccharides 5. Solvents and reaction media for the modification of cellulose. BioResources. 2008;3:576–601.
46. Ebner G, Schiehser S, Potthast A, Rosenau T. Side reaction of cellulose with common 1-alkyl-3-methylimidazolium-based ionic liquids. Tetrahedron Lett. 2008;49:7322–4.
47. Karatzos S, Edye L, Wellard R. The undesirable acetylation of cellulose by the acetate ion of 1-ethyl-3-methylimidazolium acetate. Cellulose. 2012;19:307–12.
48. Rodriguez H, Gurau G, Holbrey JD, Rogers RD. Reaction of elemental chalcogens with imidazolium acetates to yield imidazole-2-chalcogenones: direct evidence for ionic liquids as proto-carbenes. Chem Commun. 2011;47:3222–4.
49. King AWT, Parviainen A, Karhunen P, Matikainen J, Hauru LKJ, Sixta H, Kilpeläinen I. Relative and inherent reactivities of imidazolium-based ionic liquids: the implications for lignocellulose processing applications. RSC Adv. 2012;2:8020–6.
50. Çetinkol ÖP, Dibble DC, Cheng G, Kent MS, Knierim B, Auer M, Wemmer DE, Pelton JG, Melnichenko YB, Ralph J, Simmons BA, Holmes BM. Understanding the impact of ionic liquid pretreatment on eucalyptus. Biofuels. 2010;1:33–46.
51. Gericke M, Liebert T, Seoud OAE, Heinze T. Tailored media for homogeneous cellulose chemistry: ionic liquid/Co-solvent mixtures. Macromol Mater Eng. 2011;296:483–93.
52. Rinaldi R. Instantaneous dissolution of cellulose in organic electrolyte solutions. Chem Commun. 2011;47:511–3.

53. Qu C, Kishimoto T, Hamada T, Nakajima N. Dissolution and acetylation of ball-milled lignocellulosic biomass in ionic liquids at room temperature: application to nuclear magnetic resonance analysis of cell-wall components. Holzforschung. 2012;67:25–32.
54. Xie H, Shen H, Gong Z, Wang Q, Zhao ZK, Bai F. Enzymatic hydrolysates of corn stover pretreated by a N-methylpyrrolidone-ionic liquid solution for microbial lipid production. Green Chem. 2012;14:1202–10.
55. Zhang Z, O'Hara IM, Doherty WOS. Effects of pH on pretreatment of sugarcane bagasse using aqueous imidazolium ionic liquids. Green Chem. 2013;15:431–8.
56. Brandt A, Ray MJ, To TQ, Leak DJ, Murphy RJ, Welton T. Ionic liquid pretreatment of lignocellulosic biomass with ionic liquid-water mixtures. Green Chem. 2011;13:2489–99.
57. Zhang Y, Du H, Qian X, Chen EY. Ionic liquid-water mixtures: enhanced Kw for efficient cellulosic biomass conversion. Energy Fuel. 2010;24:2410–7.
58. Fu D, Mazza G. Aqueous ionic liquid pretreatment of straw. Bioresour Technol. 2011;102:7008–11.
59. Zoia L, King AWT, Argyropoulos DS. Molecular weight distributions and linkages in lignocellulosic materials derivatized from ionic liquid media. J Agric Food Chem. 2011;59:829–38.
60. Casas A, Alonso MV, Oliet M, Rojo E, Rodríguez F. FTIR analysis of lignin regenerated from Pinus radiata and Eucalyptus globulus woods dissolved in imidazolium-based ionic liquids. J Chem Technol Biotechnol. 2012;87:472–80.
61. Emmel A, Mathias AL, Wypych F, Ramos LP. Fractionation of Eucalyptus grandis chips by dilute acid-catalysed steam explosion. Bioresour Technol. 2003;86:105–15.
62. Cantow MJR, editor. Polymer fractionation. New York: Academic; 1967. p. 527.
63. Yanagisawa M, Shibata I, Isogai A. SEC-MALLS analysis of softwood kraft pulp using LiCl/1,3-dimethyl-2-imidazolidinone as an eluent. Cellulose. 2005;12:151–8.
64. Schult T, Hjerde T, Optun OI, Kleppe PJ, Moe S. Characterization of cellulose by SEC-MALLS. Cellulose. 2002;9:149–58.
65. Jacobs A, Dahlman O. Characterization of the molar masses of hemicelluloses from wood and pulps employing size exclusion chromatography and matrix-assisted laser desorption ionization time-of-flight mass spectrometry. Biomacromolecules. 2001;2:894–905.
66. Lundqvist J, Teleman A, Junel L, Zacchi G, Dahlman O, Tjerneld F, Stålbrand H. Isolation and characterization of galactoglucomannan from spruce (Picea abies). Carbohydr Polym. 2002;48:29–39.
67. Guerra A, Filpponen I, Lucia LA, Saquing C, Baumberger S, Argyropoulos DS. Toward a better understanding of the lignin isolation process from wood. J Agric Food Chem. 2006;54:5939–47.
68. Lateef H, Grimes S, Kewcharoenwong P, Feinberg B. Separation and recovery of cellulose and lignin using ionic liquids: a process for recovery from paper-based waste. J Chem Technol Biotechnol. 2009;84:1818–27.
69. Lawoko M, Henriksson G, Gellerstedt G. Characterisation of lignin-carbohydrate complexes (LCCs) of spruce wood (Picea abies L.) isolated with two methods. Holzforschung. 2006;60:156–61.
70. Åkerholm M, Salmen L. Interactions between wood polymers studied by dynamic FT-IR spectroscopy. Polymer. 2001;42:963–9.
71. Stevanic J, Salmén L. Orientation of the wood polymers in the cell wall of spruce wood fibres. Holzforschung. 2009;63:497–503.
72. Schwanninger M, Rodrigues JC, Pereira H, Hinterstoisser B. Effects of short-time vibratory ball milling on the shape of FT-IR spectra of wood and cellulose. Vib Spectrosc. 2004;36:23–40.
73. Ebringerova A, Hromadkova Z, Hribalova V, Xuc C, Holmbom B, Sundberg A, Willför S. Norway spruce galactoglucomannans exhibiting immunomodulating and radical-scavenging activities. Int J Biol Macromol. 2008;42:1–5.
74. Wu J, Zhang J, Zhang H, He J, Ren Q, Guo M. Homogeneous acetylation of cellulose in a new ionic liquid. Biomacromolecules. 2004;5:266–8.
75. Dence CW, Lin SY, editors. Methods in lignin chemistry. Heidelberg/Berlin: Springer; 1992. p. 578.

Part III
Production of Biofuels and Chemicals in Ionic Liquids

Chapter 7
Biofuel Production with Ionic Liquids

Haibo Xie and Zongbao Kent Zhao

Abstract In consideration of unique properties of ionic liquids, the research into using ionic liquids as solvents and catalysts for lipids extraction, biodiesel production and purification, as well as bioalcohol extraction from fermentation booth have been investigated to develop clean and cost-competitive new technologies. This review summarizes up-to-date progress in these areas and analyzes examples with the aim to provide an in-depth understanding of how to integrate ionic liquids-based technologies into traditional biofuel production processes.

Keywords Ionic liquids • Lipids • Extraction • Transesterification • Biodiesel • Bioethanol • Biobutanol • Catalysis

7.1 Introduction

The search for alternative resources for transportation fuel production is driven by the increasing concerns on global warming and fossil resources depletion. Biomass refers to all organic matters derived from the process of photosynthesis. It is produced in large quantity, with an estimated 140 billion metric tons per year. Thus, sugar, starch, vegetable oils, agricultural wastes, forest residues, and dedicated industrial materials from plants, all belong to biomass by definition. Lignocellulose, is mainly consisted of cellulose, hemicellulose and lignin, is by far the most abundant form of biomass. Biomass has been considered as the most important alternative feedstock for the production of fuels, chemicals and materials [1].

Ionic liquids (ILs) are specifically referred to salts that melt below 100 °C. ILs are usually organic salts comprised of cations and anions. Some ILs exist as liquids

H. Xie (✉) • Z.K. Zhao (✉)
Division of Biotechnology, Dalian National Laboratory for Clean Energy, Dalian Institute of Chemical Physics, Chinese Academy of Sciences (CAS), 457 Zhongshan Road, Dalian 116023, People's Republic of China
e-mail: hbxie@dicp.ac.cn; zhaozb@dicp.ac.cn

Z. Fang et al. (eds.), *Production of Biofuels and Chemicals with Ionic Liquids*, Biofuels and Biorefineries 1, DOI 10.1007/978-94-007-7711-8_7,

at relatively low temperatures, even below room temperature, so they are usually called as room temperature ILs. Due to their unique structural properties, ILs may have some of the following properties, such as non-detectable volatile under atmosphere pressure, good solubility to most of inorganic and organic compounds, high thermal stability, high ionic conductivity, tunable miscibility with traditional solvents, etc. Notably, the chemical and physical properties of ILs can be tuned by combining different cation and anion [2]. These unique properties suggest their important roles in green chemical processes, especially in the topic of biomass processing and conversion towards a foundation of sustainable biorefinery process [3, 4]. Up to date, the use of ILs for biomass processing and conversion mainly focuses on the following subjects:

- Dissolution, derivation, and regeneration of biopolymers
- Catalytic conversion of biopolymers and their monomers into platform chemicals
- Biomass pretreatment, components fractionation and structural elucidation
- Separation, production of biofuels

Currently, biodiesel, bio-alcohols (e.g. ethanol, butanol) and bio-oil from pyrolysis are among the most extensively studied biofuels [5]. Biodiesel is produced mainly from vegetable oils and animal fats *via* transesterification process. These kinds of oils (lipids) are regarded as important energetic biorenewable resources and chemical raw materials [6, 7]. Bio-alcohols are normally made by microbial fermentation of sugars produced from carbohydrates or starch crops such as sugarcane or corn [8, 9]. With the significant progress in the conversion and utilization of biomass, it is well-recognized that clean and efficient technologies are in high demand to develop a cost-effective bioenergy production system, such as extraction of lipids and bio-alcohols, biodiesel preparation and purification [10].

The applications of ILs as a solvent and/or catalyst in conversion of biomass in a wide perspective have been covered in other chapters of this book or being reviewed elsewhere [11–13]. Especially, with ILs-based pretreatment technology, lignocellulosic biomass is also being developed as a feedstock for the production of bio-alcohols and biodiesel [14, 15]. This chapter focuses on those areas, in which ILs are used for the production of biofuel molecules in a more direct way. Specifically, the chapter will provide an up-to-date overview on how to apply ILs in the following areas: lipids separation, biodiesel production, bio-alcohols extraction, and bio-oil production through pyrolysis and upgrading.

7.2 Biodiesel

Biodiesel, chemically defined as monoalkyl esters of long chain fatty acids, are derived from renewable feedstocks like vegetable oils and animal fats [16]. Recently, the production of biodiesel from lipid produced by oleaginous yeasts and algae has been obtained much attention [15–19]. Biodiesel has potentials

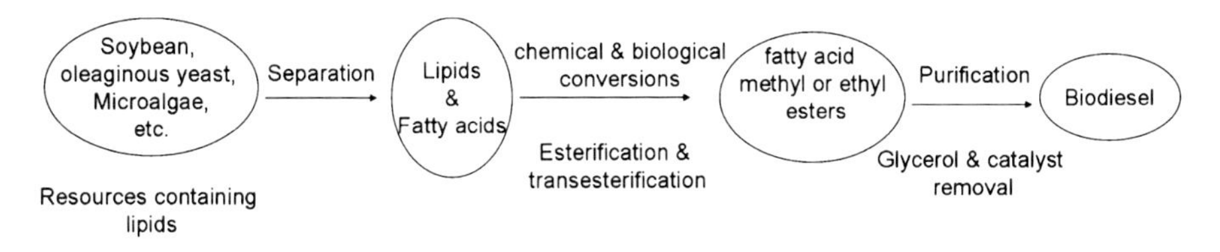

Fig. 7.1 Key process of biodiesel production from lipids

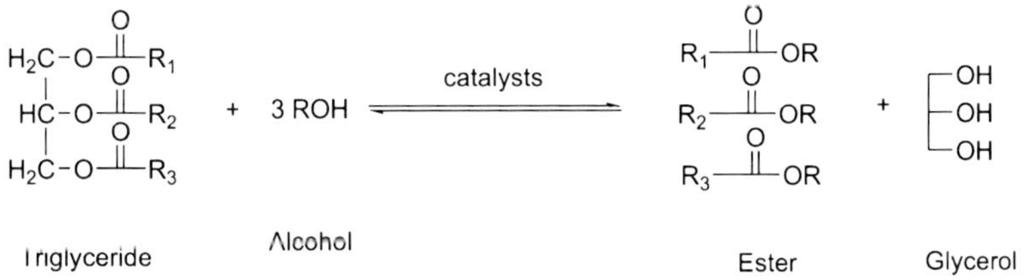

$$\begin{matrix} H_2C{-}O{-}C(=O){-}R_1 \\ HC{-}O{-}C(=O){-}R_2 \\ H_2C{-}O{-}C(=O){-}R_3 \end{matrix} + 3\,ROH \xrightleftharpoons{\text{catalysts}} \begin{matrix} R_1{-}C(=O){-}OR \\ R_2{-}C(=O){-}OR \\ R_3{-}C(=O){-}OR \end{matrix} + \begin{matrix} {-}OH \\ {-}OH \\ {-}OH \end{matrix}$$

Triglyceride Alcohol Ester Glycerol

Scheme 7.1 Catalytic transesterification of triacylglycerides for biodiesel production

as an alternative to petroleum diesel. Emission of carbon dioxide into the atmosphere can be reduced by substituting diesel fuel with biodiesel [20]. The production of biodiesel from lipids include some key steps which are (1) lipids extraction from oily materials, such as soybean, sunflower seeds, and cell mass of oleaginous microorganisms, (2) esterification or transesterification of fatty acids or lipids, and (3) purification of fatty acid esters (Fig. 7.1). Biodiesel can be produced by either chemical or enzymatic conversion of lipids or fatty acids with a monohydric alcohol in the presence of acid or base catalysts, or lipases (Scheme 7.1). These procedures require a large quantity of organic solvents and corrosive acidic or basic catalysts. Downstream processing costs and environmental problems associated with biodiesel production and byproducts recovery have stimulated the search for alternative production methods and alternative substrates. In consideration of the unique properties of ILs comparing to traditional solvents and acidic or basic catalysts, a lot of efforts by using ILs as solvents and catalysts have been devoted into developing a more clean and efficient process for biodiesel production.

7.2.1 Lipids Extraction

The lipids extraction from various resources is essential for production of biodiesel. Usually, the lipids extraction process uses large amounts of organic solvents, such as hexane, $CHCl_3$, etc. which also results in significant losses and energy consumption during the solvent recycling process. Hexanes and related hydrocarbon extractants are also becoming an environmental and health concerns. Therefore, exploration of new extraction technologies has received much attention [21].

A new class of “switchable solvents” has been proposed [22, 23], which are based on an exothermic transformation from an organic base, an alcohol and an acid gas (e.g. CO_2). These solvents are capable of changing composition reversibly under mild conditions to shift between molecular liquid and ionic liquid, in associated with switching properties, such as polarity and viscosity [24–27]. Thus, ‘switchable solvents’ have been tested to extract lipids from soybean flakes [28]. It was found that the combination of an amidine and excess water gave superior solvent/oil separation, adequate oil extraction. The contamination levels of residual amidine in the soy oil are very low. This method takes advantage of the fact that amidines can be made to switch their hydrophilicity by application or removal of CO_2 in the system. However, the extraction efficiency is lower than those of traditional hexane, and ethanol.

Microalgae are one of the most important emerging resources for lipids. In 2010, Samori, et al. extended the switchable solvent to the lipid extraction from water-suspended and dried microalgae *Botryococcus braunii*. It was found that DBU/octanol exhibited the highest yields of extracted hydrocarbons from both freeze-dried and liquid algal samples (16 and 8.2 % respectively against 7.8 and 5.6 % with n-hexane) [29]. Their follow-up research demonstrated that a new switchable system based on the reversible reaction of *N,N*-dimethylcyclohexylamine (DMCHA) with water showed better performance in lipid extraction of wet algal samples or cultures (Fig. 7.2) [30].

The lipid extraction efficiency of the system applied to both of wet and dry biomass, was higher than that obtained through a typical extraction procedure with $CHCl_3$–MeOH (Tables 7.1, 7.2). The FAMEs yield was very good for all of the tested algae, independent of the biomass/DMCHA ratio. The higher extraction content may be because DMCHA had access to structural lipids which are resistant to extraction with $CHCl_3$–MeOH [30].

In 2010, Young et al. reported the ability of a co-solvent system composed of a hydrophilic IL 1-ethyl-3-methyl imidazolium methyl sulfate and methanol at a mass ratio of 1.2:1 to extract and auto partition lipids from various biomass. The extraction yields of lipids were summarized in Table 7.3. The results suggest that the ILs–methanol co-solvent is successful in complete extraction of the lipids from these biomass sources. The proposed IL–methanol co-solvent system differs from traditional organic co-solvent systems which both dissolve and solubilize the extracted lipid and thus suffer extraction efficiencies limited by the solvent’s carrying capacity [31].

Considering the main components (e.g. polysaccharide, protein, lipids) of algae, Teixeira investigated the use of traditional ILs to extract lipids and produced chemical feedstocks from algae without acids, bases or other catalysts, which is based on the fact that a dissolution and hydrolysis of wet alga biomass in ILs. Deconstruction reached completion in <50 min regardless of algae species, at 100–140 °C and atmospheric pressure. The fast rate of hydrolysis without acids or bases suggests the ionic liquid itself is acting as both a solvent and a catalyst. Depolymerization of algae cell wall polysaccharides could result in the deconstruction of the cell wall, including the phospholipid membrane, and creation of a cell-free mixture that can be separated into constituent fractions, and result in a full separation of lipids from algae. This work presented a strategy of full utilization of algae biomass [32].

Fig. 7.2 Extraction of lipids from algal wet biomass with DMCHA (50 mg/mL extraction system): (**a**) DMCHA containing algal lipid (*green* layer) after 24 h of extraction (algal biomass was removed by centrifugation); (**b**) on the top DMCHA, on the bottom H_2O; (**c**) CO_2 bubbling; (**d**) after formation of $DMCHAH^+$ HCO_3^-, lipids (*green* layer and drops) float on the surface of the system (Reprinted with permission from [30]. Copyright © 2012 The Royal Society of Chemistry)

Table 7.1 TLS and FAMEs content expressed on algal dry weight basis (means ± standard deviation, n = 3), obtained through $CHCl_3$-MeOH hot extraction of dried samples and DMCHA extraction of wet samples (50 mg/mL, 24 h extraction) (Reprinted with permission from [30]. Copyright © 2012 The Royal Society of Chemistry)

	TLs (wt%)		FAMEs (wt%)	
	$CHCl_3$-MeOH	DMCHA	$CHCl_3$-MeOH	DMCHA
D. communis	17.8 ± 0.1	29.2 ± 0.9	6.0 ± 0.1	6.1 ± 0.7
N. gaditana	45.1 ± 0.9	57.9 ± 1.3	10.6 ± 0.1	11.0 ± 0.9
T. suecica	25.4 ± 2.6	31.9 ± 1.5	4.5 ± 0.5	5.4 ± 0.6

Table 7.2 Lipids fractionation of oils obtained through extraction of dried algae samples ($CHCl_3$-MeOH) and wet algal biomass (DMCHA), expressed on algal dry weight basis (Reprinted with permission from [30]. Copyright © 2012 The Royal Society of Chemistry)

	$CHCl_3$-MeOH			DMCHA		
Lipid (wt%)	*D. communis*	*N. gaditana*	*T. suecica*	*D. communis*	*N. gaditana*	*T. suecica*
NL	1.6	7.2	2.8	4.1	5.2	3.5
GL	10.3	12.6	12.7	14.3	29.5	14.7
PL	5.9	25.3	9.9	10.8	23.2	13.7

NL neutral lipids, *GL* glycolipids, *PL* Phospholipids

Table 7.3 Lipid extraction results with ionic liquids co-solvent (Reprinted with permission from [31]. Copyright © 2010 Elsevier Limited)

Biomass type	Sample preparation	Reported yield (%)	Experimental yield (%)
Duniella microalgae	Freeze dried	11.1	8.6
Chlorella microalgae	Freeze dried	11–23	38
Canola oil seed	Oven dried	42.9	44
Jatropha oil seed	Removal of shell and husk	55–58	50
Kamani oil seed	Removal of shell and husk	49	38
Pongamia oil seed	Removal of shell and husk	30–40	11

7.2.2 Biodiesel Production by Chemical Catalysis

Currently, most commercial biodiesel is produced from plant lipids using homogeneous basic catalysts such as NaOH or KOH [20]. Nevertheless, these catalytic systems have a number of drawbacks: (I) catalysts cannot be reused and have to be neutralized which produces wastewater; (II) formation of stable emulsions that makes FAMEs separation difficult; (III) glycerol is obtained as an aqueous solution with low purity; (IV) the process is sensitive to residual water and free fatty acids [33]. ILs are recognized as green solvents due to their special properties comparing with traditional organic solvents, such as tunability, non-detectable vapor point and performance benefits over molecular solvents. Their properties can be designed to suit a particular need by changing the structures of the cation, anion or both acidic or basic for special synthesis [34, 35]. The principle was widely used for biodiesel production from lipids [11–13].

In consideration of good solubility of ILs to inorganic and organic compounds, and tunable miscibility with organic solvents, the simplest way is to use ILs for biodiesel production as solvents to immobilize traditional acidic or basic catalysts, such as K_2CO_3, NaOH, hydroxide salts of ammonium cations, sodium methoxide, lithium diisopropylamide, and H_2SO_4 [36]. Usually, a two-phase system (a glycerol-methanol-ILs-catalyst phase and biodiesel phase) forms due to the immiscibility of biodiesel with ILs after the reaction is done. The catalytic system can be reused after decanting the biodiesel directly. For example, under basic conditions, the combination of 1-n-butyl-3-methylimidazolium bis(trifluoromethylsulfonyl)imide ($BMI \cdot NTf_2$), alcohols, and K_2CO_3 (40 mol%) results in production of biodiesel from soybean oil in high yield and purity. H_2SO_4 immobilized in $BMI \cdot NTf_2$ efficiently promotes the transesterification reaction of soybean oil and various primary and secondary alcohols. In this multiphase process the acid is almost completely retained in the IL phase, while the biodiesel forms a separate phase. The recovered IL containing the acid catalyst could be reused for six times without significant yield or selectivity loss [36].

It is known that the acidity and basicity of ILs can be tuned by changing the composition of cationic and anionic species. Some acidic or basic ILs have been used as both catalysts and solvents for the synthesis of biodiesel (Scheme 7.2) These ILs can be synthesized by introduction of acidic functional groups into either the cation or anion, or adding a Lewis acid catalysts in ILs to form a catalytic active Lewis acid ILs [12, 37]. No matter the use of ILs as solvents or catalysts, the processes are usually efficient and facile for biodiesel production (Table 7.4). Inexpensive materials such as non-edible oils and waste cooking oils contain high free fatty acid contents, which are not suitable for base-catalyzed biodiesel production process. Therefore, free fatty acids should be converted into FAMEs, for which acidic ILs have been better than traditional mineral acids. For example, the dicationic IL N,N,N,N-tetramethyl-N,N-dipropanesulfonic acid ethylenediammonium hydrogen sulfate could be used as efficient and recyclable catalyst for the synthesis of biodiesel from long-chain free fatty acids or their mixtures [44]. The reaction was accomplished in a monophase at 70 °C for 6 h, while the products were separated by liquid/liquid biphase separation at room temperature with yields of 93–96 %. The work-up process was simple, and the catalysts could be reused for six times with little activity loss. This novel and clean procedure offered advantages including short reaction time, high yield, operational simplicity, and environmental friendliness. To achieve a better catalyst separation, acidic ILs-based catalysts have been covalently immobilized onto SBA-15. The immobilized catalysts displayed relatively high activity in esterification of oleic acid with short-chain alcohols because of the synergistic effects of both Lewis and Brønsted acidic sites. Under the optimal reaction conditions (molar ratio of methanol to oleic acid 6:1, 5 wt% catalyst loading, and 363 K for 3 h), the conversion of oleic acid reached 87.7 %. It was found that some metal chloride-based ILs could efficiently convert un-pretreated *Jatropha oil* with high-acid value (13.8 mg KOH/g) to biodiesel. For example, when $FeCl_3$ was added to [1-butyl-3-methyl-imidazolium][CH_3SO_3], a biodiesel yield of 99.7 % was achieved at 120 °C [53].

The basic ILs can be designed by introducing a strong basic anion or an organic basic moiety. The principle was used widely to synthesize ILs for the transesterification of lipids with methanol and ethanol. Most recently, three novel dicationic basic ILs were prepared for synthesis of biodiesel from soybean oil. Among them, 1,2-bis(3-methylimidazolium-1-yl)ethylene imidazolide showed the highest biodiesel yield of 99.6 %. When the acidity of soybean oil was 0.49 mg KOH/g, the yield of biodiesel was 99.6 %. However, when it was 1 mg KOH/g, the yield of biodiesel dropped to 82.5 %. Thus, basic ILs had limited capacity to use high acidity feedstock for biodiesel production [54].

It is well recognized that organic bases (e.g. 1,8-Diazabicyclo[5.4.0]undec-7-ene (DBU), guanidine) are important catalysts for the transesterification of lipid with alcohols [27, 55–57]. As both organic bases and alcohols are important components in the reaction, a novel phase-switching homogeneous catalysis was devised for clean production of biodiesel and glycerol (Scheme 7.3). It was found that the FAMEs can be decanted from the system and the yields were up to 95.2 % [27].

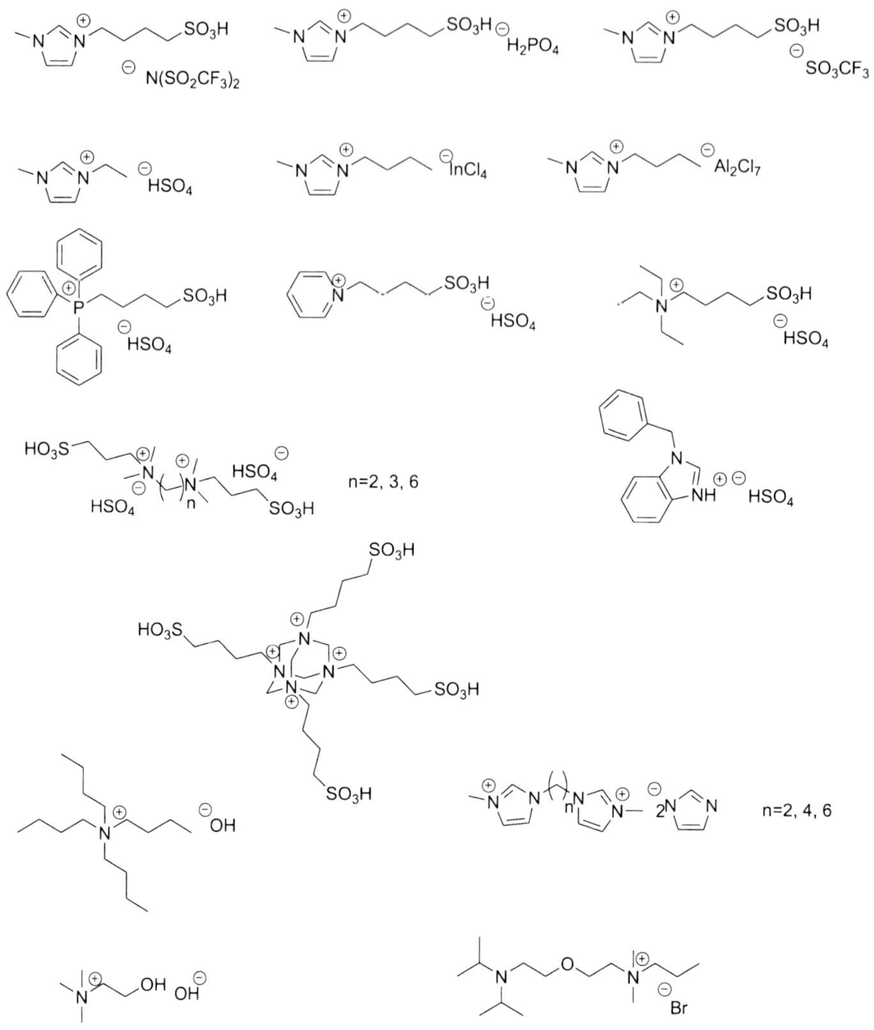

Scheme 7.2 Typical acidic and basic ILs for biodiesel synthesis

The produced glycerol was extracted from FAMEs completely by the "switchable solvents", and recovered with high purity after recycling DBU by another extraction process. This system has been tested for integrated production of biodiesel from cell mass of the oleaginous yeast *Rhodosporidium toruloides* Y4. While intracellular lipid was successfully extracted, only about 21.9 % of the lipid was converted into FAMEs. Nonetheless, such systems offered significant advantages.

Table 7.4 Ionic liquids as catalysts and solvents for biodiesel production

Raw materials[a]	Catalyst	Ionic liquids	Condit on	Biodiesel yield (%)	References
Soybean oil	Acid/base-catalyzed	[Bmim][NTf_2]	H_2SO_4,K_2CO_3 (40 mol%)	>98	[36]
Waste oil	Brønsted acidic IL	[$(CH_2)_4SO_3HPy$]HSO_4	170 °C for 4 h; methnol:oils: catalyst 12:1:0.06 (molar ratios)	>93	[38]
Soybean oil	Chloroaluminate IL	[Et_3NH]Cl–$AlCl_3$ (x($AlCl_3$) = 0.7),	70 °C, 9 h	98	[39]
Long-chain fatty acids	Brønsted acidic ionic liquid	[NMP][CH_3SO_3]	70 °C, 8 h	Up to 95	[40]
Rapeseed oil or free fat acid	Brønsted acidic ionic liquid	Zwitterion IL	70 °C, 7 h	98	[41]
Soybean oil,	Choline chloride · x$ZnCl_2$ ILs	Choline chloride · x$ZnCl_2$ ILs	70 °C, 72 h	54	[42]
Crude palm oil	KOH	[Bmim][HSO_4]	1.0 % KOH, 50 min, 60 °C	98	[43]
Oleic acid/EtOH	[TMEDAPS][HSO_4]	[TMEDAPS][HSO_4]	70 °C, 6 h	96	[44]
Stearic acid/EtOH	[TMEDAPS][HSO_4]	[TMEDAPS][HSO_4]	70 °C, 6 h	94	[44]
Myristic acid/EtOH	[TMEDAPS][HSO_4]	[TMEDAPS][HSO_4]	70 °C, 6 h	94	[44]
Lauric acid	p-toluenesulfonic acid	Quaternary ammonium methanesulfonate salts	60 °C, 2 h	97	[45]
Tung oil	Brønsted acidic ILs	Brønsted acidic ILs	70 °C, from methanol to butanol	93–96	[46]
Soybean	Brønsted acidic ILs	Brønsted acidic ILs	70 °C, methanol	94	[46]
Rapeseed oil	A multi –SO_3H functionalized Brønsted acidic IL	A multi –SO_3H functionalized Brønsted acidic IL	70 °C, 7 h	98	[40]
Cottonseed oil	1-(4-sulfonic acid) butylpyridinium hydrogen sulfate	1-(4-sulfonic acid) butylpyridinium hydrogen sulfate	70 °C, 5 h	92	[47]

(continued)

Table 7.4 (continued)

Raw materials[a]	Catalyst	Ionic liquids	Condition	Biodiesel yield (%)	References
Jatropha oil	[Bmim][CH_3SO_3]	$FeCl_3$	120 °C, 5 h	99	[48]
Canola oil	[3,3′-(hexane-1,6-diyl)bis (6-sulfo-1-(4-sulfobenzyl)-1Hk-benzimidazolium) hydrogensulfate]	[3,3′-(hexane-1,6-diyl)bis(6-sulfo-1-(4-sulfobenzyl)-1H-benzimidazolium) hydrogensulfate]	5 h	95	[49]
Glycerol trioleate	[Bmim][OH]	[Bmim][OH]	120 °C, 8 h, methanol	87	[50]
Oleates	1-(4-sulphonic acid) butyl-3-methy-limidazolium hydrogen sulphate	1-(4-sulphonic acid) butyl-3-methylimidazolium hydrogen sulphate		99	[51]
Soybean oil	3-(N,N,N-triethylamino)-1-propanesulfonic hydrogen sulfate	3-(N,N,N-triethylamino)-1-propanesulfonic hydrogen sulfate	80 °C, 60 min, ultrasound-assisted	98	[52]

[a] MeOH was another raw material

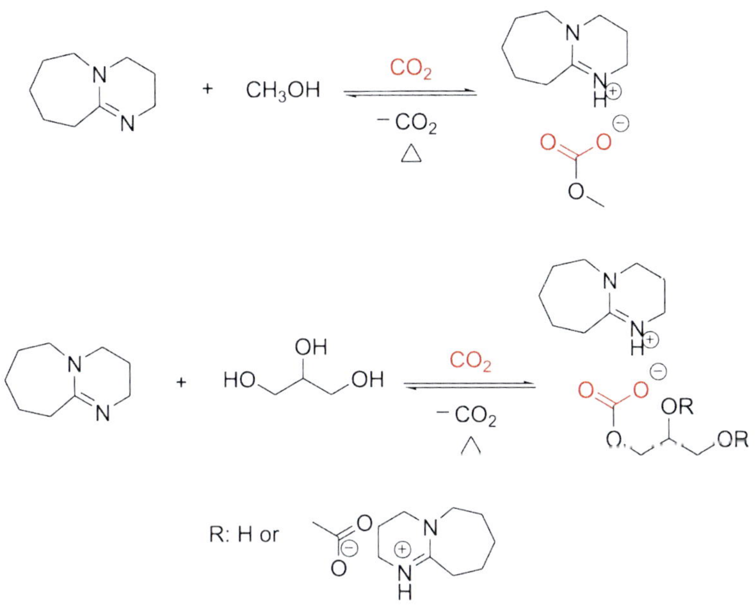

Scheme 7.3 Reversible chemical absorption of CO_2 by methanol and glycerol in the presence of DBU

7.2.3 Biodiesel Production by Biochemical Catalysis

The recovery of glycerol and the removal of inorganic salts remain cost in-efficient in traditional biodiesel production process catalyzed by acid or base catalysts [58]. Biocatalytic processes by lipases, especially immobilized biocatalysts, offer a promising route to improve the greenness of biodiesel production, as these processes can be done with high activity and selectivity under mild reaction conditions [59]. However, the traditional use of immobilized lipases also has several disadvantages, because lipids and methanol constitute a two-phase system which severely impedes mass transfer and enzyme catalysis. Some ILs with good enzyme compatibility were also good solvents for biocatalytic conversion [60, 61]. The design and application of lipase-compatible ILs have been investigated [62–71]. The main task of the introduction of ILs into these areas was to balance enzymatic compatibility and miscibility of ILs with lipids and methanol, thus to facilitate catalyst recycling and reuse, and product separation and purification.

For example, lipases from *Candida Antarctica* and *Pseudomonas cepacia* have been successfully immobilized in different ILs. The methanolysis of lipids was performed at room temperature, and product separation was realized by simple decantation, resulting in a facile reuse of the ILs/enzyme catalytic system. However, it was also found that many hydrophobic ILs have poor capability in dissolving lipids, while hydrophilic ILs tended to cause enzyme inactivation [62, 72]. A new type of ether-functionalized ILs carrying anions of acetate or formate was prepared. These ILs dissolved oils at 50 °C, at which temperature lipases maintained high catalytic activities even in the presence of high concentrations of methanol (up to 50 % v/v). High conversions of miglyol oil were observed in mixtures of ILs and methanol (70/30, v/v) when the reaction was catalyzed by a variety of lipases and different enzyme preparations (free and immobilized). The preliminary study on the transesterification of soybean oil in ILs/methanol mixtures further confirmed the potential of using oil-dissolving and lipase-stabilizing ILs in biodiesel production [69].

Homogeneous systems involving ILs have also been known for biocatalytic production of biodiesel [65]. Three hydrophobic ILs capable of dissolving triolein-methanol mixtures were devised. These hydrophobic ILs were based on cations attached with long chain alkyl groups. The $[C_{18}mim][NTf_2]$ was able to dissolve vegetable oil to form a monophasic system and provided an excellent microenvironment for catalysis. It has also been shown that polystyrene divinylbenzene porous matrix covalently attached with 1-decyl-2-methyimidazolium cation could be used as carrier to immobilize *C. antarctica* lipase B. The immobilized lipase was applied for methanolysis of triolein in both *tert*-butanol and supercritical (sc) CO_2 as reaction media. It was found that the use of modified supports with low ionic-liquid loading led to the highest yields (up to 95 %) and operational stability (85 % biodiesel yield after 45 cycles) in $scCO_2$ at 45 °C, 18 MPa. The presence of *tert*-butanol as an inert co-solvent in the $scCO_2$ phase at the same concentration as triolein was key to avoid poisoning the biocatalyst through the blockage of its active sites by the polar byproduct (glycerol) [65]. Additionally, the anion also played an important role on the efficiency, which was improved by increasing its hydrophobicity (i.e., $[NTf_2] > [PF_6] > [BF_4] > [Cl]$). More hydrophobic microenvironment provided by NTf_2 on the surface could facilitate a more efficient enzyme conformation as well as a higher accessibility of the substrate to the enzyme active site, leading to an increase in the observed activity. A large alkyl chain in the cation resulted in a clear improvement of the efficiency for the biotransformation in monophasic liquid systems [73]. Many other progresses are summarized in Table 7.5.

7.2.4 Biodiesel Purification

The purification of biodiesel is an essential process towards the production of high quality fuel. The main task of which is to remove glycerol and residual catalyst. Glycerol has low solubility in FAMEs and can be separated by settling or centrifugation. The presence of residual glycerides can cause deposition of biodiesel in internal combustion engine injectors (carbon residue) [78]. In addition, residual glycerol can initiate settling problems in the engine and, on the long term, affect human or animal health by the emission of hazardous acrolein into the environment. The presence of catalysts in biodiesel can form deposits (carbon residue) in fuel injection system, poison the emission control system, and weaken the engine [79].

Table 7.5 Ionic liquids as solvents for biodiesel production by enzymes

Raw materials[a]	Catalyst	Ionic liquids	Condition	By-product	Biodiesel yield (%)	References
Soybean oil	Immobilized Candida antarctica Lipase-catalyzed	[C_2mim][TfO]	50 °C, 12 h	<20 %	80	[62]
Soybean oil	Pseudomonas cepacia lipase	[C_4mim][NTf_2]	Room temperature	N.C.[a]	96	[72]
Soybean oil	Lipase-producing filamentous fungi immobilized on biomass support particles	[C_2mim][BF_4] or [C_4mim][BF_4]	24 h in biphasic systems	N.C.[a]	60	[63]
Soybean oil	Fungus whole-cell Biocatalysts	[C_4mim][BF_4]	72 h	N.C.[a]	60	[63]
Triolein	Novozym 435	[C_{18}mim][NTf_2]	60 °C, 6 h	N.C.[a]	96	[65]
Triolein or Waste Canola Oil	Novozym 435	[C_4mim][PF_6]	48 °C	Triacylglycerol	72	[67]
Triolein	Lipase	[C_4mim][PF_6]	48–55 °C	Triacylglycerol	80	[74]
Miglyol oil	Novozym 435	[$Me(OEt)_3$-Et_3N][OAc]	50 °C, 96 h	N.C.[b]	98	[70]
Triolein	Novozym 435	[C_{16}mim][NTf_2]	60 °C, 24 h	N.C.[b]	99	[64]
Olive oil	Novozym 435	[C_{16}mim][NTf_2]	60 °C, 24 h	N.C.[b]	93	[64]
Sunflower oil	Novozym 435	[C_{16}mim][NTf_2]	60 °C, 24 h	N.C.[b]	92	[64]
Palm oil	Novozym 435	[C_{16}mim][NTf_2]	60 °C, 24 h	N.C.[b]	94	[64]
Cooking waste oil	Novozym 435	[C_{16}mim][NTf_2]	60 °C, 24 h	N.C.[b]	96	[64]
Refined corn oil	Penicillium expansum lipase	[C_4mim][PF_6]	40 °C, 20 h	N.C.[b]	86	[73]
Miglyol oil	Novozym 435	Choline acetate	40 °C, 3 h, choline acetate/glycerol (1:1.5 molar ratio)	N.C.[b]	97	[69]

(continued)

Table 7.5 (continued)

Raw materials[a]	Catalyst	Ionic liquids	Condition	By-product	Biodiesel yield (%)	References
Microalgal oil	Novozym 435	1-butyl-3-methylimidazolium hexafluorophosphate/tert-butanol	48 h, 50 °C		90	[75]
Soybean oil	*Burkholderia cepacia* lipase	[OmPy][BF_4]	40 °C, 12 h		83	[76]
Cooking oil	Novozym 435	1-ethyl-3-methylimidazolium trifluoromethanesulfonate	40 °C, 24 h		99	[77]

[a] MeOH was another raw material
[b] N.C. = not characterized
Ionic liquids' abbreviation and full name: [C_2mim][TfO]: 1-Ethyl-3-methylimidazolium trifluoromethanesulfonate; [C_4mim][NTf_2]: 1-*n*-butyl-3-methylimidazolium *N*-bistrifluoromethanesulfonyl)imidate; [C_2mim][BF_4]: 1-ethyl-3-methylimidazolium tetrafluoroborate; [C_4mim][BF_4]: 1-butyl-3-methylimidazolium tetrafluoroborate; [C_{18}mim][NTf_2]: 1-methyl-3-octadecylimidazolium *bis*(trifluoromethylsulfonyl)imide; [C_{16}mim][NTf_2]: 1-hexadecyl-3-methylimidazolium *bis* (trifluoromethylsulfonyl)imide; [C_4mim][PF_6]: 1-butyl-3-methylimidazolium hexafluorophosphate; [C_8mPy][BF_4]: 1-octyl-3-methyl-pyrdininium tetrafluoroborate

As glycerol and methanol are highly soluble in water, water washing was widely used to remove excess contaminations (e.g. glycerol, alcohols, residual metal salts, soaps, fatty acids). However, the presence of water brings many disadvantages, including increased costs and production time, and generation of waste water [80]. Traditionally, several other methods have been used to remove glycerol from biodiesel, such as adsorption over silica, membrane reactors, and the addition of lime and phosphoric acid. Yet, technical problems remain for biodiesel production at an industrial scale [81]. To develop better process for byproduct removal, some classes of deep eutectic solvents based on mixtures of quaternary ammonium salts and compounds with hydrogen bond-donating group, have been applied in biodiesel production from rapeseed and soybean oil [82]. Deep eutectic solvents are inexpensive, non-toxic, and environmentally benign. While pure quaternary ammonium salts alone were inefficient, the quaternary ammonium salt–glycerol mixture solvents were successful for extraction of glycerol from biodiesel production mixtures, and a glycerol/salt molar ratio of 1.1 was found most effective. Of those salts studied, choline chloride, $ClEtNMe_3Cl$ and $EtNH_3Cl$ showed high efficiency for glycerol removal.

Deep eutectic solvents have also been used to extract glycerol from palm oil-based biodiesel production in order to meet the EN 14214, and ASTM D6751 standards. The extraction process involved different compositions of a quaternary ammonium salt to glycerol as the solvent, and a ratio of 1:1 was found most efficient. Moreover, the ratio of biodiesel to deep eutectic solvent was more important than the ratio of quaternary ammonium salt to glycerol. The used solvent can be recovered by crystallization [83].

Deep eutectic solvents based on methyl triphenyl phosphonium bromide and different hydrogen bond donors (e.g. glycerol, ethylene glycol, and triethylene glycol) were employed to remove glycerol from palm-oil-based biodiesel [84]. It was found that the solvents including ethylene glycol or triethylene glycol were successful in removing free glycerol to below the ASTM standards. These solvents were able to reduce the content of monoacylglycerides (MGs) and diacylglycerides (DGs), but DGs were removed more effectively than MGs. Choline chloride and methyltriphenylphosphonium bromide based deep eutectic solvents could also be used to remove residual KOH efficiently from palm oil-based biodiesel [85].

7.3 Bioalcohols Separation

Bioethanol and biobutanol are two important energy additives and chemicals, while most of them are produced so far by fermentation using sugars as carbon source. Separation of these alcohols from their fermentation broth requires up to 6 % of the energetic value of the compounds themselves [86]. Conventionally, alcohols and water were separated by distillation or membrane technology, but these technologies are energetically costly, or are not mature for large scale application. The tenability of miscibility and solubility of ILs to water, and other organic compounds offers significant advantages for separation of alcohols from water [87].

In an early study [88], ILs were tested for extraction of BuOH from aqueous solutions. It was found that BuOH distribution coefficient in 1-Butyl-3-methylimidazolium hexafluorophosphate ([BMIM][PF_6])–water system was 0.85, which was very close to the predicted value. 1-octyl-3-methylimidazolium hexafluorophosphate ([OMIM][PF_6]) had increased molecular dimensions of the alkylimidazolium cation gave lower mutual solubility of the ionic liquid and water, resulting in higher extractive selectivity for BuOH. Pervaporative BuOH recovery from 1 wt% aqueous solution and [OMIM][PF_6] was investigated using commercial polydimethylsiloxane membrane MEM-100. Although the viscosity of [OMIM][PF_6]–water–BuOH solution was about 100 fold higher than the viscosity of the BuOH–water mixture, the flux rate through the membrane was only 0.6 fold lower at higher selectivity. BuOH–water ratio in the permeation was close to that in ionic liquid feed, suggesting that the membrane did not improve the separation. Distillation is thought to be more economical for BuOH recovery from ILs [89].

It was demonstrated that the solubility of water in the hydrophobic IL 1-alkyl-3-methylimidazolium hexafluorophosphates could be significantly increased in the presence of ethanol as a so-solute. It was found that 1-hexyl-3-methylimidazolium hexafluorophosphate and 1-octyl-3-methylimidazolium hexafluorophosphate are completely miscible with ethanol, and immiscible with water, whereas 1-butyl-3-methylimidazolium hexafluorophosphate is totally miscible with aqueous ethanol only between 0.5 and 0.9 mole fraction ethanol at 25 °C. At higher and lower mole fraction of ethanol, the aqueous and IL components were only partially miscible and a biphasic system was obtained upon mixing equal volumes of the IL and aqueous ethanol. These observations indicated that ILs may be exploited as an extraction solvent for bioethanol recovery from fermentation broth.

As the hexafluorophosphate anion are partially hydrolyzed in the presence of water and thus generating corrosive HF [90], following up research in this area was moved to ILs with anions bis(trifluoromethylsulfonyl)imide and tetracyanoborate. These ILs are more stable in water and are more hydrophobic than those of hexafluorophosphate anion based ILs. Studies have led to the development of ternary diagrams, separation coefficients of 1-hexyl-3-methylimidazolium bis (trifluoromethylsulfonyl)imide/ethanol/water, and 1-hexyl-3-methylimidazolium bis(trifluoromethylsulfonyl)imide/1-butanol/water. It was found that 1-hexyl-3-methylimidazolium bis(trifluoromethylsulfonyl)imide can be used successfully to separate 1-butanol from water. Although it can also be used for ethanol separation, the solvent/feed ratio has to be unreasonably high [91, 92].

The tetracyanoborate-based ILs, 1-hexyl-3-methylimidazolium tetracyanoborate, 1-decyl-3-methylimidazolium tetracyanoborate, and trihexyltetradecylphosphonium tetracyanoborate, have also been studied. A complete miscibility in the binary liquid systems of 1-butanol with these ILs was observed. The presence of imidazolium cation gave lower selectivity and distribution ratio than those with phosphonium cation. The ILs with the longer alkyl chains at the cation showed higher selectivity and distribution ratio. Regarding to performance of the imidazolium based ILs, the choice of anion was shown to have a large impact on the upper critical solution temperatures of the system. The relative alcohol affinity for the different anions was $(CN)_2N > CF_3SO_3 > (CF_3SO_2)_2N > BF_4 > PF_6$ [93, 94].

Liquid–liquid extraction of 1-butanol from water employing nonfluorinated task-specific ILs (TSILs) has also been described recently [95]. Tetraoctylammonium 2-methyl-1-naphthoate [TOAMNaph] was identified as the best IL which had butanol distribution coefficient of 21 and selectivity of 274. These data were substantially better than those of the benchmark solvent oleyl alcohol, which had butanol distribution coefficient of 3.4 and selectivity of 192. The conceptual design study showed that butanol extraction with [TOAMNaph] requires 5.65 MJ/kg BuOH, which is 73 % less than that by conventional distillation.

To establish a more economic separation technology, studies were carried out to get in depth understanding of the vapor–liquid equilibrium of ethanol–water–ILs [96], 1-butanol–water–ILs, and high pressure CO_2-induced phase changes [97]. The results suggested that ILs were capable of breaking the binary azeotrope ethanol–water, opening a new possibility as entrainer for this system, while the 1-hexyl-3-methyl imidazolium chloride moved the azeotrope composition to a smaller fraction of ethanol. In addition, hexane assisted ILs extraction of ethanol [98], and phosphonium and ammonium ionic liquid-based supported liquid membranes have also been investigated [99].

7.4 Ionic Liquids for Bio-Oil Production and Upgrading

Bio-oil is a renewable liquid fuel, having negligible contents of sulfur, nitrogen, and ash, and is widely recognized as one of the most promising renewable fuels that may one day replace fossil fuels. Fast pyrolysis of biomass technologies for the production of bio-oil have been developed extensively in recent years, which is usually carried out by the rapid (in a few seconds) raising of temperature to around 450–550 °C under atmospheric pressure and anaerobic conditions. The bio-oils are of high oxygen content of high viscosity, thermal instability, corrosiveness, and chemical complexity. These characteristics limit the applications of bio-oils, precluding it from being used directly as a liquid fuel [100, 101]. Therefore, bio-oils need to be upgraded to improve its fuel properties. In order to develop a mild pyrolysis process with higher selectivity to favored compounds, the ILs-based technology was also introduced into this area. Several reports have demonstrated that ILs could be used as solvents or catalysts for this purpose. For example, Sheldrake et al. reported that dicationic molten salts were used as solvents for the controlled pyrolysis of cellulose to anhydrosugars [102]. It was demonstrated that the use of serials of dicationic ionic liquids for the pyrolysis of cellulose gave levoglucosenone as the dominant anhydrosugar product at 180 °C. An acidic dicationic IL were prepared and used as the catalyst to upgrade bio-oil through the esterification reaction of organic acids and ethanol at room temperature [103]. It was found that no coke and deactivation of the catalyst were observed. The yield of upgraded oil reached 49 %, and its properties were significantly improved with higher heating value of 24.6 MJ/kg, an increase of pH value to 5.1, and a decrease of moisture content to 8.2 wt%. The data showed that organic acids could be successfully converted into esters and that the dicationic IL can facilitate the esterification to upgrade bio-oil. It is also found that microware irradiation could promote the pyrolysis of rice straw and sawdust with 1-butyl-3-methylimidazolium chloride and 1-butyl-3-methylimidazolium tetrafluoroborate ILs as catalysts, and the bio-oil yield from rice straw reached 38 % and that from sawdust reached 34 % [104]. However, due to the high cost of ILs, and thermo stability during the pyrolysis, the use of ILs for the pyrolysis of biomass will not be the right direction.

7.5 Conclusion and Prospects

Biodiesel and bioalcohols are major biofuels that will offer many advantages over traditional fossil fuels and chemicals. In consideration of the unique properties of ILs and the key issues of biodiesel production from lipids, and bioalcohols separation from fermentation process, ILs including functionalized acidic and basic ILs, switchable ILs and deep eutectic solvents have been used for more efficient production of biodiesel and bioalcohols. Although satisfactory results have been achieved in terms of lipids extraction, catalytic conversion of lipids and fatty acids, biodiesel purification, and bioalcohols separation, major challenges remain in this area in terms of lowering the costs, improving recyclability and environmental compatibility of ILs. In the future, the effect of possible residual ILs on the quality of biofuel products and downstream application need to be addressed. Bearing all of this in mind, new switchable ionic liquid systems may have great potential in application because of their unique properties, such as easy preparation and good recyclability. It is expected that ILs will be applied in a wider and more integrated way for biofuel production from various raw materials.

References

1. Ragauskas A, Williams C, Davison B, Britovsek G, Cairney J, Eckert C, Frederick W, Hallett J, Leak D, Liotta C, Mielenz J, Murphy R, Templer R, Tschaplinski T. The path forward for biofuels and biomaterials. Science. 2006;311(5760):484–9.
2. Rogers R, Seddon K. Ionic liquids-solvents of the future? Science. 2003;302(5646):792–3.
3. Pinkert A, Marsh K, Pang S, Staiger M. Ionic liquids and their interaction with cellulose. Chem Rev. 2009;109(12):6712–28.
4. Xie H, Liu W, Beadham I, Gathergood N. Biorefinery with ionic liquids (chapter 3). In: Xie H, Gathergood N, editors. The role of green chemistry for the biomass processing and conversion. Hoboken, New Jersey: Wiley; 2012. p. 75–133. ISBN 978-0-470-64410-2.

5. Srivastava A, Prasad R. Triglycerides-based diesel fuels. Renew Sust Energy Rev. 2000;4 (2):111–33.
6. Galbe M, Zacchi G. A review of the production of ethanol from softwood. Appl Microbiol Biotechnol. 2002;59(6):618–28.
7. Fukuda H, Kondo A, Noda H. Biodiesel fuel production by transesterification of oils. J Biosci Bioeng. 2001;92(5):405–16.
8. Sun Y, Cheng J. Hydrolysis of lignocellulosic materials for ethanol production: a review. Bioresour Technol. 2002;83(1):1–11.
9. Mascal M. Chemicals from biobutanol: technologies and markets. Biofuels Bioprod Bioref. 2012;6(4):483–93.
10. Clark J, Deswarte F, Farmer T. The integration of green chemistry into future biorefineries. Biofuels Bioprod Bioref. 2009;3(1):72–90.
11. Zhao H, Baker G. Ionic liquids and deep eutectic solvents for biodiesel synthesis: a review. J Chem Technol Biotechnol. 2013;88(1):3–12.
12. Earle M, Plechkova N, Seddon K. Green synthesis of biodiesel using ionic liquids. Pure Appl Chem. 2009;81(11):2045–57.
13. Fauzi A, Amin N. An overview of ionic liquids as solvents in biodiesel synthesis. Renew Sust Energy Rev. 2012;16(8):5770–86.
14. Zhu Z, Zhu M, Wu Z. Pretreatment of sugarcane bagasse with NH_4OH-H_2O_2 and ionic liquid for efficient hydrolysis and bioethanol production. Bioresour Technol. 2012;119:199–207.
15. Xie H, Shen H, Gong Z, Wang Q, Zhao Z, Bai F. Enzymatic hydrolysates of corn stover pretreated by a N-methylpyrrolidone-ionic liquid solution for microbial lipid production. Green Chem. 2012;14(4):1202–10.
16. Pinto A, Guarieiro L, Rezende M, Ribeiro N, Torres E, Lopes W, Pereira P, de Andrade J. Biodiesel: an overview. J Braz Chem Soc. 2005;16(6B):1313–30.
17. Ahmad A, Yasin N, Derek C, Lim J. Microalgae as a sustainable energy source for biodiesel production: a review. Renew Sust Energy Rev. 2011;15(1):584–93.
18. Li Y, Zhao Z, Bai F. High-density cultivation of oleaginous yeast Rhodosporidium toruloides Y4 in fed-batch culture. Enzyme Microb Technol. 2007;41(3):312–7.
19. Chisti Y. Biodiesel from microalgae. Biotechnol Adv. 2007;25(3):294–306.
20. Shahid E, Jamal Y. Production of biodiesel: a technical review. Renew Sust Energy Rev. 2011;15(9):4732–45.
21. Erikson D. Practical handbook of soybean processing and utilization. St. Louis: AOCS Press/ United Soybean Board; 1995.
22. Jessop P, Heldebrant D, Li X, Eckert C, Liotta C. Green chemistry – reversible nonpolar-to-polar solvent. Nature. 2005;436(7054):1102.
23. Jessop P. Searching for green solvents. Green Chem. 2011;13(6):1391–8.
24. Fadhel A, Pollet P, Liotta C, Eckert C. Novel solvents for sustainable production of specialty chemicals. Annual Rev Chem Biomol Eng (Ed.: J. M. Prausnitz). 2011; 2:189–210.
25. Subramaniam B. Gas-expanded liquids for sustainable catalysis and novel materials: recent advances. Coord Chem Rev. 2010;254(15–16):1843–53.
26. Wang J, Su X, Jessop P, Feng Y. CO_2 Switchable solvents, solutes and surfactants: state of the art. Prog Chem. 2010;22(11):2099–105.
27. Cao X, Xie H, Wu Z, Shen H, Jing B. Phase-switching homogeneous catalysis for clean production of biodiesel and glycerol from soybean and microbial lipids. ChemCatChem. 2012;4(9):1272–8.
28. Phan L, Brown H, White J, Hodgson A, Jessop P. Soybean oil extraction and separation using switchable or expanded solvents. Green Chem. 2009;11(1):53–9.
29. Samori C, Torri C, Samori G, Fabbri D, Galletti P, Guerrini F, Pistocchi R, Tagliavini E. Extraction of hydrocarbons from microalga Botryococcus braunii with switchable solvents. Bioresour Technol. 2010;101(9):3274–9.
30. Samori C, Barreiro D, Vet R, Pezzolesi L, Brilman D, Galletti P, Tagliavini E. Effective lipid extraction from algae cultures using switchable solvents. Green Chem. 2013;15(2):353–6.

31. Young G, Nippgen F, Titterbrandt S, Cooney M. Lipid extraction from biomass using co-solvent mixtures of ionic liquids and polar covalent molecules. Sep Purif Technol. 2010;72(1):118–21.
32. Teixeira R. Energy-efficient extraction of fuel and chemical feedstocks from algae. Green Chem. 2012;14(2):419–27.
33. Azocar L, Ciudad G, Heipieper H, Navia R. Biotechnological processes for biodiesel production using alternative oils. Appl Microbiol Biotechnol. 2010;88(3):621–36.
34. Hallett J, Welton T. Room-temperature ionic liquids: solvents for synthesis and catalysis, 2. Chem Rev. 2011;111(5):3508–76.
35. Cole A, Jensen J, Ntai I, Tran K, Weaver K, Forbes D, Davis J. Novel Bronsted acidic ionic liquids and their use as dual solvent-catalysts. J Am Chem Soc. 2002;124(21):5962–3.
36. Lapis A, de Oliveira L, Neto B, Dupont J. Ionic liquid supported acid/base-catalyzed production of biodiesel. ChemSusChem. 2008;1(8–9):759–62.
37. Crocker W. Washing with eutectic solvents cleans up biodiesel and produces glycerol – mild green ionic liquids. J Mater Chem. 2007;17(22):T41–1.
38. Han M, Yi W, Wu Q, Liu Y, Hong Y, Wang D. Preparation of biodiesel from waste oils catalyzed by a Bronsted acidic ionic liquid. Bioresour Technol. 2009;100(7):2308–10.
39. Liang X, Gong G, Wu H, Yang J. Highly efficient procedure for the synthesis of biodiesel from soybean oil using chloroaluminate ionic liquid as catalyst. Fuel. 2009;88(4):613–6.
40. Zhang L, Xian M, He Y, Li L, Yang J, Yu S, Xu X. A Bronsted acidic ionic liquid as an efficient and environmentally benign catalyst for biodiesel synthesis from free fatty acids and alcohols. Bioresour Technol. 2009;100(19):4368–73.
41. Liang X, Yang J. Synthesis of a novel multi -SO_3H functionalized ionic liquid and its catalytic activities for biodiesel synthesis. Green Chem. 2010;12(2):201–4.
42. Long T, Deng Y, Gan S, Chen J. Application of choline chloride center dot xZnCl(2) ionic liquids for preparation of biodiesel. Chin J Chem Eng. 2010;18(2):322–7.
43. Elsheikh Y, Man Z, Bustam M, Yusup S, Wilfred C. Bronsted imidazolium ionic liquids: synthesis and comparison of their catalytic activities as pre-catalyst for biodiesel production through two stage process. Energy Convers Manag. 2011;52(2):804–9.
44. Fang D, Yang J, Jiao C. Dicationic ionic liquids as environmentally benign catalysts for biodiesel synthesis. ACS Catal. 2011;1(1):42–7.
45. De Santi V, Cardellini F, Brinchi L, Germani R. Novel Bronsted acidic deep eutectic solvent as reaction media for esterification of carboxylic acid with alcohols. Tetrahedron Lett. 2012;53(38):5151–5.
46. Zhou J, Lu Y, Huang B, Huo Y, Zhang K. Preparation of biodiesel from Tung oil catalyzed by sulfonic-functional Bronsted acidic ionic liquids. Adv Manuf Technol. 2011;Pts 1–3, (Ed.: J. Gao), (314–316):1459–62.
47. Li K, Chen L, Yan Z, Wang H. Application of pyridinium ionic liquid as a recyclable catalyst for acid-catalyzed transesterification of Jatropha oil. Catal Lett. 2010;139(3–4):151–6.
48. Guo F, Fang Z, Tian X, Long Y, Jiang L. One-step production of biodiesel from Jatropha oil with high-acid value in ionic liquids. Bioresour Technol. 2011;102(11):6469–72.
49. Ghiaci M, Aghabarari B, Habibollahi S, Gil A. Highly efficient Bronsted acidic ionic liquid-based catalysts for biodiesel synthesis from vegetable oils. Bioresour Technol. 2011;102 (2):1200–4.
50. Zhou S, Liu L, Wang B, Xu F, Sun R. Biodiesel preparation from transesterification of glycerol trioleate catalyzed by basic ionic liquids. Chin Chem Lett. 2012;23(4):379–82.
51. Zhou S, Liu L, Wang B, Xu F, Sun R. Facile biodiesel synthesis from esterification of free fatty acids catalyzed by SO_3H-functionalized ionic liquid. Asian J Chem. 2013;25(1):240–4.
52. Guo W, Li H, Ji G, Zhang G. Ultrasound-assisted production of biodiesel from soybean oil using Bronsted acidic ionic liquid as catalyst. Bioresour Technol. 2012;125:332–4.
53. Zhang L, Cui Y, Zhang C, Wang L, Wan H, Guan G. Biodiesel production by esterification of oleic acid over Bronsted acidic ionic liquid supported onto Fe-incorporated SBA-15. Ind Eng Chem Res. 2012;51(51):16590–6.

54. Fan M, Yang J, Jiang P, Zhang P, Li S. Synthesis of novel dicationic basic ionic liquids and its catalytic activities for biodiesel production. RSC Adv. 2013;3(3):752–6.
55. Karavalakis G, Anastopoulos G, Stournas S. Tetramethylguanidine as an efficient catalyst for transesterification of waste frying oils. Appl Energy. 2011;88(11):3645–50.
56. Lohmeijer B, Pratt R, Leibfarth F, Logan J, Long D, Dove A, Nederberg F, Choi J, Wade C, Waymouth R, Hedrick J. Guanidine and amidine organocatalysts for ring-opening polymerization of cyclic esters. Macromolecules. 2006;39(25):8574–83.
57. Balbino J, de Menezes E, Benvenutti E, Cataluna R, Ebeling G, Dupont J. Silica-supported guanidine catalyst for continuous flow biodiesel production. Green Chem. 2011;13(11):3111–6.
58. Salvi B, Panwar N. Biodiesel resources and production technologies – a review. Renew Sust Energy Rev. 2012;16(6):3680–9.
59. Demirbas A. Current technologies in biodiesel production. In: Biodiesel. London: Springer; 2008. pp. 161–173.
60. van Rantwijk F, Sheldon R. Biocatalysis in ionic liquids. Chem Rev. 2007;107(6):2757–85.
61. van Rantwijk F, Lau R, Sheldon R. Biocatalytic transformations in ionic liquids. Trends Biotechnol. 2003;21(3):131–8.
62. Ha S, Lan M, Lee S, Hwang S, Koo Y. Lipase-catalyzed biodiesel production from soybean oil in ionic liquids. Enzyme Microb Technol. 2007;41(4):480–3.
63. Arai S, Nakashima K, Tanino T, Ogino C, Kondo A, Fukuda H. Production of biodiesel fuel from soybean oil catalyzed by fungus whole-cell biocatalysts in ionic liquids. Enzyme Microb Technol. 2010;46(1):51–5.
64. De Diego T, Manjon A, Lozano P, Vaultier M, Iborra J. An efficient activity ionic liquid-enzyme system for biodiesel production. Green Chem. 2011;13(2):444–51.
65. Lozano P, Garcia-Verdugo E, Bernal J, Izquierdo D, Isabel Burguete M, Sanchez-Gomez G, Luis S. Immobilised lipase on structured supports containing covalently attached ionic liquids for the continuous synthesis of biodiesel in $scCO_2$. ChemSusChem. 2012;5(4):790–8.
66. Nakashima K, Arai S, Tanino T, Ogino C, Kondo A, Fukuda H. Production of biodiesel fuel in ionic liquids catalyzed by whole-cell biocatalysts. J Biosci Bioeng. 2009;108:S43–3.
67. Ruzich N, Bassi A. Investigation of lipase-catalyzed biodiesel production using ionic liquid BMIM PF_6 as a co-solvent in 500 mL jacketed conical and shake flask reactors using triolein or waste canola oil as substrates. Energy Fuels. 2010;24:3214–22.
68. Zhang K, Lai J, Huang Z, Yang Z. Penicillium expansum lipase-catalyzed production of biodiesel in ionic liquids. Bioresour Technol. 2011;102(3):2767–72.
69. Zhao H, Baker G, Holmes S. New eutectic ionic liquids for lipase activation and enzymatic preparation of biodiesel. Org Biomol Chem. 2011;9(6):1908–16.
70. Zhao H, Song Z, Olubajo O, Cowins J. New ether-functionalized ionic liquids for lipase-catalyzed synthesis of biodiesel. Appl Biochem Biotechnol. 2010;162(1):13–23.
71. Hernández-Fernández F, de los Ríos A, Lozano-Blanco L, Godínez C. Biocatalytic ester synthesis in ionic liquid media. J Chem Technol Biotechnol. 2010;85(11):1423–35.
72. Gamba M, Lapis A, Dupont J. Supported ionic liquid enzymatic catalysis for the production of biodiesel. Adv Synth Catal. 2008;350(1):160–4.
73. Lozano P, Bernal J, Vaultier M. Towards continuous sustainable processes for enzymatic synthesis of biodiesel in hydrophobic ionic liquids/supercritical carbon dioxide biphasic systems. Fuel. 2011;90(11):3461–7.
74. Ruzich N, Bassi A. Investigation of enzymatic biodiesel production using ionic liquids as a co-solvent. Can J Chem Eng. 2010;88(2):277–82.
75. Lai J, Hu Z, Wang P, Yang Z. Enzymatic production of microalgal biodiesel in ionic liquid Bmim PF_6. Fuel. 2012;95(1):329–33.
76. Liu Y, Chen D, Yan Y, Peng C, Xu L. Biodiesel synthesis and conformation of lipase from Burkholderia cepacia in room temperature ionic liquids and organic solvents. Bioresour Technol. 2011;102(22):10414–8.
77. Liu Y, Wu H, Yan Y, Dong L, Zhu M, Liang P. Lipase-catalyzed transesterification for biodiesel production in ionic liquid EmimTfO. Int J Green Energy. 2013;10(1):63–71.

78. Ma F, Hanna M. Biodiesel production: a review. Bioresour Technol. 1999;70(1):1–15.
79. Mudge S, Pereira G. Stimulating the biodegradation of crude oil with biodiesel preliminary results. Spill Sci Technol Bull. 1999;5(5–6):353–5.
80. Berrios M, Skelton R. Comparison of purification methods for biodiesel. Chem Eng J. 2008;144(3):459–65.
81. Dube M, Tremblay A, Liu J. Biodiesel production using a membrane reactor. Bioresour Technol. 2007;98(3):639–47.
82. Abbott A, Cullis P, Gibson M, Harris R, Raven E. Extraction of glycerol from biodiesel into a eutectic based ionic liquid. Green Chem. 2007;9(8):868–72.
83. Hayyan M, Mjalli F, Hashim M, AlNashef I. A novel technique for separating glycerine from palm oil-based biodiesel using ionic liquids. Fuel Process Technol. 2010;91(1):116–20.
84. Shahbaz K, Mjalli F, Hashim M, AlNashef I. Using deep eutectic solvents based on methyl triphenyl phosphonium bromide for the removal of glycerol from palm-oil-based biodiesel. Energy Fuels. 2011;25(6):2671–8.
85. Shahbaz K, Mjalli F, Hashim M, AlNasheff I. Eutectic solvents for the removal of residual palm oil-based biodiesel catalyst. Sep Purif Technol. 2011;81(2):216–22.
86. Roffler S, Blanch H, Wilke C. *In situ* recovery of fermentation products. Trends Biotechnol. 1984;2(5):129–36.
87. Kertes A, King C. Extraction chemistry of low molecular weight aliphatic alcohols. Chem Rev. 1987;87(4):687–710.
88. Fadeev A, Meagher M. Opportunities for ionic liquids in recovery of biofuels. Chem Commun. 2001;3:295–6.
89. Swatloski R, Visser A, Reichert W, Broker G, Farina L, Holbrey J, Rogers R. On the solubilization of water with ethanol in hydrophobic hexafluorophosphate ionic liquids. Green Chem. 2002;4(2):81–7.
90. Freire M, Neves C, Marrucho I, Coutinho J, Fernandes A. Hydrolysis of tetrafluoroborate and hexafluorophosphate counter ions in imidazolium-based ionic liquids. J Phys Chem A. 2010;114(11):3744–9.
91. Chapeaux A, Simoni L, Ronan T, Stadtherr M, Brennecke J. Extraction of alcohols from water with 1-hexyl-3-methylimidazolium bis(trifluoromethylsulfonyl)imide. Green Chem. 2008;10(12):1301–6.
92. Chapeaux A, Simoni L, Stadtherr M, Brennecke J. Liquid phase behavior of ionic liquids with water and 1-octanol and modeling of 1-octanol/water partition coefficients. J Chem Eng Data. 2007;52(6):2462–7.
93. Crosthwaite J, Aki S, Maginn E, Brennecke J. Liquid phase behavior of imidazolium-based ionic liquids with alcohols. J Phys Chem B. 2004;108(16):5113–9.
94. Crosthwaite J, Aki S, Maginn E, Brennecke J. Liquid phase behavior of imidazolium-based ionic liquids with alcohols: effect of hydrogen bonding and non-polar interactions. Fluid Phase Equilib. 2005;228:303–9.
95. Garcia-Chavez L, Garsia C, Schuur B, de Haan A. Biobutanol recovery using nonfluorinated task-specific ionic liquids. Ind Eng Chem Res. 2012;51(24):8293–301.
96. Calvar N, Gomez E, Gonzalez B, Dominguez A. Experimental vapor–liquid equilibria for the ternary system ethanol + water + 1-ethyl-3-methylpyridinium ethylsulfate and the corresponding binary systems at 101.3 kPa: Study of the effect of the cation. J Chem Eng Data. 2010;55(8):2786–91.
97. Najdanovic-Visak V, Rebelo L, da Ponte M. Liquid-liquid behaviour of ionic liquid-1-butanol-water and high pressure CO_2-induced phase changes. Green Chem. 2005;7 (6):443–50.
98. Pereiro A, Rodriguez A. Effective extraction in packed column of ethanol from the azeotropic mixture ethanol plus hexane with an ionic liquid as solvent. Chem Eng J. 2009;153 (1–3):80–5.

99. Cascon H, Choudhari S. 1-Butanol pervaporation performance and intrinsic stability of phosphonium and ammonium ionic liquid-based supported liquid membranes. J Membr Sci. 2013;429:214–24.
100. Furimsky E. Catalytic hydrodeoxygenation. Appl Catal Gen. 2000;199(2):147–90.
101. Mohan D, Pittman C, Steele P. Pyrolysis of wood/biomass for bio-oil: a critical review. Energy Fuels. 2006;20(3):848–89.
102. Sheldrake G, Schleck D. Dicationic molten salts (ionic liquids) as reusable media for the controlled pyrolysis of cellulose to anhydrosugars. Green Chem. 2007;9(10):1044–6.
103. Xiong W, Zhu M, Deng L, Fu Y, Guo Q. Esterification of organic acid in bio-oil using acidic ionic liquid catalysts. Energy Fuels. 2009;23:2278–83.
104. Du J, Liu P, Liu Z, Sun D, Tao C. Fast pyrolysis of biomass for bio-oil with ionic liquid and microwave irradiation. J Fuel Chem Technol. 2010;38(5):554–9.

Chapter 8
Catalytic Transformation of Biomass in Ionic Liquids

Blair J. Cox and John G. Ekerdt

Abstract This chapter focuses on a number of developing technologies based on catalytic transformations of biomass in ionic liquids. As an introduction, an overview of biomass and ionic liquids is given. The chapter continues with a description of catalysis of monosaccharides and polysaccharides in ionic liquids, covering saccharification, depolymerization, isomerization, dehydration into 5-hydroxymethylfurfural, and further processing. The derivatization of mono- and polysaccharides is also discussed. Because fermentation of biomass is an important technology that is widely used and continuing to grow, a section is devoted to the use of ionic liquids in pretreatment of biomass for saccharification and fermentation into ethanol. Extraction and depolymerization of lignin model compounds and the whole lignin polymer in ionic liquids are discussed both for pretreatment and use of lignin fragments as a source of fuel and chemicals. A discussion of deoxygenation and hydrogenation of lignin fragments is also given, followed by a concluding section outlining the advantages, challenges, and prospects for catalytic processing of biomass in ionic liquids.

Keywords Biomass • Ionic liquid • Cellulose • Saccharides • Lignin • Catalysis • Carbohydrates

B.J. Cox
Department of Chemical Engineering, The University of Texas at Austin, 200 E. Dean Keeton St. Stop C0400, Austin, TX 78712-1589, USA

UT Dallas Venture Development Center, Cyclewood Solutions Inc., Richardson, TX 75080, USA
e-mail: blair.cox@cyclewood.com

J.G. Ekerdt (✉)
Department of Chemical Engineering, The University of Texas at Austin, 200 E. Dean Keeton St. Stop C0400, Austin, TX 78712-1589, USA
e-mail: Ekerdt@che.utexas.edu

Z. Fang et al. (eds.), *Production of Biofuels and Chemicals with Ionic Liquids*, Biofuels and Biorefineries 1, DOI 10.1007/978-94-007-7711-8_8,

8.1 Introduction

Biomass is already the single largest source of renewable energy in the United States and has great potential for further utilization as a renewable resource [1]. Globally, the potential for sustainable biomass derived energy is 100 EJ/a, which is 30 % of the 2003 global energy consumption [2]. The majority of biomass falls into the category of lignocellulosic biomass, so named because it is composed of three biopolymers: cellulose, hemicellulose and lignin. Agricultural lands in the United States can produce nearly 1 billion dry tons of biomass annually while still meeting food, feed, and export demands, while US forest resources can produce an additional 368 million dry tons [1]. Other forms of biomass, such as corn starch (840 million tons in 2010 worldwide [3] and 216 million tons corn annually in the US [4]), or simple sugars such as glucose, fructose, or sucrose may also prove to be important resources.

The utilization of biomass as a source of fuels and chemicals has increased in recent years. Production of ethanol from biomass has seen rapid growth with the US and Brazil leading the world in bio-ethanol production. Corn based ethanol production in the US has reached 13.9 million gal while Brazil produces 5.6 million gal of ethanol from the fermentation of sugar cane annually [5]. In the US, the Energy Independence and Security Act of 2007 mandated production and blending of ethanol as a biofuel, which has led to the large scale production of corn based ethanol [6]. Based on the availability of resources, other substrates for fermentation can be used as is the case with sugar cane in Brazil [7]. Cellulose is looked to as the next generation of substrates for ethanol production using feed stocks such as switch grass, sugarcane bagasse, or corn stover as a cellulose source [8]. In order for the cellulose to ethanol conversion to work, biomass sources must be pretreated to make the structural carbohydrates accessible to saccharification in preparation for conventional fermentation into ethanol [9]. The pretreatment step has been the subject of considerable research. Steam explosion, ammonia treatment, dilute acid treatment, milling, and even treatment with ionic liquids have been explored as methods for preparing biomass for saccharification [10–14].

While fermentation into ethanol is one option for converting biomass into fuel, other catalytic processes have been investigated and developed for the utilization of biomass. Using algae as a means of production for both bio-oil and carbohydrates has been looked to as a next generation source of biomass products due the algae's high energy yield per cultivation area and ability to thrive in a wide range of locations [15]. Conversion of biomass into bio-oil has received considerable attention. Both fast pyrolysis and syn-gas processes hold considerable potential for biofuel production [16, 17]. Catalyst development and application of petrochemical technology is also an important subject in the field of biofuels [18]. Even the less technologically advanced method of burning biomass provides a significant source of energy. Residue from processing of biomass into food or consumer products and biomass harvested specifically for fuel are a significant source of energy and have a high sustainable potential that has not yet been realized

[2]. Further discussion of the current state of the utilization of biomass for the production of ethanol, bio-oil, commodity chemicals, and other products is covered in a number of articles [2, 15, 17, 19, 20].

One of the challenges in utilizing biomass in chemical processing for fuels or other products is that, in most cases, the biomass is insoluble in commonly used solvents. Ionic liquids (ILs) are a class of compounds that are composed completely of anions and cations and melt at temperatures below 100 °C. Recently, it has been found that some ILs are effective for dissolution of many kinds of biomass. Some can even completely dissolve lignocellulose up to 25 % by weight without chemical modification of the biomass occurring [21]. Based on this discovery, the research on the catalytic transformation of biomass in ionic liquids has increased markedly in recent years. The hope of this research is that the unique solvent properties of ionic liquids coupled with the potential of biomass as a renewable resource will lead to advances in the next generation of fuel and chemical production.

8.1.1 Biomass

The most common biomass source, lignocellulose, is the principal component of plant matter and is the largest renewable resource available [1]. Lignocellulose is composed primarily of three biopolymers: cellulose, hemicellulose, and lignin. Cellulose, which is the most abundant biopolymer in lignocellulosic biomass, is composed of glucose monomers linked together through 1–4 glycosidic linkages. As shown in Fig. 8.1, these chains of glucose hydrogen bond with the hydroxyl groups of neighboring cellulose molecules, providing a stable, crystalline structure to cellulose fibers in cell walls [23]. Because cellulose is the single most abundant renewable resource available, there has been significant work in its utilization across a wide range of applications. Cellulose can be somewhat difficult to break into its component glucose units and there are a number of methods, such as enzymatic or acid catalyzed hydrolysis, that will convert cellulose into monosaccharides or short carbohydrate chains [24, 25].

Hemicellulose, which, like cellulose, is a polymer composed of monosaccharides, makes up 20–30 % of plant biomass. Unlike cellulose, however, hemicellulose is a branched carbohydrate that can be made up of multiple monosaccharides, bonded through a number of different glycosidic linkages. The structure of hemicellulose is composed of a polysaccharide backbone made from glucose, xylose, or mannose units connected through β–(1–3) or β–(1–4) glycosidic bonds. From these backbones, there are side chains of glucose, glucuronic acid, 4-O-methylglucuronic acid, mannose, xylose, arabinose, or galactose [26]. The composition of the hemicellulose is dependent on the plant species that produced it [26, 27]. Compared to cellulose, hemicellulose is significantly easier to hydrolyze into small carbohydrate chains and monosaccharides. Currently, there are a number of methods for hemicellulose extraction and degradation, including steam explosion, dilute acid treatment, and ammonia explosion [9].

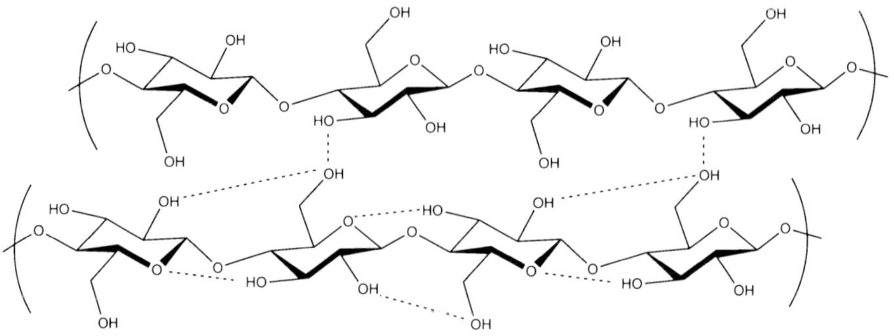

Fig. 8.1 Intra- and intermolecular hydrogen bonds in cellulose (Adapted with permission from [22]. Copyright 2009 American Chemical Society)

Lignin is the third kind of structural biopolymer, which composes 15–30 % lignocellulosic biomass by weight [28]. In the structure of the cell wall, lignin fills the space between cellulose/hemicellulose fibers. Unlike cellulose and hemicellulose, lignin is not made from carbohydrates but from phenylpropane units that are linked through enzymatic radical polymerization [29]. The monomers that plants employ to create lignin are cinnaminyl alcohol, sinapyl alcohol, and *p*-coumaryl alcohol. These monomers are bonded together through a number of different linkages that form a complex, amorphous structure. The most common of these bonds are the β–O–4, 5–5, β–5, β–1, and α–O–4 linkages (Fig. 8.2), which represent 45–50, 18–25, 9–12, 7–10, and 6–8 % of the linkages in softwood lignin, respectively [30]. Lignin is a major inhibitor of biological degradation of lignocellulose, as there are only a few species in nature that can effectively metabolize it [31, 32]. While lignin benefits living plants, it presents a significant challenge to the successful utilization of biomass in the production of fuels or other chemicals.

Currently, lignin is processed through a number of different techniques. The pulp and paper industry generally uses a process called kraft pulping, in which a strong bases and sulfur compounds depolymerize and extract lignin from wood pulp [30, 33]. Other methods, such as acid pulping, organosolve pulping, and high temperature ethanol/water have been used to degrade lignin have also been employed on an industrial scale [34–38]. In other cases, lignin depolymerization, along with disruption of cellulose crystallinity, has been used for biomass pretreatment for further processing [9, 14].

Other sources of biomass can be used as substrates. Starches and simple sugars are currently used in the production of fuel ethanol. These carbohydrates can be sourced from corn, sugar cane, beets, or as a product from the depolymerization of longer chain polysaccharides [20, 39]. Algae have also received significant attention as a source of renewable energy feed stocks [15]. In research on catalysis of biomass, simple sugars are often used as a model or substitute for more complicated carbohydrates and biomass in general. Other sources of biomass may be similar to common lignocellulose, but have unique characteristics that warrant special

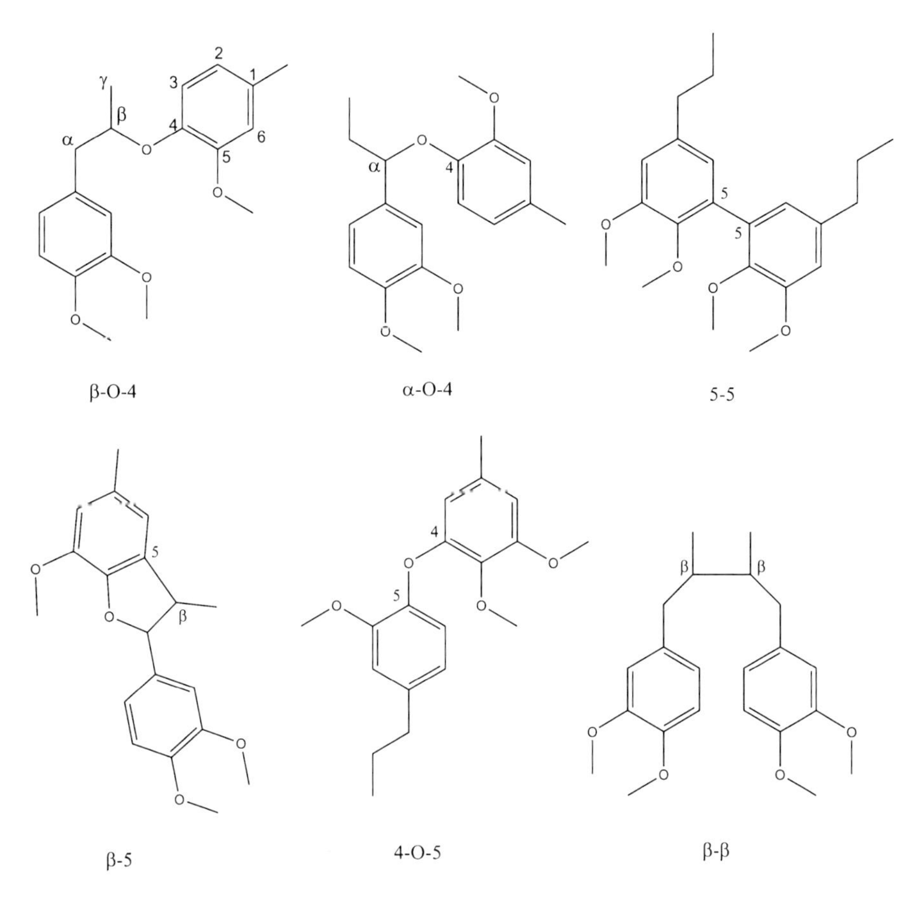

Fig. 8.2 Common linkages in softwood lignin (Adapted with permission from [30]. Copyright 2004 Elsevier)

attention. Rice hulls, for example, are coated in a layer of silica that makes effective catalytic conversion difficult [40].

8.1.2 Ionic Liquids

Ionic liquids were reported as early as 1914 when ethylammonium nitrate was shown to melt at 12 °C [41]. In recent years the study of these compounds has experienced a resurgence. An ionic liquid is defined as a chemical compound that exists as an organic anion and a cation and has a melting point below 100 °C. There are many different kinds of ILs, and many more are being developed (Fig. 8.3). The most common forms are based on dialkylimidazolium, tetraalkylammonium, alkylpyridinium, or tetraalkylphosphonium cations coupled with an inorganic

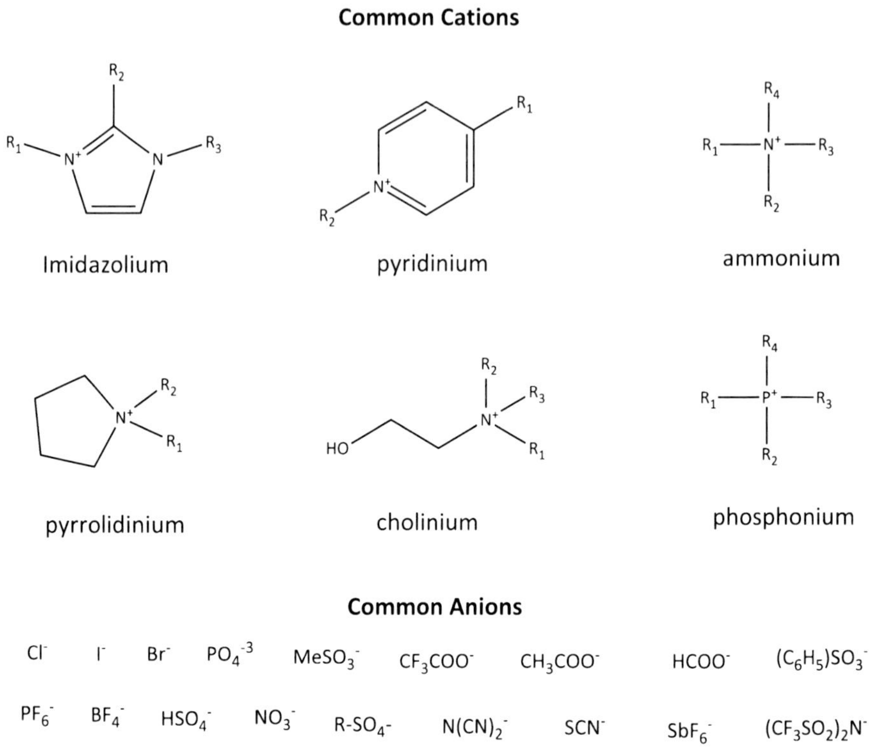

Common Cations

Imidazolium pyridinium ammonium

pyrrolidinium cholinium phosphonium

Common Anions

Cl^- I^- Br^- PO_4^{-3} $MeSO_3^-$ CF_3COO^- CH_3COO^- $HCOO^-$ $(C_6H_5)SO_3^-$

PF_6^- BF_4^- HSO_4^- NO_3^- $R\text{-}SO_4\text{-}$ $N(CN)_2^-$ SCN^- SbF_6^- $(CF_3SO_2)_2N^-$

Fig. 8.3 Common IL anions and cations

anion [42]. Due to their ionic character, ILs have essentially no vapor pressure. While it has been reported that some ILs can be distilled under the right conditions [43], in general, the vapor pressure is low enough to be neglected. Because ILs are composed of discrete anions and cations, the solvent properties, such as viscosity, melting point, and miscibility with other solvents can be tuned through the right combination and design of each ion. The ability to design ILs to specific substrates, chemistries, and situations is important, because ILs are increasingly being looked to as a medium for applications, including biomass processing [44].

Additionally, most ILs display good stability under a wide range of chemical, thermal, and electrochemical conditions. Some ILs have reported thermal stability of over 300 °C, although the stability is highly dependent on the identity of the IL's constituent ions [45]. There has been research to show that the long term thermal stability of some ILs is significantly less than that indicated by standard thermogravimetric analysis techniques [46]. This may be important in the development of biomass processing techniques in ILs, because one of the main advantages of ILs is the potential for the essentially complete recycling of solvents. ILs are also generally assumed to have good chemical stability. In many cases, strongly acidic, reducing, or oxidizing agents can be used without degradation of the ILs [47, 48].

There are, however, some exceptions to this rule [49]. Dialkylimidazolium based ILs have a mildly acidic hydrogen that undergoes hydrogen exchange in aqueous media and can even deprotonate to form a reactive carbene under basic conditions [49, 50]. Some ILs, such as halide, acetate, or formate based ILs, can form volatile acids (such as hydrochloric, acetic, or formic acid) [43, 51]. Additionally, some ILs can be designed to be reactive with the addition of acidic moieties or metal centers [47, 52, 53]. ILs have also been looked to as media for novel electrochemistry, as some of them have a wide window of electrochemical stability [54].

ILs have found a place in a number of catalytic reactions. In many cases, the solvent properties of ILs increase reaction rate and selectivity [55–57]. Additionally, post reaction separations are often made easier due to immiscibility of products with the IL phase, such as in the case of esterification in acid ILs [58, 59]. ILs also give the ability to distill volatile products and reuse the IL [60]. The coordination of ILs to metal centers has also been shown to increase the activity and recyclability of some metal catalysts [48, 52]. While ILs are often designed around specific solvent properties, some ILs are designed to work as a combined solvent and catalyst. A common method for this is to attach an acid group to the end of an alkyl chain on the cation, which has been used to depolymerize cellulose and to catalyze esterification reactions [53, 61, 62].

Recently, some ILs have been shown to be able to either partially or completely dissolve cellulose, lignin, or lignocellulosic biomass. Imidazolium-based ILs seem to be especially well suited for this application. The most common solvents that are used to dissolve biomass are alkylimidazolium chlorides, acetates, and formates, although others have been investigated and used in biomass chemistry [21, 63, 64]. This property of ILs has been exploited in the production of novel materials, such as cellulosic aerogels and films in addition to being used as a solvent for catalysis of lignocellulose [65–67].

The property of these ILs that enables them to effectively dissolve biomass is their ability to function as a hydrogen bond acceptor while only having a limited ability to act as a hydrogen bond donor. In general, it is the ability of the anion to form hydrogen bonds with the hydroxyl groups of the cellulose, disrupting the hydrogen bond crosslinking of the polysaccharide, that makes these ILs effective at solubilizing biomass [22, 68] (Fig. 8.4). Dissolution of glucose in 1,3-dimethylimidazolium chloride was studied through computer modeling to analyze the IL/saccharide interactions further. This work demonstrated the almost exclusive coordination of the chloride anion to saccharides with only minimal contributions from hydrogen bonding and van der Waal forces from the imidazolium cation [69]. The dissolution process first swells the cellulose and, in the case of lignocellulose, extracts the lignin [70]. Some ILs have even been specifically designed to dissolve carbohydrates without denaturing enzymes to allow for homogenous enzymatic catalysis of biomass [71]. Work has been done to investigate ILs using solvatochromic dyes to probe the hydrogen bonding acidity and basicity, the polarity, and dispersion forces in various ILs [72]. These properties are part of what makes ILs an attractive solvent for the processing of biomass.

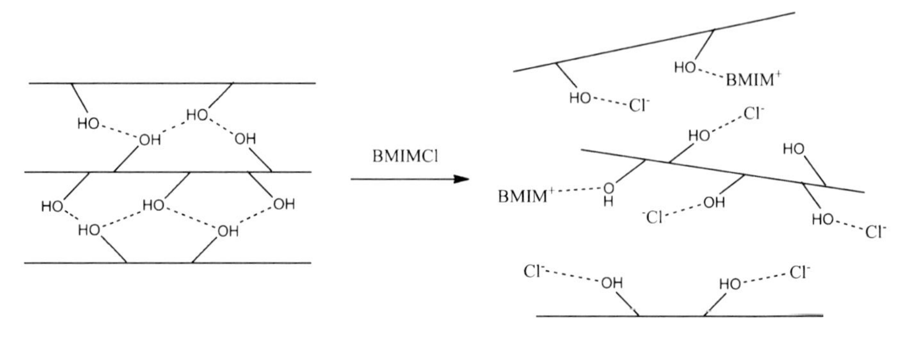

Fig. 8.4 Proposed dissolution mechanism of cellulose in 1-butyl-3-methylimidazolium chloride (BMIMCl) (Adapted with permission from [22]. Copyright 2009 American Chemical Society)

8.2 Catalysis of Carbohydrate

8.2.1 Pretreatment of Biomass

The production of ethanol through fermentation is already a common process for the utilization of biomass resources [39]. While simple sugars and starches can be easily used in this process, feed stocks of cellulose and hemicellulose would provide a source of sugars that would not compete with food and could be produced in otherwise unused land area. The conversion of these polysaccharides into simple sugars adds a significant challenge compared to the use of corn or sugarcane feed stocks [20]. In order for cellulosic biomass to be used in ethanol production, the feedstock must be pretreated and saccharified to provide a substrate suitable to the ethanol producing yeast. Pretreatment is a key step in a number of catalytic biomass processes [9], so while the pretreatment step itself may not be catalytic, it is important to an understanding of processing of biomass in ILs.

Because the structure of biomass, especially the presence of lignin, inhibits the saccharification of structural carbohydrates, pretreatment is needed to open up the structure of plant matter (Fig. 8.5). While a number of methods, such as steam explosion, dilute acid treatment, ammonia explosion, and milling have been explored, the unique solvent properties of ILs have garnered significant attention as a pretreatment option [9, 73]. What makes ILs promising for the catalytic treatment of biomass, namely their ability to make homogeneous solutions of lignocellulose, is also what make ILs a good medium for pretreatment. Because much of the recalcitrance of biomass to saccharification comes from the structure of the cell wall and the presence of lignin, when the structure is disrupted through dissolution in ILs, the carbohydrates are made available for enzymatic attack [70].

The general procedure for IL pretreatment of biomass is to dissolve or swell the biomass with an IL solvent. After treatment at a given temperature for a given time, an antisolvent, such as water, ethanol, or an acetone/water mixture, is added

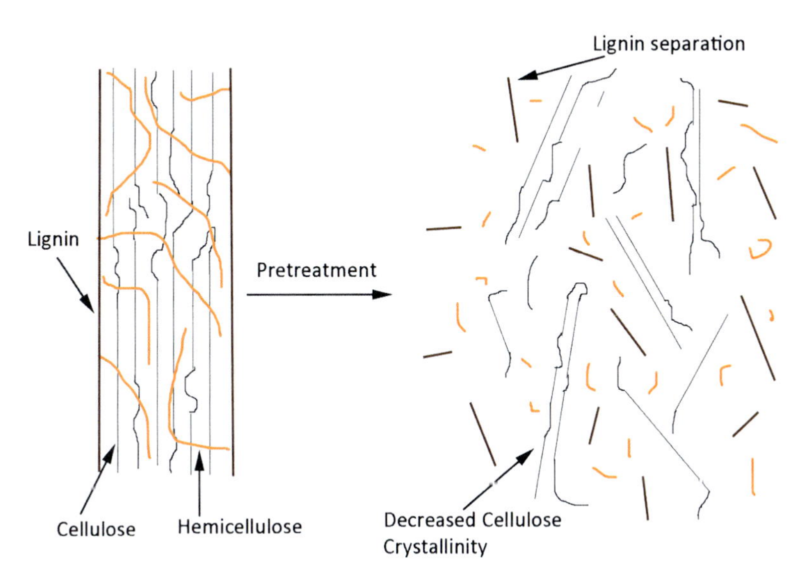

Fig. 8.5 Schematic of the role of pretreatment in the conversion of biomass to fuel (Adapted with permission from [73]. Copyright 2009 American Chemical Society)

to precipitate the biomass and wash away the IL (see Fig. 8.6). The biomass is then dried and saccharified through enzymatic or chemical methods. This process builds off of the work by Rogers in which lignocellulose was fractionated into a cellulose and a lignin rich phase through dissolution in 1-ethyl-3-methylimidazolium acetate (EMIMAc). In this work, an acetone/water solution was added to precipitate the cellulose while keeping the lignin in solution. Evaporation of the acetone precipitates the lignin after the cellulose had been filtered from the solution [74].

This method was applied to pretreatment using a variety of ILs by a number of different researchers. Lynam conducted a study to measure the effect of a few different ILs on the composition and structure of lignocellulose in which ground rice hulls pretreated with 1-ethyl-3-methylimidazolium acetate, 1-allyl-3-methylimidazolium chloride (AMIMCl), or 1-hexyl-3-methylimidazolium chloride (HMIMCl) and either ethanol or water as antisolvents for cellulose precipitation. In this study, the EMIMAc was able to completely remove the lignin and significant amounts of hemicellulose from the rice hulls, while the other ILs removed less lignin but more hemicellulose [75]. Lee et al. worked with EMIMAc and other ILs to treat wood flour for enzymatic saccharification. By removing the lignin and reducing the cellulose crystallinity, the IL treated wood flour was 95 % digestible to cellulase enzymes from *Trichoderma viride* [76]. It has been demonstrated that IL pretreatment works even at a biomass loading well above the solubility limit of the biomass in the IL [77]. Other ILs, such as 1-ethyl-3-methylimidazolium diethyl phosphate, alkyloxyalkyl substituted imidazolium acetate, or alkyloxyalkyl substituted ammonium acetate have also been investigated [78, 79]. The use of

Fig. 8.6 SEM micrographs of (**a**) untreated and (**b**) ionic liquid pretreated and recovered fibers from switchgrass (Reprinted with permission from [70]. Copyright 2009 John Wiley and Sons)

ILs as a pretreatment strategy has been compared to a more common method, namely pretreatment with dilute acid. In this study, it was found that IL pretreated samples produced a higher yield of monosaccharides and in a shorter time than samples pretreated with dilute acid [14]. Some studies have employed variations on dissolution and washing, such as the addition of an ammonia treatment step [80] or combining the pretreatment and saccharification into one step with aqueous ILs and enzymes [78, 81, 82].

One of the more difficult challenges facing pretreatment of biomass with ILs or saccharification in ILs is the separation of the ILs and carbohydrates after the treatment is completed. Ideally, the products will be insoluble in the IL or be precipitated with an antisolvent. These options may not be sufficient, such as when monosaccharides must be extracted from the IL. Brennan and coworkers developed a liquid-liquid extraction procedure for the removal of sugars from an IL phase using organic soluble boronic acids that have an affinity for sugars [83]. Another method relies on the use of kosmotropic, or water-structuring, salts to induce a biphasic system with water and an IL [84]. This effect has been used to separate and reuse ILs after pretreatment of biomass [85].

8.2.2 Hydrolysis of Carbohydrates with Acids

Cellulase enzymes are not the only catalysts that are effective for the saccharification of polysaccharides. In a number of reaction schemes for the utilization of biomass, lignocellulose must be hydrolyzed into monosaccharides. This is important not just in fermentation of biomass into ethanol, but also if the saccharides are to undergo processing directly into commodity chemicals or fuels. Without the use of ILs, the most common methods to hydrolyze polysaccharides into monosaccharides is through enzymatic hydrolysis or acid catalyzed hydrolysis [19]. This process can be slow and expensive, in part due to the necessity of heterogeneous reactions due to the insolubility of cellulose in conventional solvents. The ability of ILs to solubilize biomass has led to a considerable research effort in the hydrolysis of cellulose. These studies have focused on a wide range of catalysts, from conventional acid catalysts, both solid and homogeneous, to novel metal catalysts and ILs that are designed to be both solvent and catalyst.

Acid hydrolysis of lignocellulose is well understood and has been used to quantitatively saccharify biomass for decades [24, 86]. The specific reaction occurring in the hydrolysis of cellulose begins with the formation of a conjugate acid leading to the cleavage of the glycosidic linkage as a water molecule is added and a H^+ ion is released [87]. Because many ILs are stable under acidic conditions, coupling the saccharification ability of acids with the dissolution ability of ILs is a natural choice. The acidity in any IL can only be as high as the conjugate acid of the anion of the IL. If HCl is added to an acetate based ILs, for example, acetic acid will be formed and the acetate anion from the IL will be effectively replaced by a chloride anion. The pH scale is not an appropriate measure of the acidity of an IL, as pH is defined in dilute aqueous solutions. For this reason, the Hammett acidity, as measured by nitroanilines with known pKa values, is used to determine the acidity in ILs [88, 89].

A number of studies have been conducted in which biomass, acid, and an IL have been mixed to hydrolyze the polysaccharides into mono- or oligosaccharides. Proof of this concept was demonstrated by dissolving various cellulose sources, including spruce wood, in BMIMCl and adding HCl, sulfuric acid, nitric acid, or phosphoric acid with heat and stirring. With this method, glucose yields as high as 43 % and total reducing sugars as high as 77 % could be obtained after 9 h [90]. Sievers et al. were able to depolymerize pure cellulose and the cellulose and hemicellulose in loblolly pine in 1-butyl-3-methylimidazolium chloride (BMIMCl) using 0.2 wt% trifluoroacetic acid as a catalyst. In the case of pure cellulose, 97 % could be transformed into soluble mono- or oligosaccharides after 2 h at 120 °C, while 62 % of the pine wood could be converted into soluble products (representing 97 % of the carbohydrate content of the wood) [91]. Other experiments have been done with AMIMCl and added HCl to hydrolyze eucalyptus, pine, and spruce thermomechanical pulps. Higher HCl loadings and longer reaction times resulted in higher degrees of hydrolysis and yielded products consistent with lignin depolymerization [92].

A study by Li et al. demonstrated the hydrolysis of corn stalk, rice straw, pine wood, and sugarcane bagasse using combinations of the ILs 1-butyl-3-methylimidazolium bromide (BMIMBr), AMIMCl, 1-hexyl-3-methylimidazolium chloride (C_6MIMCl), 1-butyl-3-methylimidazolium hydrogensulfate ($BMIMHSO_4$), and 1-(4-sulfobutyl)-3-methylimidazolium hydrogensulfate ($SBMIMHSO_4$) with HCl, sulfuric acid, nitric acid, phosphoric acid, and maleic acid. BMIMCl coupled with HCl was found to be the most effective system. Interestingly, sulfuric acid was less effective than HCl, possibly due to interaction between the sulfuric acid and lignin [93]. While this study successfully hydrolyzed cellulose in ILs as other researchers had demonstrated, it also worked with the naturally acidic ILs $BMIMHSO_4$ and $SBMIMHSO_4$. Other groups have worked with acidic ILs as catalysts in a number of situations [53, 58, 61]. Amarasekara et al. also utilized this novel acidic IL to depolymerize cellulose and found that 1-(1-propylsulfonic)-3-methylimidazolium chloride and 1-(1-butylsulfonic)-3-methylimidazolium chloride could dissolve cellulose up to a loading of 20 g per 100 g of IL at room temperature. The cellulose could then be hydrolyzed with the addition of water and mild heating (70 °C) to produce up to 62 % yield of reducing sugars and 14 % yield of glucose [62]. In one interesting study by Zhang et al., a simple EMIMCl/water system without added acid was used to hydrolyze cellulose at temperatures between 90 and 140 °C with a total reducing sugar yield of up to 97 %. This study of EMIMCl/water systems also demonstrated the ability of "neutral" ILs such dialkyl- and trialkylimidazolium chlorides to lower the pH of an aqueous solution, enhancing cellulose hydrolysis [94].

Solid acid catalysts have also been successfully employed to hydrolyze the polysaccharides of biomass. A pair of studies by Rinaldi and coworkers demonstrates the ability of H-type ion exchange resins and zeolites to catalyze the depolymerization of cellulose. This process showed continuously decreasing degree of polymerization of cellulose along with a continuously increasing yield of reducing sugars, reaching 13 % reducing sugar yield after 5 h using Amberlyst 15 to depolymerize cellulose [95]. Further study examined the mechanisms of Amberlist 15dry as an acid catalyst and determined the effects of catalyst concentration, substrate concentration, temperature, and impurities have on the reaction. Additionally, they found that the catalyst releases H^+ ions into solution, which subsequently catalyze cellulose depolymerization instead of acting as a true heterogeneous catalyst [96]. Zhang and Zhao also studied H-form zeolites and H-type ion exchange resins in ILs as a method for depolymerizing cellulose but with the addition of microwave irradiation to effect the reaction. The combination of high surface area H-form zeolites and microwave irradiation produced a much quicker reaction, yielding 37 % glucose after only 8 min [97], as compared to the 13 % reducing sugar yield obtained by Rinaldi et al. after 5 h. In other work, silica was modified with tethered sulfonic acid functionalized ionic liquids and subsequently used to hydrolyze cellulose in BMIMCl [98]. When using solid catalysts, especially ion exchange resins, with ILs, it is important to note that ion exchange between the catalyst and IL will most likely occur. This makes the catalysis homogeneous and alters the composition of the IL.

8.2.3 Hydrolysis of Carbohydrates with Metal Catalysts

Recently, Su et al. demonstrated the ability to use paired metal chlorides for the depolymerization of cellulose into monosaccharides and other products. In this work, $CuCl_2$, $CrCl_2$, $CrCl_3$, $PdCl_2$, and $FeCl_3$ were initially tested as catalysts, but none were effective on their own. It was then discovered that by pairing $CuCl_2$ and $PdCl_2$, high yields of monosaccharides could be obtained. The total yield of products (including glucose, cellobiose, 5-hydroxymethylfurfural, and other) from this reaction system was found to be as high as 70 % with glucose yields up to roughly 45 %. This yield is much higher and occurs much faster than when using sulfuric acid under the same conditions [99].

8.2.4 5-Hydroxymethylfurfural and Other Products

5-Hydroxymethylfurfural (HMF) can be produced from poly- and monohexoses and is a valuable platform chemical that can be used to make polymers, fuels, and commodity chemicals (Fig. 8.7). HMF is the product of the dehydration of 6-carbon sugars such as fructose, glucose, and mannose. Because ketoses are furanoses when in their cyclic form, they are much easier to convert into HMF than aldoses. Polymers of six carbon sugars can also be used to produce HMF. The following procedure is generally used in the conventional production of HMF: (1) hydrolyze polyhexoses into monomers, (2) isomerize aldoses into ketoses, and (3) use an acid catalyst to dehydrate ketoses into HMF [100]. Once HMF is produced, it can be processed into fuels, resins, solvents, alkanes, fuel additives, or polymers as a replacement for petroleum resources. The utility of HMF as a renewable biomass based platform chemical has led to significant research in the production of HMF in ILs, although the technology has still not matured into an industrial process [101, 102].

Acid catalyzed dehydration of a ketose, such as fructose, is the easiest method for production of HMF. This method has been implemented successfully using ILs as solvents. Lansalot-Matras and Moreau demonstrated up to 80 % yield of HMF from fructose in the ionic liquids 1-butyl-3-methylimidazolium tetrafluoroborate ($BMIMBF_4$) and 1-butyl-3-methylimidazolium hexafluorophosphate ($BMIMPF_6$) with added DMSO to solubilize fructose. The advantage of the IL was demonstrated in a reaction without an added catalyst. An HMF yield of 36 % was obtained after 32 h in a DMSO/IL solution while only trace amounts of HMF were detected in pure DMSO after 44 h [103]. Later, the same research group demonstrated the use of the acidic IL 1-H-3-methylimidazolium chloride as a combination solvent and catalyst by producing 92 % yield of HMF from fructose and nearly quantitative amounts of HMF and glucose from sucrose. It was also noted in this article that there was no observed degradation of the HMF after the conclusion of the reaction [104]. Qi and coworkers tested sulfuric acid, HCl, phosphoric acid, acetic acid,

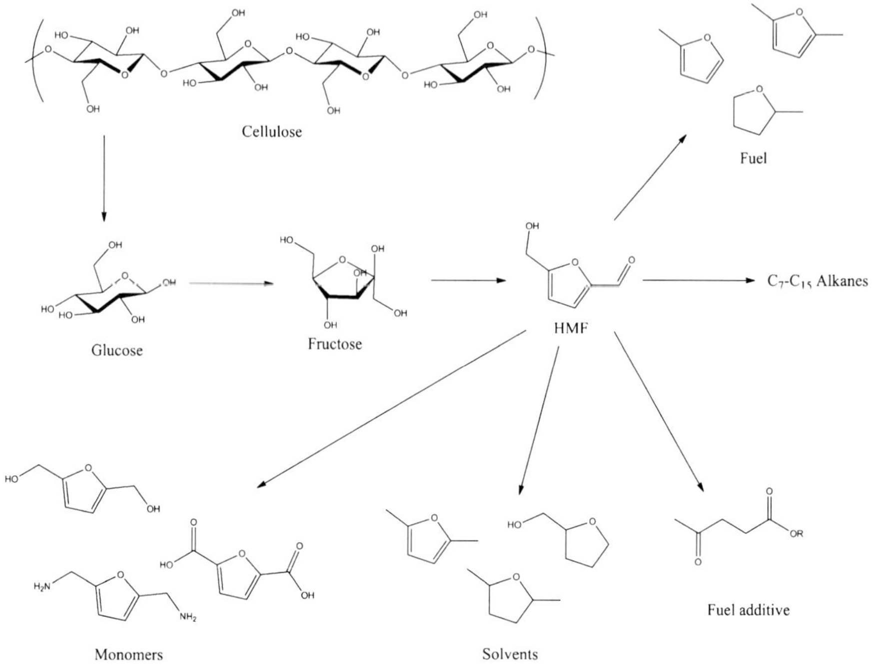

Fig. 8.7 The synthesis of HMF from carbohydrates and its further derivatization to important chemicals (Adapted with permission from [100]. Copyright 2011 American Chemical Society. Adapted with permission from [101]. Copyright 2011 John Wiley and Sons)

$CuCl_2$, $PdCl_2$, Dowex resin, and Amberlyst 15 as catalysts for production of HMF from fructose in BMIMCl. The Amberlyst catalyst was the best of the catalysts, producing an 83.3 % yield of HMF after only 10 min [105]. Others have taken the idea of acid catalyzed dehydration of fructose in ILs and worked to make it more environmentally friendly by using ILs made from renewable materials. Hu et al. tested a number of ILs and found that choline chloride coupled with citric acid was the most effective system for the fructose to HMF conversion, achieving over 90 % yield [106]. Acid catalyzed dehydration was tested on a number of different 6-carbon sugars by Sievers et al. through the use of added sulfuric acid in BMIMCl. Since glucose and mannose can both be isomerized into fructose, both sugars should be viable feed stocks for HMF production. As has been demonstrated previously, fructose gave high yields of HMF (>90 %). Glucose only produced up to 12 % HMF yield while only very small amounts of HMF were detected when mannose was used as a substrate. Additionally, xylose, a 5-carbon sugar, was shown to undergo an analogous reaction to form furfural with up to a 13 % yield [107].

Recently, it was discovered that some metal chlorides can catalyze the conversion of aldoses such as glucose and mannose into HMF in ILs. Zhao et al. demonstrated the conversion of glucose to HMF using $CrCl_2$ in 1-ethyl-3-methylimidazolium chloride with a yield of almost 70 %. Both $CrCl_2$ and $CrCl_3$ were found to be effective in this system, although $CrCl_2$ demonstrated the highest catalytic activity [108]. This discovery has led to many studies investigating metal chloride promoted production of HMF in ILs. Pidko et al. performed work using a combination of experimental techniques and computational modeling to show that $CrCl_2$ and $CrCl_3$ effect the dehydration of glucose into HMF through ring opening and hydrogen shift catalyzed by $CrCl_4$ ions in solution [109, 110]. Binder et al. worked to elucidate the mechanism further through the use of glucose, mannose, galactose, lactose, tagatose, psicose and sorbose along with isotopic labeling. In this study, it was demonstrated that the chromium catalyst causes a 1,2-hydride shift which leads to a furanose that can be dehydrated [111]. In these studies, the efficacy of the $CrCl_2$/IL system could be enhanced through the use of microwave irradiation and by supporting the $CrCl_3$. The addition of microwave irradiation instead of simple heating in an oil bath allowed HMF to be recovered with up to a 40.2 % yield after only 2.5 min while conventionally heated reactions obtained 32.5 % yield after 32 min [112]. In addition to $CrCl_2$ and $CrCl_3$, it has been shown that $SnCl_4$ can catalyze the dehydration of sugars, inulin, and starch into HMF in $EMIMBF_4$. In this system, and HMF yield of between 40 and 65 % could be obtained, depending on the substrate [113].

While production of HMF from monosaccharides is a useful technology, by producing HMF directly from lignocellulose, the saccharification step of biomass processing would be removed. Su et al. demonstrated the use of coupled metal chlorides to produce HMF from cellulose in a single step. By using $CuCl_2$ and $CrCl_2$ in EMIMCl, cellulose could be converted to HMF with a yield of 55.4 % that stayed constant over several recycles of the IL/catalyst system. Additionally, the metal chlorides were shown to work in a synergistic manner, with almost no HMF production with either metal chloride on its own [114]. Binder and Raines demonstrated a similar system in which $CrCl_2$ or $CrCl_3$ in a solution of N, N-dimethylformamide (DMA) with LiCl and the IL EMIMCl. With this system, fructose, glucose, cellulose and lignocellulose from corn stover and pine wood could all be converted into HMF. The yields were dependent on conditions and substrate, although even the lignocellulose produced up to a 48 % yield [115]. In a study by Zhang et al. that was previously mentioned in the section on cellulose hydrolysis, the EMIMCl/water mixtures that could be used to hydrolyze cellulose were also demonstrated to be effective at HMF production when a $CrCl_2$ catalyst was added [94]. By incorporating the use of microwave irradiation with the $CrCl_3$/IL system, glucose and cellulose could be converted to HMF with 90 and 60 % yields, respectively [116]. Expanding the use of microwave treatments, lignocellulose from corn stover, rice straw and pine wood could be converted to HMF and furfural with yields of 45–52 and 23–31 %, respectively, with $CrCl_3$ in BMIMCl in 3 min or less [117].

8.2.5 Pyrolysis and Deoxygenate Production in ILs

There has been some research into deoxygenation and pyrolysis of carbohydrates in ILs. Generally, the thermal stability of ILs precludes their use under the conditions needed for deoxygenation catalysts and pyrolysis. In some cases, however, ILs have been shown to support these reactions. Sheldrake et al. demonstrated the use of dicationic ILs as a medium for pyrolysis of cellulose to dehydration products of glucose. The ILs used were composed of two imidazolium rings connected with a 4-, 6-, or 9-carbon alkyl chain. Generally, pyrolysis occurs at temperatures above 300 °C, but the use of these dicationic ILs allowed the production of levoglucosenone, 1-(2-furanyl)-2-hydroxyethanone, and HMF along with trace amounts of levoglucosan at 180 °C. Only 5.5 % combined yield was obtained for these products. Monocationic imidazolium ILs and dicationic pyridinium ILs were not thermally stable at this temperature and no products, other than IL decomposition products were detected [60].

Chidambaram and coworkers demonstrated the production of HMF and subsequent deoxygenation into 2,5-dimethylfuran in EMIMCl. While a number of acids were tested as catalysts, it was found that heteropoly acids gave a better combination of conversion and selectivity than any of the simple acids tested (such as sulfuric, fluoroacidic, nitric, HCl, and phosphoric). The best of the heteropoly acids, $H_3PMo_{12}O_{40}$ (12-MPA), produced a 71 % conversion of glucose and a 89 % HMF selectivity. To investigate the next step of processing HMF in ILs, metal catalysts were used to deoxygenate the HMF into 2,4-dimethylfurfural in the IL. Palladium, platinum, ruthenium, and rhodium catalysts supported by carbon were added to IL/HMF solutions at 120 °C under 62 bar of H_2. The most effective of these catalysts was Pd/C which gave 19 % conversion and 51 % selectivity. When a small amount of acetonitrile was added to prevent the formation of humins, a conversion of 47 % with 36 % selectivity was obtained [118]. Other work has been done on the hydrodeoxygenation of lignin specifically and will be discussed later in this chapter.

8.2.6 Derivatization of Biomass in ILs

Biopolymers are an important area of research as a replacement for petroleum derived products. In the US, 331 million barrels, or 4.6 % of the total US petroleum consumption, were used to make polymers in 2006 (329 million as feedstock, 2 million for energy) [119]. By displacing the need for petroleum, production of biopolymers could lessen global oil demand. While cellulose, starches, and other naturally occurring biopolymers can be difficult to work with due to their chemical and physical properties, chemical modification of these naturally occurring polymers allows for a wide range of properties to be achieved [120]. See Table 8.1 for representative examples. Additionally, modification of biopolymers can be used to aid in analytical methods by making otherwise insoluble polymers such as cellulose soluble in a wide range of solvents [130]. Smaller molecules of interest can also be manufactured through the chemical modification of monosaccharides. In many cases, it is even possible to couple the use of ILs and enzymes to effect a biocatalytic change while maintaining the advantages of an IL system [131]. Because ILs have the ability to solubilize unmodified biomass, they are

Table 8.1 Biomass modification reactions in ionic liquids

Name	Reaction	References
Trans-esterification	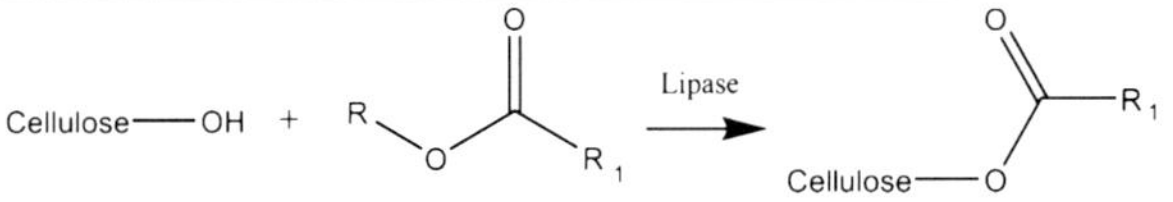 Cellulose—OH + R O O R1 —Lipase→ O R1 Cellulose—O	[121–124]
Acetylation	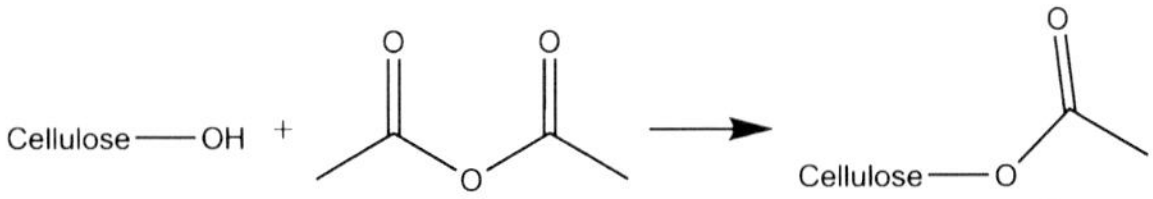 Cellulose—OH + O O O → O Cellulose—O	[125, 126]
Acylation	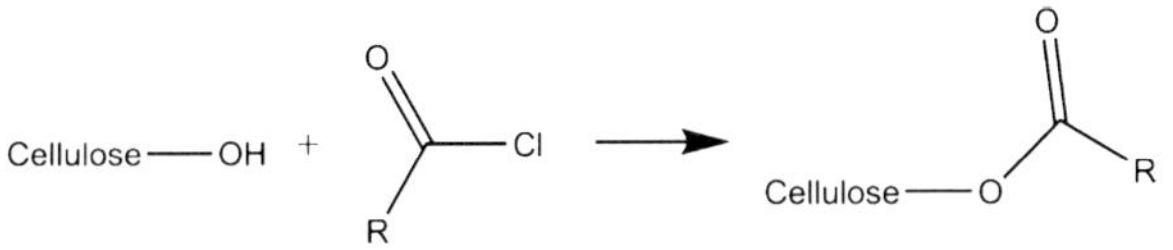 Cellulose—OH + O Cl R → O R Cellulose—O	[127, 128]
Carbanilation	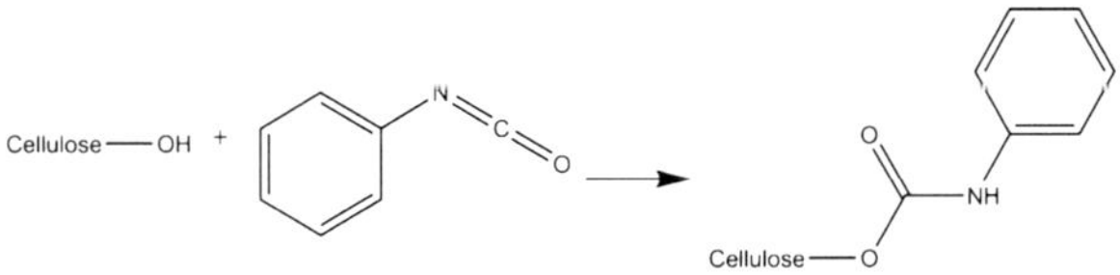 Cellulose—OH + N C O → O NH Cellulose—O	[125, 128]
Succination	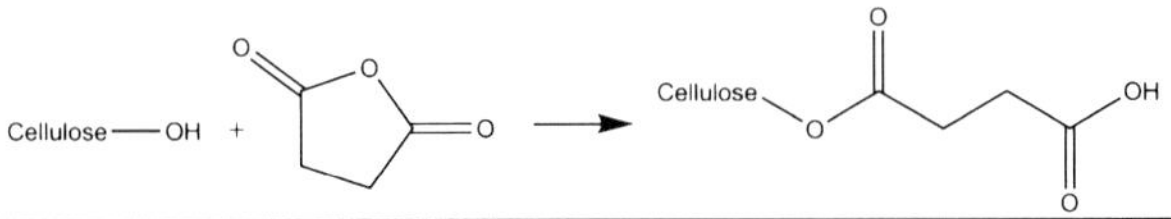 Cellulose—OH + O O O → O Cellulose O OH O	[129]

well suited to the task of chemical modification of lignocellulose for a wide variety of applications. ILs are particularly well suited to chemical modification of carbohydrates because, generally, the sugars or biopolymers are more hydrophilic while the reactants for derivatization are more hydrophobic. The ILs are often able to solubilize both reactants and, in the case of smaller molecules, the amphiphilic product. Research into modification of biomass in ILs is also important because there is significant work focusing on using enzymes in IL based systems, which could lead to other enzymatic processing of biomass in ILs.

The modification of monosaccharides with laurates is a common method for the production of surfactants. Glucose modification with vinyl laurates in ILs has been studied by Lee et al. using 1-butyl-3-methylimidazolium trifluoromethanesulfonate (BMIMTfO) and 1-butyl-3-methylimidazolium bis(trifluorosulfonyl)imide ($BMIMTf_2N$). Using lipase enzymes, Lee and coworkers demonstrated that a super saturated solution of glucose in a mixture of BMIMTfO and $BMIMTf_2N$, along with ultrasound treatment produces a better conversion and yield in less time than subsaturation solutions in a pure IL without the ultrasound [121–123]. This reaction, producing sugar esters with lipase enzymes, has been demonstrated in other ILs, such as $BMIMBF_4$ and $BMIMPF_6$ [124].

Modification of larger saccharide chains requires ILs that are better suited to biomass solvation. Unfortunately, the ILs that solvate cellulose and lignocellulose are also destructive to enzymes [132]. Consequently, most of the research in the modification of cellulose and lignocellulose uses non-enzymatic catalysts or no catalysts at all. Success has been seen in acetylation, carbanilation, sulfation, succinilation (Table 8.1), and benzoylation of cellulose chains in imidazolium based chlorides and bromides along with choline chloride ILs [125–129]. There has been work in combining the enzymes and ILs that will dissolve biomass. Zhao and coworkers developed ILs with $Me(EtO)_n-$ substituted imidazolium and alkylammonium (where n = 2–7) cations coupled with an acetate anion that are both effective for solubilizing cellulose and as a solvent for lipase catalyzed esterification. These ILs can dissolve cellulose up to 10 wt%, and support enzymatic esterification with methylmethacrylate with yield up to 66 % [71].

8.3 Catalysis of Lignin Conversion

While catalysis of lignin conversion in ILs has not received the same attention as cellulose and monosaccharides, there is growing interest in the application of ILs to the catalysis of lignin conversion. The main goal of much of the research relating to lignin in ILs have been for the purpose of pretreatment [76, 80, 85, 133]. Sun et al. demonstrated the ability to fractionate wood into cellulose rich and lignin rich samples using EMIMAc and acetone/water as an antisolvent for cellulose followed by evaporation of acetone to precipitate the lignin [74]. Other groups then used this discovery as a stepping stone to pretreatment of biomass. While some use ILs as a path towards delignification for pretreatment [76], others have shown that a loss of cellulose crystallinity is also a source of IL pretreatment efficacy [14, 77, 79]. The other work in catalysis of lignin conversion has covered thermal and chemical depolymerization and hydrodeoxygenation. As lignin is a complex, amorphous polymer, most studies on the catalysis of lignin conversion work with lignin model compounds as a way to test a process while keeping analytical complications to a minimum.

One of the simplest treatments of lignin in ILs is to simply dissolve lignin and heat the IL/lignin mixture for a period of time. Kubo et al. performed a series of experiments with the lignin model compound guaiacylglycerol-β-guaiacyl ether (GG) mixed with BMIMCl, EMIMAc, or AMIMCl at 120 °C. This model compound simulates the β–O–4 ether linkage, which is the most common structure in lignin. The main product of this reaction was 3-(4-hydroxy-3methoxyphenyl)-2-(2-methoxyphenoxy)2-propenol, which is an enol ether (EE). EE is the dehydration product of GG, and an analogous process has been implicated as an intermediate in the depolymerization of lignin under both acidic and alkaline conditions [134]. This intermediate has been detected in other studies involving lignin model compounds.

The cleavage of the β–O–4 ether linkage is a possible pathway both to general delignification of biomass and to the utilization of lignin as a feedstock for fuel and chemical production [135, 136]. A number of methods have been explored to sever this bond. In one study, N-bases were used in 1-butyl-2,3-dimethylimidazolium chloride (BDMIMCl). It was demonstrated that the base 1,3,5-triazabicyclo[4.4.0]dec-5-ene (TBD) was effective at cleaving the β–O–4 ether linkage in GG with a yield of up to 23 %. As in the study by Kubo et al., EE was observed as an intermediate. Other N-bases, including 7-methyl-1,3,5-triazabicyclo[4.4.0]dec-5-ene (MTBD), did not show the same activity as TBD, indicating a unique functionality for TBD. It was suggested that TBD could act as a combination base and nucleophile to break down lignin using the same mechanism as kraft pulping, although the process was not shown to be catalytic [137]. The trialkylimidazolium IL was utilized in this study as opposed to the more common dialkylimidazolium ILs because the hydrogen at the 2 position on the dialkylimidazolium ring can be extracted under basic conditions to form a reactive carbene [49]. Another method, using metal chlorides as catalysts, was demonstrated to be effective at the hydrolysis of both phenolic and non-phenolic lignin model compounds. In this study, GG and veratrylglycerol-β-guaiacyl ether (VG) were used to model the β–O–4 ether linkage in lignin. $FeCl_3$, $CuCl_2$, and $AlCl_3$ were shown to be effective at catalyzing the hydrolysis of the ether linkage with $AlCl_3$ showing the highest yield of 80 % for GG and 75 % for VG. The metal chlorides most likely acted as acid catalysts to break the bond through the same mechanism of other acid promoted systems [138].

The task-specific acidic IL HMIMCl has been shown to catalytically hydrolyze the ether bonds in both GG and VG up to a 71.5 % yield. The mechanism for this hydrolysis starts with dehydration into an enol ether structure, which is then susceptible to acidic attack of the β–O–4 ether linkages. This process, as shown in Fig. 8.8, occurs both in the individual model compounds and dimers of the model compound that form under reaction conditions [139]. This method was extended to a number of other acidic ILs. ILs composed of 1-H-methylimidazolium cations and chloride, bromide, hydrogensulfate, and tetrafluoroborate anions, along with $BMIMHSO_4$ were used to hydrolyze the β–O–4 ether linkage in GG and VG. HMIMCl was found to be the most effective of these ILs. The Hammett acidity of each of these ILs was measured using UV–vis measurements to determine protonation of 3-nitroanaline added to the ILs, but the acidity of the IL did not correlate with the yield of hydrolysis products. The efficacy of acid catalyzed hydrolysis in these ILs was determined by the ability of the anion to hydrogen bond with the hydroxyl groups on the lignin model compounds [88]. Further study demonstrated the ability of HMIMCl to depolymerize lignin through acid catalyzed hydrolysis of the β–O–4 ether linkage. The lignin used was extracted from oak wood using EMIMAc. Treated lignin was demonstrated to be reduced in size from the untreated lignin and the disappearance of the ether structures was observed through NMR and IR spectroscopy [136].

Other transformations of lignin have been demonstrated in ILs as well. Binder et al. performed work with many catalysts in EMIMCl and 1-ethyl-3-methylimidazolium triflate (EMIMOTf). While a number of catalysts were able to dealkylate the lignin model compound eugenol, these catalysts failed to produce monomeric products from organosolv lignin [140]. Further use of metals for catalysis of lignin in ILs was demonstrated by Jiang and Ragauskas. This study

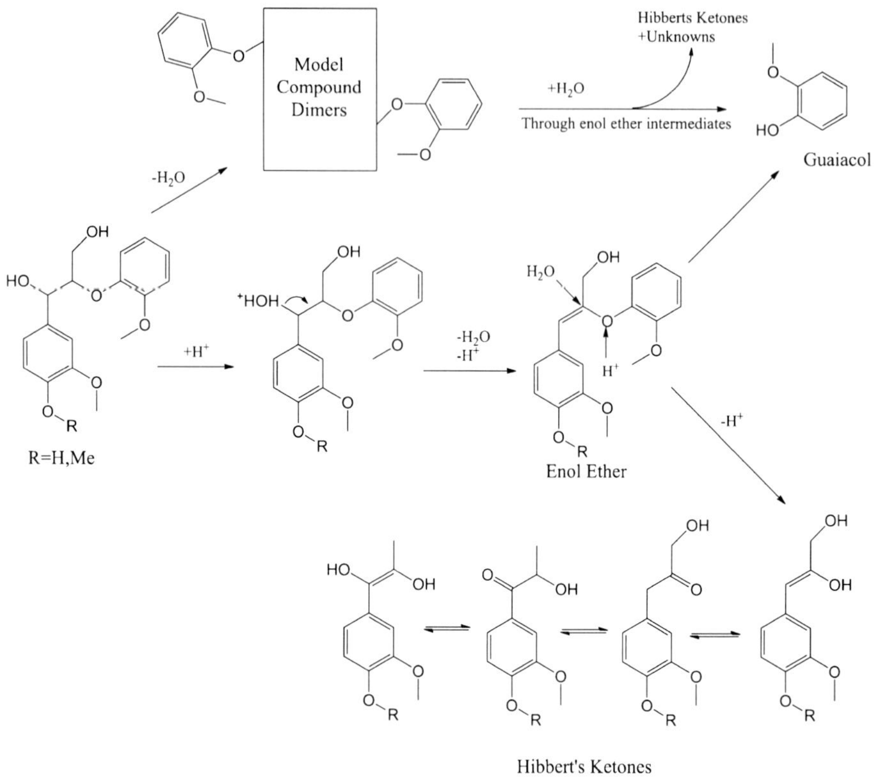

Fig. 8.8 Proposed acid-catalyzed mechanism for hydrolysis of β-O-4 bonds lignin model compounds (Adapted with permission from [138]. Copyright 2010 John Wiley and Sons)

dealt with the use of vanadyl acetylacetonate in $BMIMPF_6$ along with Cu(II) or Cu(I) co-catalysts to selectively oxidize aromatic alcohols into carbonyl or carboxylic acid groups. While most of this work focused on a wide variety of alcohols, 3,4-dimethoxybenzyl alcohol and 1-(3,4-dimethoxyphenyl)ethanol were specifically noted as being model compounds for lignin [141]. Other work has been performed with metal catalysts in ILs for the purpose of deoxygenation of lignin model compounds. In a study by Yan and coworkers, cyclohexanol was dehydrated into cyclohexene with Brønsted acidic ILs. Then, by combining the acidic ILs with Ru, Rh, or Pt nanoparticles, phenolic lignin model compounds were hydrogenated and deoxygenated to non-aromatic hexane species [142]. Other work on hydrodeoxygenation of lignin in ILs have been limited because the temperature at which traditional hydrodeoxygenation catalysts function exceeds the stability limit of IL, especially of those that have the ability to solubilize biomass.

8.4 Current and Future Research

While research in the area of catalysis of biomass in ionic liquids has been the subject of significant work in recent years, there is still much to do. The potential of these biomass/IL systems has been proven at the laboratory scale, but the use of ILs in biomass processing has not reached its potential as an industrially viable technology yet. One of the main inhibitors of industrial adoption is the cost of ILs. While increased adoption of IL technology will increase production and thus drive costs down, improved IL synthesis and processing methods would be beneficial to any process that relies on ILs.

Separation and recovery of ILs is one of the most important aspects of any biomass processing scheme due to the high cost of ILs and the potential for detrimental effects from ILs remaining in later processing steps. Separation of soluble products such as sugars from ILs and separation of ILs from aqueous solutions have received some attention, although there is still room for significant advances in this area [83–85]. Simply evaporating water from ILs is too energy intensive to be useful in most processes. Until satisfactory separation techniques are developed for recovery of both ILs and biomass products, the use of ILs will not gain wider use in the industrial world.

There are some areas that have been the subject of initial investigation that could use further research. As has been discussed in this chapter, metals and metal salts have been shown to be effective in a number of different catalytic systems. Since ILs provide a unique environment for metal complexes [143], there may still be better catalysts that take advantage of ILs for the catalysis of biomass. Supported metal catalysts could also use further study in the catalysis of biomass as long as the catalysts remain truly heterogeneous throughout the reaction processes. Additionally, the use of microwave irradiation to enhance rates and yields of reactions in ILs has received some attention from a limited number of researchers and could use further research.

Enzymatic catalysis of biomass in ionic liquids also has room to expand. Some work has already been done in making lipase-compatible ILs that also dissolve biomass [71], but more work could be done to expand on this idea. Cellulase-compatible ILs would allow biomass to be pretreated and saccharified in a one-pot process without the use of harsh acids or metal catalysts. Cellulase that is tolerant of ILs is already the subject of some research [144]. Because many ILs tend to inhibit enzymatic and microbial activity, there is significant room to develop ways to make ILs and bio-processes compatible.

8.5 Conclusions

ILs present a unique set of challenges as well as a unique set of advantages as solvents for processing biomass. ILs provide lignocellulose solvation, enhanced catalyst activity, recyclability and, in some cases simple separations making them a promising avenue of research and potential candidate as a technology in the next generation of biorefineries. In addition to the basic research needed to find the optimize the IL/catalyst/substrate combination, overcoming the unique challenges of ILs must be thoroughly investigated [145]. When applying the work done with compounds such as monosaccharides and lignin model compounds to lignocellulose, care must be taken to select ILs that will work for the process and accommodate the realities of a more difficult lignocellulosic substrate. Processes designed with ILs will need to take these challenges into account, along with possible health effects and corrosion caused by highly ionic media [146].

Even with the remaining challenges of developing new industrial processes, the promise of homogenous conversion of biomass into fuels, commodity chemicals, and polymers is a strong motivator. While ILs have worked their way into some pilot scale and industrial processes [147], the technology for IL use in biomass processing on an industrial scale may still be somewhat immature. More research aimed at optimizing recent discoveries, developing separations and recycling processes, and discovering new uses for IL/biomass systems has the potential to make IL systems practicable for industrial biomass processing.

References

1. Perlack RD, Wright LL, Turhollow AF, et al. Biomass as feedstock for a bioenergy and bioproducts industry: the technical feasibility of a billion-ton annual supply. Agriculture. 2005;1–78.
2. Parikka M. Global biomass fuel resources. Biomass Bioenergy. 2004;27:613–20.
3. United Nations. Maize production quantity. Rome: Food and Agriculture Organization of the United Nations; 2010. p. 1.
4. Graham RL, Nelson R, Sheehan J, et al. Current and potential U.S. corn stover supplies. Agron J. 2007;99:1.
5. Renewable Fuels Association. Accelerating industry innovations – 2012 ethanol industry outlook. Washington, DC: Renewable Fuels Association; 2012. p. 1–38.
6. Bang G. Energy security and climate change concerns: triggers for energy policy change in the United States? Energy Policy. 2010;38:1645–53.
7. Goldemberg J, Coelho ST, Nastari PM, Lucon O. Ethanol learning curve—the Brazilian experience. Biomass Bioenergy. 2004;26:301–4.
8. Solomon BD, Barnes JR, Halvorsen KE. Grain and cellulosic ethanol: history, economics, and energy policy. Biomass Bioenergy. 2007;31:416–25.
9. Yang B, Wyman CE. Pretreatment: the key to unlocking low-cost cellulosic ethanol. Biofuels Bioprod Biorefin. 2008;2(1):26–40.
10. Kaar WE, Gutierrez CV, Kinoshita CM. Steam explosion of sugarcane bagasse as a pretreatment for conversion to ethanol. Biomass Bioenergy. 1998;14:277–87.
11. Kim TH, Kim JS, Sunwoo C, Lee Y. Pretreatment of corn stover by aqueous ammonia. Bioresour Technol. 2003;90:39–47.
12. Saha BC, Iten LB, Cotta MA, Wu YV. Dilute acid pretreatment, enzymatic saccharification and fermentation of wheat straw to ethanol. Process Biochem. 2005;40:3693–700.
13. Hideno A, Inoue H, Tsukahara K, et al. Wet disk milling pretreatment without sulfuric acid for enzymatic hydrolysis of rice straw. Bioresour Technol. 2009;100:2706–11.

14. Li C, Knierim B, Manisseri C, et al. Comparison of dilute acid and ionic liquid pretreatment of switchgrass: biomass recalcitrance, delignification and enzymatic saccharification. Bioresour Technol. 2010;101:4900–6.
15. Singh J, Gu S. Commercialization potential of microalgae for biofuels production. Renew Sust Energ Rev. 2010;14:2596–610.
16. Sutton D, Kelleher B, Ross JRH. Review of literature on catalysts for biomass gasification. Fuel Process Technol. 2001;73:155–73.
17. Mohan D, Pittman CU, Steele PH. Pyrolysis of wood/biomass for bio-oil: a critical review. Energy Fuel. 2006;20:848–89.
18. Elliott DC. Historical developments in hydroprocessing bio-oils. Energy Fuel. 2007;21:1792–815.
19. Demirbas MF. Biorefineries for biofuel upgrading: a critical review. Appl Energy. 2009;86: S151–61.
20. Lin Y, Tanaka S. Ethanol fermentation from biomass resources: current state and prospects. Appl Microbiol. 2006;69:627–42.
21. Zhu S, Wu Y, Chen Q, et al. Dissolution of cellulose with ionic liquids and its application: a mini-review. Green Chem. 2006;8:325.
22. Pinkert A, Marsh KN, Pang S, Staiger MP. Ionic liquids and their interaction with cellulose. Chem Rev. 2009;109:6712–28.
23. Foyle T, Jennings L, Mulcahy P. Compositional analysis of lignocellulosic materials: evaluation of methods used for sugar analysis of waste paper and straw. Bioresour Technol. 2007;98:3026–36.
24. Sluiter A, Hames B, Ruiz R, et al. Determination of structural carbohydrates and lignin in biomass. Golden: National Renewable Energy Laboratory. Technical report NREL/TP-510-42618. 2011;1–18.
25. Brethauer S, Wyman CE. Review: continuous hydrolysis and fermentation for cellulosic ethanol production. Bioresour Technol. 2010;101:4862–74.
26. Ebringerová A. Structural diversity and application potential of hemicelluloses. Macromol Symp. 2006;232:1–12.
27. Whistler RL, Gaillard BDE. Comparison of xylans from several annual plants. Arch Biochem Biophys. 1961;93:332–4.
28. Holladay JE, White JF, Bozell JJ, Johnson D. Top value-added chemicals from biomass volume II—results of screening for potential candidates from biorefinery lignin. Richland: Pacific Northwest National Laboratory. Report PNNL-16983. 2007; II:1–79.
29. Weng J-K, Li X, Bonawitz ND, Chapple C. Emerging strategies of lignin engineering and degradation for cellulosic biofuel production. Curr Opin Biotechnol. 2008;19:166–72.
30. Chakar FS, Ragauskas AJ. Review of current and future softwood kraft lignin process chemistry. Ind Crop Prod. 2004;20:131–41.
31. Tien M, Kirk TK. Lignin-degrading enzyme from the Hymenomycete Phanerochaete chrysosporium Burds. Science. 1984;81:2280–4.
32. Martinez D, Challacombe J, Morgenstern I, et al. Genome, transcriptome, and secretome analysis of wood decay fungus Postia placenta supports unique mechanisms of lignocellulose conversion. Proc Natl Acad Sci U S A. 2009;106:1954–9.
33. Gierer J. Chemical aspects of kraft pulping. Wood Sci Technol. 1980;266:241–66.
34. Pye EK, Lora JH. The Alcell process: a proven alternative to kraft pulping. Tappi J. 1991;74 (3):113–7.
35. McDonough TJ. The chemistry of organosolv delignification. Atlanta: Institute for Paper Science and Technology. IPST technical paper series number 455; 1992.
36. Xu F, Sun J, Sun R, et al. Comparative study of organosolv lignins from wheat straw. Ind Crop Prod. 2006;23:180–93.
37. Adler E. Lignin chemistry-past, present and future. Wood Sci. 1977;8:169–218.

38. Yokoyama T, Matsumoto Y. Revisiting the mechanism of β-O-4 bond cleavage during acidolysis of lignin. Part 1: kinetics of the formation of enol ether from non-phenolic C 6 -C 2 type model compounds. Holzforschung. 2008;62:164–8.
39. Pimentel D, Patzek TW. Ethanol production using corn, switchgrass, and wood; biodiesel production using soybean and sunflower. Nat Resour Res. 2005;14:65–76.
40. Saha BC, Cotta MA. Enzymatic saccharification and fermentation of alkaline peroxide pretreated rice hulls to ethanol. Enzyme Microb Technol. 2007;41:528–32.
41. Welton T. Room-temperature ionic liquids. Solvents for synthesis and catalysis. Chem Rev. 1999;99:2071–84.
42. Olivier-Bourbigou H. Ionic liquids: perspectives for organic and catalytic reactions. J Mol Catal A Chem. 2002;182–183:419–37.
43. Earle MJ, Esperança JMSS, Gilea MA. The distillation and volatility of ionic liquids. Nature. 2006;439:831–4.
44. Wishart JF. Energy applications of ionic liquids. Energy Environ Sci. 2009;2:956.
45. Kulkarni PS, Branco LC, Crespo JG, et al. Comparison of physicochemical properties of new ionic liquids based on imidazolium, quaternary ammonium, and guanidinium cations. Chem Eur J. 2007;13:8478–88.
46. Kosmulski M. Thermal stability of low temperature ionic liquids revisited. Thermochim Acta. 2004;412:47–53.
47. Zhao D, Wu M, Kou Y, Min E. Ionic liquids: applications in catalysis. Catal Today. 2002;74:157–89.
48. Miao W, Chan TH. Ionic-liquid-supported synthesis: a novel liquid-phase strategy for organic synthesis. Acc Chem Res. 2006;39:897–908.
49. Chowdhury S, Mohan RS, Scott JL. Reactivity of ionic liquids. Tetrahedron. 2007;63:2363–89.
50. Arduengo AJ, Krafczyk R, Schmutzler R, et al. Imidazolylidenes, imidazolinylidenes and imidazolidines. Tetrahedron. 1999;55:14523–34.
51. Yoshizawa M, Xu W, Angell CA. Ionic liquids by proton transfer: vapor pressure, conductivity, and the relevance of DeltapKa from aqueous solutions. J Am Chem Soc. 2003;125:15411–9.
52. Gordon C. New developments in catalysis using ionic liquids. Appl Catal Gen. 2001;222:101–17.
53. Cole AC, Jensen JL, Ntai I, et al. Novel Brønsted acidic ionic liquids and their use as dual solvent-catalysts. J Am Chem Soc. 2002;124:5962–3.
54. Greaves TL, Drummond CJ. Protic ionic liquids: properties and applications. Chem Rev. 2008;108:206–37.
55. Sheldon R. Catalytic reactions in ionic liquids. Chem Commun 2001;23:2399–407.
56. Kim DW, Hong DJ, Seo JW, et al. Hydroxylation of alkyl halides with water in ionic liquid: significantly enhanced nucleophilicity of water. J Org Chem. 2004;69:3186–9.
57. Yang Z, Pan W. Ionic liquids: green solvents for nonaqueous biocatalysis. Enzyme Microb Technol. 2005;37:19–28.
58. Zhu H-P, Yang F, Tang J, He M-Y. Brønsted acidic ionic liquid 1-methylimidazolium tetrafluoroborate: a green catalyst and recyclable medium for esterification. Green Chem. 2003;5:38.
59. Gui J, Cong Z, Liu D, et al. Novel Brønsted acidic ionic liquid as efficient and reusable catalyst system for esterification. Catal Commun. 2004;5:473–7.
60. Sheldrake GN, Schleck D. Dicationic molten salts (ionic liquids) as re-usable media for the controlled pyrolysis of cellulose to anhydrosugars. Green Chem. 2007;9:1044.
61. Fei Z, Zhao D, Geldbach TJ, et al. Brønsted acidic ionic liquids and their zwitterions: synthesis, characterization and pKa determination. Chem Eur J. 2004;10:4886–93.
62. Amarasekara AS, Owereh OS. Hydrolysis and decomposition of cellulose in Brönsted acidic ionic liquids Under mild conditions. Ind Eng Chem Res. 2009;48:10152–5.

63. Zavrel M, Bross D, Funke M, et al. High-throughput screening for ionic liquids dissolving (ligno-)cellulose. Bioresour Technol. 2009;100:2580–7.
64. Mäki-Arvela P, Anugwom I, Virtanen P, et al. Dissolution of lignocellulosic materials and its constituents using ionic liquids—a review. Ind Crop Prod. 2010;32:175–201.
65. Aaltonen O, Jauhiainen O. The preparation of lignocellulosic aerogels from ionic liquid solutions. Carbohydr Polym. 2009;75:125–9.
66. Turner MB, Spear SK, Holbrey JD, Rogers RD. Production of bioactive cellulose films reconstituted from ionic liquids. Biomacromolecules. 2004;5:1379–84.
67. Su S, Tan Y, Macfarlane DR. Ionic liquids in biomass processing. Top Curr Chem. 2009;290:311–39.
68. Remsing RC, Swatloski RP, Rogers RD, Moyna G. Mechanism of cellulose dissolution in the ionic liquid 1-n-butyl-3-methylimidazolium chloride: a 13C and 35/37Cl NMR relaxation study on model systems. Chem Commun (Camb). 2006;(12):1271–3.
69. Youngs TG, Hardacre C, Holbrey JD. Glucose solvation by the ionic liquid 1,3-dimethylimidazolium chloride: a simulation study. J Phys Chem B. 2007;111:13765–74.
70. Singh S, Simmons BA, Vogel KP. Visualization of biomass solubilization and cellulose regeneration during ionic liquid pretreatment of switchgrass. Biotechnol Bioeng. 2009;104:68–75.
71. Zhao H, Baker GA, Song Z, et al. Designing enzyme-compatible ionic liquids that can dissolve carbohydrates. Green Chem. 2008;10:696.
72. Anderson JL, Ding J, Welton T, Armstrong DW. Characterizing ionic liquids on the basis of multiple solvation interactions. J Am Chem Soc. 2002;124:14247–54.
73. Kumar P, Barrett DM, Delwiche MJ, Stroeve P. Methods for pretreatment of lignocellulosic biomass for efficient hydrolysis and biofuel production. Ind Eng Chem Res. 2009;48(8):3713–29.
74. Sun N, Rahman M, Qin Y, et al. Complete dissolution and partial delignification of wood in the ionic liquid 1-ethyl-3-methylimidazolium acetate. Green Chem. 2009;11:646–55.
75. Lynam JG, Reza MT, Vasquez VR, Coronella CJ. Pretreatment of rice hulls by ionic liquid dissolution. Bioresour Technol. 2012;114:629–36.
76. Lee SH, Doherty TV, Linhardt RJ, Dordick JS. Ionic liquid-mediated selective extraction of lignin from wood leading to enhanced enzymatic cellulose hydrolysis. Biotechnol Bioeng. 2009;102:1368–76.
77. Wu H, Mora-Pale M, Miao J, et al. Facile pretreatment of lignocellulosic biomass at high loadings in room temperature ionic liquids. Biotechnol Bioeng. 2011;108:2865–75.
78. Kamiya N, Matsushita Y, Hanaki M, et al. Enzymatic in situ saccharification of cellulose in aqueous-ionic liquid media. Biotechnol Lett. 2008;30:1037–40.
79. Zhao H, Baker GA, Cowins JV. Fast enzymatic saccharification of switchgrass after pretreatment with ionic liquids. Biotechnol Prog. 2009;26:127–33.
80. Nguyen T-AD, Kim K-R, Han SJ, et al. Pretreatment of rice straw with ammonia and ionic liquid for lignocellulose conversion to fermentable sugars. Bioresour Technol. 2010;101:7432–8.
81. Fu D, Mazza G. Aqueous ionic liquid pretreatment of straw. Bioresour Technol. 2011;102:7008–11.
82. Fu D, Mazza G. Optimization of processing conditions for the pretreatment of wheat straw using aqueous ionic liquid. Bioresour Technol. 2011;102:8003–10.
83. Brennan TCR, Datta S, Blanch HW, et al. Recovery of sugars from ionic liquid biomass liquor by solvent extraction. BioEnergy Res. 2010;3:123–33.
84. Gutowski KE, Broker GA, Willauer HD, et al. Controlling the aqueous miscibility of ionic liquids: aqueous biphasic systems of water-miscible ionic liquids and water-structuring salts for recycle, metathesis, and separations. J Am Chem Soc. 2003;125:6632–3.
85. Shill K, Padmanabhan S, Xin Q, et al. Ionic liquid pretreatment of cellulosic biomass: enzymatic hydrolysis and ionic liquid recycle. Biotechnol Bioeng. 2011;108:511–20.
86. Saeman JF, Bubl JL, Harris EE. Quantitative saccharification of wood and cellulose. Ind Eng Chem Anal Ed. 1945;17:35–7.

87. Jacobsen SE, Wyman CE. Cellulose and hemicellulose hydrolysis models for application to current and novel pretreatment processes. Appl Biochem Biotechnol. 2000;84–86:81–96.
88. Cox BJ, Jia S, Zhang ZC, Ekerdt JG. Catalytic degradation of lignin model compounds in acidic imidazolium based ionic liquids: Hammett acidity and anion effects. Polym Degrad Stab. 2011;96:426–31.
89. Thomazeau C, Olivier-Bourbigou H, Magna L, et al. Determination of an acidic scale in room temperature ionic liquids. J Am Chem Soc. 2003;125:5264–5.
90. Li C, Zhao ZK. Efficient acid-catalyzed hydrolysis of cellulose in ionic liquid. Adv Synth Catal. 2007;349:1847–50.
91. Sievers C, Valenzuela-Olarte MB, Marzialetti T, et al. Ionic-liquid-phase hydrolysis of pine wood. Ind Eng Chem Res. 2009;48:1277–86.
92. Li B, Filpponen I, Argyropoulos DS. Acidolysis of wood in ionic liquids. Ind Eng Chem Res. 2010;49:3126–36.
93. Li C, Wang Q, Zhao ZK. Acid in ionic liquid: an efficient system for hydrolysis of lignocellulose. Green Chem. 2008;10:177.
94. Zhang Y, Du H, Qian X, Chen EY-X. Ionic liquid–water mixtures: enhanced K w for efficient cellulosic biomass conversion. Energy Fuel. 2010;24:2410–7.
95. Rinaldi R, Palkovits R, Schüth F. Depolymerization of cellulose using solid catalysts in ionic liquids. Angew Chem Int Ed Engl. 2008;47:8047–50.
96. Rinaldi R, Meine N, Vom Stein J, et al. Which controls the depolymerization of cellulose in ionic liuqids: the solid acid catalyst or the cellulose? ChemSusChem. 2010;3:266–76.
97. Zhang Z, Zhao ZK. Solid acid and microwave-assisted hydrolysis of cellulose in ionic liquid. Carbohydr Res. 2009;344:2069–72.
98. Amarasekara AS, Owereh OS. Synthesis of a sulfonic acid functionalized acidic ionic liquid modified silica catalyst and applications in the hydrolysis of cellulose. Catal Commun. 2010;11:1072–5.
99. Su Y, Brown HM, Li G, et al. Accelerated cellulose depolymerization catalyzed by paired metal chlorides in ionic liquid solvent. Appl Catal Gen. 2011;391:436–42.
100. Kuster BFM. 5-hydroxymethylfurfural (HMF). A review focussing on its manufacture. Starch—Stärke. 1990;42:314–21.
101. Zakrzewska ME, Bogel-Łukasik E, Bogel-Łukasik R. Ionic liquid-mediated formation of 5-hydroxymethylfurfural-a promising biomass-derived building block. Chem Rev. 2011;111:397–417.
102. Ståhlberg T, Fu W, Woodley JM, Riisager A. Synthesis of 5-(hydroxymethyl)furfural in ionic liquids: paving the way to renewable chemicals. ChemSusChem. 2011;4:451–8.
103. Lansalot-Matras C, Moreau C. Dehydration of fructose into 5-hydroxymethylfurfural in the presence of ionic liquids. Catal Commun. 2003;4:517–20.
104. Moreau C, Finiels A, Vanoye L. Dehydration of fructose and sucrose into 5-hydroxymethylfurfural in the presence of 1-H-3-methyl imidazolium chloride acting both as solvent and catalyst. J Mol Catal A Chem. 2006;253:165–9.
105. Qi X, Watanabe M, Aida TM, Smith JRL. Efficient process for conversion of fructose to 5-hydroxymethylfurfural with ionic liquids. Green Chem. 2009;11:1327.
106. Hu S, Zhang Z, Zhou Y, et al. Conversion of fructose to 5-hydroxymethylfurfural using ionic liquids prepared from renewable materials. Green Chem. 2008;10:1280.
107. Sievers C, Musin I, Marzialetti T, et al. Acid-catalyzed conversion of sugars and furfurals in an ionic-liquid phase. ChemSusChem. 2009;2:665–71.
108. Zhao H, Holladay JE, Brown H, Zhang ZC. Metal chlorides in ionic liquid solvents convert sugars to 5-hydroxymethylfurfural. Science. 2007;316:1597–600.
109. Zhang Y, Pidko EA, Hensen EJM. Molecular aspects of glucose dehydration by chromium chlorides in ionic liquids. Chem Eur J. 2011;17:5281–8.
110. Pidko EA, Degirmenci V, Van Santen RA, Hensen EJM. Coordination properties of ionic liquid-mediated chromium(II) and copper(II) chlorides and their complexes with glucose. Inorg Chem. 2010;49:10081–91.

111. Binder JB, Cefali AV, Blank JJ, Raines RT. Mechanistic insights on the conversion of sugars into 5-hydroxymethylfurfural. Energy Environ Sci. 2010;3:765.
112. Zhang Z, Zhao ZK. Production of 5-hydroxymethylfurfural from glucose catalyzed by hydroxyapatite supported chromium chloride. Bioresour Technol. 2011;102:3970–2.
113. Hu S, Zhang Z, Song J, et al. Efficient conversion of glucose into 5-hydroxymethylfurfural catalyzed by a common Lewis acid SnCl4 in an ionic liquid. Green Chem. 2009;11:1746.
114. Su Y, Brown HM, Huang X, et al. Single-step conversion of cellulose to 5-hydroxymethylfurfural (HMF), a versatile platform chemical. Appl Catal Gen. 2009;361:117–22.
115. Binder JB, Raines RT. Simple chemical transformation of lignocellulosic biomass into furans for fuels and chemicals. J Am Chem Soc. 2009;131:1979–85.
116. Li C, Zhang Z, Zhao ZK. Direct conversion of glucose and cellulose to 5-hydroxymethylfurfural in ionic liquid under microwave irradiation. Tetrahedron Lett. 2009;50:5403–5.
117. Zhang Z, Zhao ZK. Microwave-assisted conversion of lignocellulosic biomass into furans in ionic liquid. Bioresour Technol. 2010;101:1111–4.
118. Chidambaram M, Bell AT. A two-step approach for the catalytic conversion of glucose to 2,5-dimethylfuran in ionic liquids. Green Chem. 2010;12:1253.
119. U.S. Energy Information Administration. How much oil is used to make plastics? Washington, DC: U.S. Energy Information Administration; 2012. http://www.eia.gov/tools/faqs/faq.cfm?id=34&t=6
120. Heinze T, Liebert T. Unconventional methods in cellulose functionalization. Prog Polym Sci. 2001;26:1689–762.
121. Lee SH, Nguyen HM, Koo Y-M, Ha SH. Ultrasound-enhanced lipase activity in the synthesis of sugar ester using ionic liquids. Process Biochem. 2008;43:1009–12.
122. Lee SH, Dang DT, Ha SH, et al. Lipase-catalyzed synthesis of fatty acid sugar ester using extremely supersaturated sugar solution in ionic liquids. Biotechnol Bioeng. 2008;99:1–8.
123. Lee SH, Ha SH, Hiep NM, et al. Lipase-catalyzed synthesis of glucose fatty acid ester using ionic liquids mixtures. J Biotechnol. 2008;133:486–9.
124. Ganske F, Bornscheuer UT. Lipase-catalyzed glucose fatty acid ester synthesis in ionic liquids. Org Lett. 2005;7:3097–8.
125. Wu J, Zhang J, Zhang H, et al. Homogeneous acetylation of cellulose in a new ionic liquid. Biomacromolecules. 2004;5:266–8.
126. Abbott AP, Bell TJ, Handa S, Stoddart B. O-Acetylation of cellulose and monosaccharides using a zinc based ionic liquid. Green Chem. 2005;7:705.
127. Xie H, King A, Kilpelainen I, et al. Thorough chemical modification of wood-based lignocellulosic materials in ionic liquids. Biomacromolecules. 2007;8:3740–8.
128. Barthel S, Heinze T. Acylation and carbanilation of cellulose in ionic liquids. Green Chem. 2006;8:301.
129. Liu CF, Sun RC, Zhang AP, et al. Homogeneous modification of sugarcane bagasse cellulose with succinic anhydride using a ionic liquid as reaction medium. Carbohydr Res. 2007;342:919–26.
130. Eremeeva T. Size-exclusion chromatography of enzymatically treated cellulose and related polysaccharides: a review. J Biochem Biophys Methods. 2003;56:253–64.
131. Van Rantwijk F, Madeira Lau R, Sheldon RA. Biocatalytic transformations in ionic liquids. Trends Biotechnol. 2003;21:131–8.
132. Park S, Kazlauskas RJ. Biocatalysis in ionic liquids—advantages beyond green technology. Curr Opin Biotechnol. 2003;14:432–7.
133. Mora-Pale M, Meli L, Doherty TV, et al. Room temperature ionic liquids as emerging solvents for the pretreatment of lignocellulosic biomass. Biotechnol Bioeng. 2011;108:1229–45.
134. Kubo S, Hashida K, Yamada T, et al. A characteristic reaction of lignin in ionic liquids; glycelol type enol-ether as the primary decomposition product of β-O-4 model compound. J Wood Chem Technol. 2008;28:84–96.

135. Okuda K, Ohara S, Umetsu M, et al. Disassembly of lignin and chemical recovery in supercritical water and p-cresol mixture. Studies on lignin model compounds. Bioresour Technol. 2008;99:1846–52.
136. Cox BJ, Ekerdt JG. Depolymerization of oak wood lignin under mild conditions using the acidic ionic liquid 1-H-3-methylimidazolium chloride as both solvent and catalyst. Bioresour Technol. 2012;118:584–8.
137. Jia S, Cox BJ, Guo X, et al. Decomposition of a phenolic lignin model compound over organic N-bases in an ionic liquid. Holzforschung. 2010;64:577–80.
138. Jia S, Cox BJ, Guo X, et al. Hydrolytic cleavage of -O-4 ether bonds of lignin model compounds in an ionic liquid with metal chlorides. Ind Eng Chem Res. 2011;50:849–55.
139. Jia S, Cox BJ, Guo X, et al. Cleaving the β–O–4 bonds of lignin model compounds in an acidic ionic liquid, 1-H-3-methylimidazolium chloride: an optional strategy for the degradation of lignin. ChemSusChem. 2010;3:1078–84.
140. Binder JB, Gray MJ, White JF, et al. Reactions of lignin model compounds in ionic liquids. Biomass Bioenergy. 2009;33:1122–30.
141. Jiang N, Ragauskas AJ. Selective aerobic oxidation of activated alcohols into acids or aldehydes in ionic liquids. J Org Chem. 2007;72:7030–3.
142. Yan N, Yuan Y, Dykeman R, et al. Hydrodeoxygenation of lignin-derived phenols into alkanes by using nanoparticle catalysts combined with Brønsted acidic ionic liquids. Angew Chem Int Ed Engl. 2010;49(32):5549–53.
143. Zhang ZC. Catalysis in ionic liquids. Adv Catal. 2006;49:153–237.
144. Park JI, Steen EJ, Burd H, et al. A thermophilic ionic liquid-tolerant cellulase cocktail for the production of cellulosic biofuels. PLoS One. 2012;7:e37010.
145. Stark A. Ionic liquids in the biorefinery: a critical assessment of their potential. Energy Environ Sci. 2011;4:19.
146. Sun N, Rodríguez H, Rahman M, Rogers RD. Where are ionic liquid strategies most suited in the pursuit of chemicals and energy from lignocellulosic biomass? Chem Commun. 2011;47:1405–21.
147. Olivier-Bourbigou H, Magna L, Morvan D. Ionic liquids and catalysis: recent progress from knowledge to applications. Appl Catal Gen. 2010;373:1–56.

Chapter 9
Production of Versatile Platform Chemical 5-Hydroxymethylfurfural from Biomass in Ionic Liquids

Xinhua Qi, Richard L. Smith Jr., and Zhen Fang

Abstract The furan derivative, 5-hydroxymethylfurfural (5-HMF), can replace many petroleum-derived monomers and intermediates presently used in the manufacture of plastics and fine chemicals. Ionic liquid solvents provide a sustainable path for 5-HMF production since they can dissolve crude biomass and allow conversion of polysaccharide fractions to 5-HMF with high selectivity. This chapter presents current progress in the synthesis of 5-HMF with ionic liquid solvents and considers conversion of saccharide substrates such as fructose, glucose, inulin and cellulose under catalytic reaction conditions. Challenges for 5-HMF production with ionic liquids are addressed and interesting aspects that still need to be explored for developing practical systems are highlighted.

Keywords Biomass • Ionic liquid • Cellulose • Hydroxymethylfurfural • Glucose • Fructose • Catalysis

9.1 Introduction

Biomass represents a possible sustainable resource for production of fuels and valuable chemicals [1]. However, the overabundance of oxygen within the molecular structures of carbohydrates limit the application of biomass as a feedstock for

X. Qi (✉)
College of Environmental Science and Engineering, Nankai University, Tianjin, China
e-mail: qixinhua@nankai.edu.cn

R.L. Smith Jr.
Graduate School of Environmental Studies, Research Center of Supercritical Fluid Technology, Tohoku University, Sendai, Japan
e-mail: smith@scf.che.tohoku.ac.jp

Z. Fang
Xishuangbanna Tropical Botanical Garden, Chinese Academy of Sciences, Kunming, China
e-mail: zhenfang@xtbg.ac.cn

Z. Fang et al. (eds.), *Production of Biofuels and Chemicals with Ionic Liquids*, Biofuels and Biorefineries 1, DOI 10.1007/978-94-007-7711-8_9,

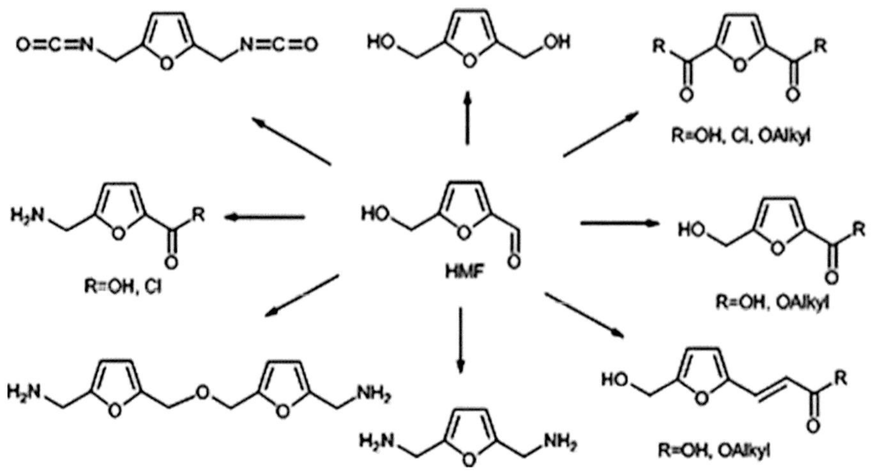

Fig. 9.1 A selection of monomers derived from 5-hydroxymethylfurfural (Reproduced with permission from [5]. Copyright © 2010 Royal Society of Chemistry)

substitutes of traditional fossil fuels [2]. Removal of water from carbohydrates by dehydration is one of the main ways for reducing their oxygen content and obtaining commodity compounds [2]. Among the possible compounds that can be derived from dehydration of carbohydrates, 5-hydroxymethylfurfural (5-HMF) is a highly versatile intermediate, since it can be used for furan-based fine chemicals and polymers [3–5] (see Fig. 9.1). In the past few years, water [6–11], organic solvents [9, 12–15] and organic-water mixtures [1, 16–18] have been broadly investigated for the production of 5-HMF from a variety of biomass-derived carbohydrates such as fructose, glucose, and other di-poly-saccharides (Fig. 9.2).

Water is a good solvent for both monosaccharides and the products and it is favored as a reaction solvent from its environmental aspects. However, as a reaction solvent, water leads to the formation of undesired side-products, especially levulinic, formic acids, and humins result from either polymerization of 5-HMF or cross-polymerization of 5-HMF and carbohydrates that is attributed to water's ionization [20–22]. When water is used as a solvent for conversion of saccharides to 5-HMF, by-product formation leads to 5-HMF yields of below 40 % and the conditions required are harsh [8, 22–24]. Therefore, it is necessary to prevent rehydration of 5-HMF in the reaction chemistry so that humin formation can be suppressed. Suppression of humin formation can be accomplished by carrying out the reactions in nonaqueous systems. Dimethylsulfoxide (DMSO) was identified early as an efficient solvent for the preparation of 5-HMF from fructose [12, 13, 25], since the furanoid form of fructose is favored in DMSO [4]. However, DMSO suffers from difficult product recovery and environmental problems in its use. Acetone is a good alternative solvent for DMSO due to some similarities in its

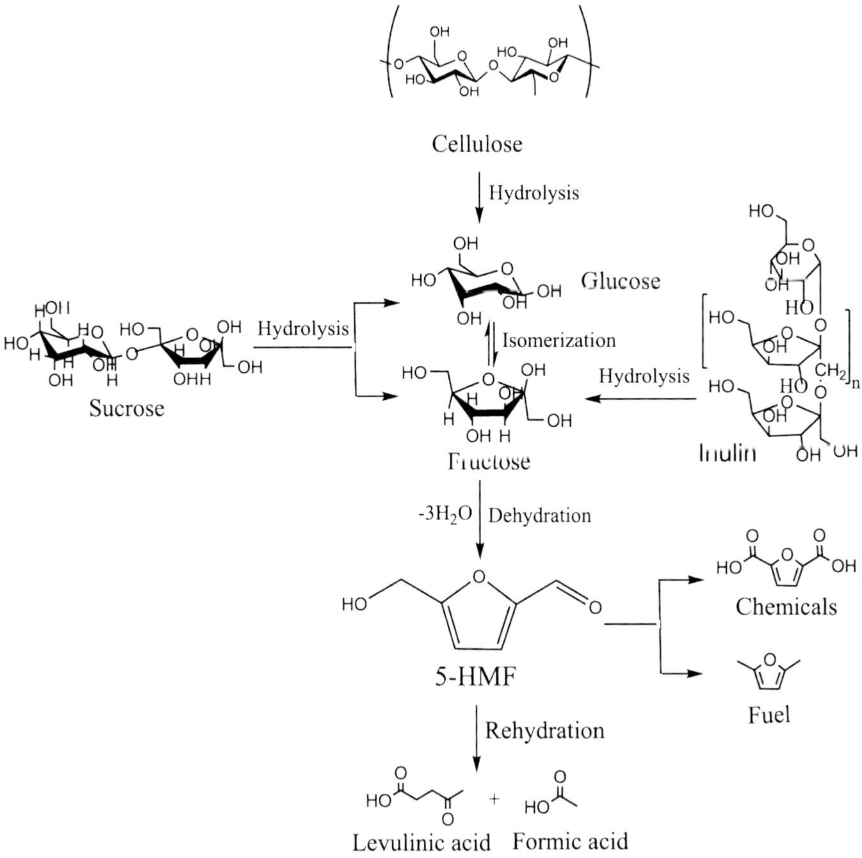

Fig. 9.2 Reaction scheme for the transformation of glucose, fructose, and other di-/polysaccharides into 5-hydroxymethylfurfural (Adapted with permission from reference [19]. Copyright © 2010 Wiley-VCH Verlag GmbH & Co. KGaA)

chemical properties, however, it has to be used with water or DMSO as cosolvent since monosaccharides have only limited solubility in pure acetone [4, 15].

Ionic liquids have favorable properties, such as nonvolatility, high thermal and chemical stability and adjustable solvent power for organic substances [26]. ILs have good attributes as solvent for producing 5-HMF from carbohydrates since they can allow reaction under relatively mild conditions and have the possibility of being used in one-pot reactions with biomass as the feedstock. This chapter provides an overview on research works that have studied the catalytic transformation of 5-HMF from a variety of carbohydrates in ionic liquids.

9.2 Catalytic Production of 5-HMF from Biomass in Ionic Liquids

9.2.1 Starting Material

9.2.1.1 Fructose

Fructose is the most common studied substrate for the preparation of 5-HMF in aqueous solutions, organic solvents, and water-organic solvent mixtures. In an early work that applied molten salts to the synthesis of 5-HMF from carbohydrates, fructose was converted to 5-HMF with a high yield of 70 % in the presence of pyridinium chloride in 1983 [27]. However, this pioneering work did not stimulate wide investigations on the 5-HMF production from biomass in melt salt solutions, until the beginning of the twenty-first century. Biomass conversions in ionic liquids became a hot topic when Lansalot-Matras and Moreau reported the acid-catalyzed dehydration of fructose in 1-butyl-3-methylimidazolium tetrafluoroborate ([BMIM][BF_4]) and 1-butyl-3-methylimidazolium hexafluorophosphate ([BMIM][PF_6]) with dimethyl sulfoxide (DMSO) as co-solvent in the presence of Amberlyst® 15 sulfonic ion-exchange resin as catalyst, to obtain a 5-HMF yield as high as 87 % at 80 °C for 32 h reaction time [28]. It was demonstrated that the addition of ionic liquids had a positive effect on the fructose dehydration to 5-HMF. They also studied acid-catalyzed dehydration of fructose in ionic liquid 1-H-3-methylimidazolium chloride ([HMIM][Cl]), which acted both as solvent and catalyst, and demonstrated a 5-HMF yield as high as 92 % [29]. According to the activation parameters calculated from the Arrhenius plots for formation and decomposition of 5-HMF, activation energies are found to be similar to those obtained in reactions catalyzed by zeolite solid catalyst. The authors attribute the high yields observed for the formation of 5-HMF in ionic liquids as solvent to be due to the differences in the preexponential factors [28]. Thereafter, many papers on the selective dehydration of fructose in ionic liquids with various catalysts, have appeared in the literature. Efficient dehydration of fructose to 5-HMF in ionic liquids was reported at much lower temperatures than in aqueous, organic solvents, and water-organic mixture systems [20, 30–34].

Qi et al. reported that 1-butyl-3-methylimidazolium chloride, [BMIM][Cl], used with a sulfonic ion-exchange resin catalyst could efficiently dehydrate fructose into 5-HMF to give a fructose conversion of 98.6 % and a 5-HMF yield of 83.3 % in 10 min reaction time at 80 °C. The reaction time could be reduced to 1 min when the temperature was increased to 120 °C which resulted in a 5-HMF yield of 82.2 % and nearly 100 % fructose conversion [20]. Comparison of the [BMIM][Cl] and [BMIM][BF_4] systems that has the same sulfonic ion-exchange resin (Amberlyst® 15) as the catalyst, indicates that the higher efficiency and selectivity observed in the [BMIM][Cl] system can be attributed to a higher tendency towards concerted catalysis due to the greater hydrogen-bonding character, nucleophilicity or basicity of the chloride ion.

Shi et al. [35] used trifluoromethanesulfonic acid (TfOH), which is an interesting catalyst, to promote the conversion of fructose into 5-HMF in imidazolium ionic liquids. They studied the reaction system for different alkyl chain length ionic liquids and used various kinds of anions. In that work, yields of 5-HMF were strongly affected by aggregation of cations and the hydrogen bonds between fructose and anions of ionic liquids. Imidazolium cationic ILs with alkyl chain lengths of the cations being shorter than four carbons were found to be suitable for 5-HMF formation. They found that the anion of an IL forms strong hydrogen bonds with fructose molecules, and thus, the acid radical leads to high reaction activity. These results not only provide evidence to explain the interaction of the structure at the molecular level in 5-HMF preparation in ionic liquids, but also provide some hints on choosing suitable ILs for 5-HMF preparation [35].

Imidazolium-based ILs provide efficient dehydration of fructose into 5-HMF, however, eutectic mixtures of choline chloride with acids have been identified as a promising catalytic system for the process. Hu et al. [36] investigated the conversion of fructose to 5-HMF in choline chloride/citric acid at 80 °C, and obtained a 5-HMF yield of 77.8 % without in situ extraction and a yield of 91.4 % when continuous extraction with ethyl acetate was used for a 1 h reaction time. The main advantage of this process was not only high 5-HMF yields obtained but also the chemical system components (fructose, choline chloride and citric acid) used that all originate from renewable sources. Ilgen et al. studied choline chloride for the purpose of using highly concentrated mixtures of fructose (up to 50 wt%). The resulting solutions had a melting region of 79–82 °C, which is much lower than the melting point of pure choline chloride (300 °C), thus processing the solutions at low temperatures was possible. The highest 5-HMF yield obtained from the choline chloride-fructose system was 67 % for which the reaction conditions were p-TsOH as catalyst, reaction temperature of 100 °C and reaction time of 0.5 h [37]. Furthermore, they made a screening study to compare the environmental impact of the choline chloride system with different conventional solvents for the conversion of carbohydrates into 5-HMF, and indicated advantages of the choline chloride systems in terms of low toxicity and reduced mobility [37]. Liu et al. [38] developed a cheap and sustainable choline chloride/CO_2 system for the dehydration of highly concentrated fructose solutions into 5-HMF with a yield of up to 72 %. In addition to the environmental benefits of this strategy, they found that in the presence of ChCl, 5-HMF is stabilized, probably due to hydrogen bonding that allows formation of a eutectic mixture between choline chloride and 5-HMF. This aspect allows fructose can be converted with a high content (up to 100 wt%) as compared to traditional procedures where 5-HMF is obtained in yields higher than 60 % only from a fructose concentration lower than 20 wt% [20].

9.2.1.2 Glucose

Glucose is an isomer of fructose and since it occurs as the monomeric unit in cellulose, it can be considered to be the most abundant monosaccharide in nature. Therefore, glucose is more appropriate than fructose as starting material for 5-HMF. Efficient routes for converting glucose to 5-HMF are an active research topic. However, glucose has been shown to be difficult to convert to 5-HMF (yields <30 %) with solvents such as water [22], organics [39] and organic-water mixtures [40]. The reason for this is apparently because glucose tends to form a stable six-membered pyranoside structure that has a low enolization rate [25]. Since enolisation rate is the rate-determining step for 5-HMF formation, glucose will react much slower than fructose. Thus, glucose is more likely to undergo cross-polymerization with reactive intermediates and 5-HMF, since it can form true oligosaccharides that contain reactive reducing groups [25]. Because of this limitation in the fundamental chemistry, there were no efficient processes for the selective dehydration of glucose into 5-HMF, until a major breakthrough came in 2007 when Zhao et al. [41] published a method for transforming glucose into 5-HMF with an ionic liquid solvent (1-ethyl-3-methylimidazolium chloride, [EMIM][Cl]) and $CrCl_2$ catalyst. In that work, a 5-HMF yield of 68 % was obtained at a temperature of 100 °C for a reaction time of 3 h. In this reaction, $CrCl_3^-$ anion is thought to not only play a key role in proton transfer that facilitates mutarotation of glucose in [EMIM][Cl], but also to play a critical role in the isomerization of glucose to fructose by a formal hydride transfer. Once fructose is formed, it is rapidly dehydrated to 5-HMF in the ionic liquid in the presence of the catalyst. Inspired by this work, a series of papers were reported for the conversion of glucose to 5-HMF using chromium chloride as catalyst [19, 42–45].

Yong et al. [45] studied the production of 5-HMF from fructose and glucose in 1-butyl-3-methylimidazolium chloride ([BMIM][Cl]) using $CrCl_2$ as catalyst, and 5-HMF yields of 96 % and 81 %, respectively for reaction conditions at 100 °C for 6 h reaction time achieved. Those authors considered that NHC/$CrCl_x$ (NHC=N-heterocyclic carbene) complexes played the key role in glucose dehydration in [BMIM][Cl]. Additionally, in the $CrCl_2$/EMIM system, a NHC/Cr complex could be formed under the reaction conditions and therefore serves as a catalyst [45]. Remarkably, 5-HMF yields were approximately 14 % higher for the reaction carried out in air than that conducted in argon. Zhang et al. [46] studied the production of 5-HMF from glucose catalyzed by hydroxyapatite supported chromium chloride in an ionic liquid (1-butyl-3-methylimidazolium chloride), and a maximum 5-HMF yield of 40 % was obtained. In the work of Zhao et al., glucose conversions and 5-HMF yields were lower when $CrCl_3$ was used instead of $CrCl_2$ [41]. However, subsequent studies suggested that there are only minor differences in the catalytic activity of bivalent and trivalent chromium salts [19, 47]. Compared with the strongly reductive Cr (II), the trivalent form, Cr (III), possesses higher stability in the environment, and Cr (III) is essential for mammals in removing glucose from the bloodstream [48]. Binder and Raines [44] made an extensive investigation on glucose conversion in dimethylacetamide (DMA) with the addition of halide salts. Addition of 10 wt% LiCl or LiBr along with $CrCl_2$, $CrCl_3$, or $CrBr_3$ resulted in 5-HMF yields up to 80 %.

A zero-valent $Cr(CO)_6$-based catalyst system was found to be effective for the conversion of glucose to 5-HMF in ionic liquid [EMIM][Cl], even at low catalyst loadings [49]. Through *in-situ*, *ex-situ*, and quantitative poisoning experiments, it was demonstrated that small, uniform Cr^0-nanoparticles, either preformed via microwave irradiation (3.6–0.7 nm) or generated *in-situ* via thermolysis (2.3–0.4 nm) during the reaction, are active species responsible for the observed catalysis when using $Cr(CO)_6$ as the precatalyst [49]. In view of some of the advantages of the $Cr(CO)_6$-derived nanoparticles catalyst system, including the relatively low cost and air-stability of the precatalyst as well as its ability to maintain high efficiency at low catalyst loadings, the results should provide a new method to develop more effective metal-nanoparticles catalysts for glucose or related biomass conversion processes.

Two analogous chromium catalysts other than chromium chloride have been reported to be effective for conversion of glucose to 5-HMF [32]. Han and co-workers [50] used 1-ethyl-3-methylimidazolium tetrafluoroborate ([EMIM][BF_4]) with $SnCl_4$ as catalyst, and obtained a 5-HMF yield of ca. 60 % at 100 °C for 3 h. They screened metal chlorides and ILs, for which only chromium(III), aluminum(III), and tin(IV) chlorides exhibited activity, and tin(IV) was concluded to be the most active catalyst. Out of the eight ILs examined, [EMIM][BF_4] was found to be most favorable. Those authors proposed that the formation of a five-membered-ring chelate complex consisting of Sn and two neighboring hydroxyl groups in glucose was a probable intermediate in the formation of 5-HMF (Fig. 9.3). Their ^{1}H-NMR measurements showed that chloride in $SnCl_4$ was transferred and interacted with hydrogen atoms, and Sn atoms interacted with oxygens to promote the formation of a straight-chain glucose required for transformation to the enol intermediate and formation of 5-HMF [50]. Stanlberg et al. [51] examined ionic liquids with lanthanide catalyst, and obtained a 5-HMF yield of 24 %. Lanthanide(III) salts have also been tried for the dehydration of glucose to 5-HMF [51]. The use of the strongest Lewis acid $Yt(OTf)_3$ resulted in 24 % of 5-HMF yield, and the catalytic effect increased with increasing atom number in the lanthanide series. Furthermore, the 5-HMF yield was observed to increase with increasing the chain length of the alkyl groups on the imidazolium cation, that is, 1-octyl-3-imidazolium chloride ([OMIM][Cl]) had a significantly higher yield than [EMIM][Cl]. This phenomenon has not been observed with other catalyst systems where [EMIM][Cl] has been superior or equivalent to other methylimidazolium chlorides [51].

Although the catalysts such as $CrCl_2$, $CrCl_3$, $Cr(CO)_6$ and $SnCl_4$ are effective for the dehydration of glucose into 5-HMF, they are poisonous, difficult to recycle and have high environmental risk. The use of inherently nonhazardous catalysts and solvent systems are needed for application in today's society. Considering that the dehydration of glucose to 5-HMF involves two steps, namely, isomerization of glucose into fructose through base catalysis that is followed by dehydration of

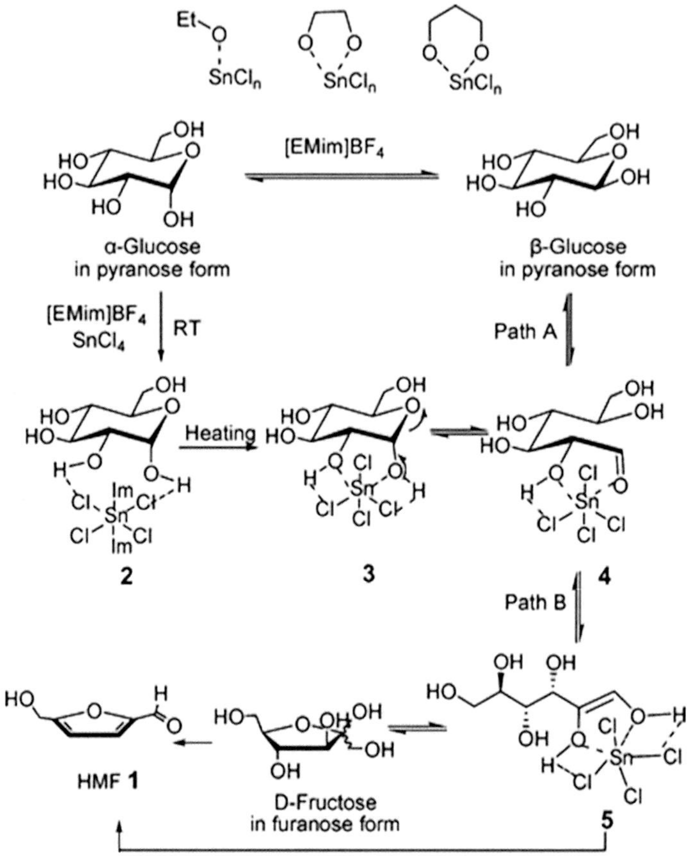

Fig. 9.3 Proposed mechanism for glucose dehydration to 5-HMF catalyzed by $SnCl_4$ in [EMIM][BF_4] (Reproduced with permission from reference [50]. Copyright © 2009 Royal Society of Chemistry)

fructose by acid catalysis to give 5-HMF [22], Qi et al. developed a method for glucose conversion into 5-HMF with ionic liquid-water mixture without using chromium-analogous catalysts [52]. They found that the addition of an appropriate amount of water into the ionic liquid has a synergistic effect on the glucose conversion to 5-HMF, and promoted the formation of 5-HMF from glucose compared with that in either pure water or in the pure ionic liquid solvent. In the proposed reaction system, a 5-HMF yield of 53 % could be obtained in 50:50 w/w% 1-hexyl-3-methyl imidazolium chloride-water mixture in 10 min reaction time at 200 °C in the presence of ZrO_2. It was confirmed that 1,3-dialkylimidazolium ILs with Cl^- and HSO_4^- anion were effective for 5-HMF formation from glucose in IL-H_2O mixture. The addition of the other protic solvents such as methanol and ethanol into the ionic liquid had a similar synergistic effect as water and promoted fructose and 5-HMF formation [52].

9.2.1.3 Sucrose

Sucrose is a disaccharide consisting of glucose and fructose moiety linked together by a glycosidic bond, and is easily hydrolyzed into fructose and glucose upon heating in ILs. Moreau et al. studied the dehydration of sucrose in 1-H-3-methyl imidazolium chloride ionic liquid, and found that it could be nearly quantitatively transformed into 5-HMF and unreacted glucose, thus 5-HMF was produced only from the fructose moiety and the glucose moiety was practically unused in the system [29]. Hu et al. investigated the conversion of sucrose in [EMIM][BF_4] in the presence of $SnCl_4$, and found that both the glucose and fructose moiety could be converted into 5-HMF (about 65 % yield) and most of the carbon in sucrose was converted [50].

Dehydration of sucrose (50 wt%) in choline chloride catalyzed by $CrCl_2$ and $CrCl_3$ resulted in 5-HMF yields of 62 and 43 %, respectively, in 1 h reaction time at 100 °C [37]. Outstanding results were reported by Lima et al. [53] who reported 5-HMF yield of 100 % when using a methyl iso-butyl ketone (MIBK) as co-solvent in the reaction system, [BMIM][Cl]/MIBK/$CrCl_3$ system. These results are surprising since yields for fructose and glucose alone in the system were only 88 and 79 %, respectively, under the same reaction conditions [53]. MIBK, however, is an undesirable additive, due to its volatility and its designation as a priority pollutant.

9.2.1.4 Cellulose

Inedible lignocellulosic biomass is a prime candidate as a starting material for 5-HMF production because it presumably would not compete directly with food sources. The development of efficient routes for converting lignocellulose biomass into 5-HMF is essential for achieving sustainable production of 5-HMF. Many research works focus on the transformation of cellulose, since it is the constituent of biomass that can be used to make 5-HMF. However, cellulose is insoluble in many conventional solvents [54–56]. The main advantage of using ionic liquids as reaction media for biomass conversion is the possibility of ionic liquids to dissolve carbohydrate polymers and subsequently form products in one-pot reactions. The conversion of cellulose to 5-HMF can be thought to involve three chemical processes: hydrolysis, isomerisation and dehydration. Although hydrolysis of cellulose in ionic liquids in the presence of mineral acids has been studied in detail [57, 58], the efficient conversion of cellulose into 5-HMF with high yield has not been realized until $CrCl_2$ was found to be active for dehydration of glucose into 5-HMF [41]. Su et al. [59] presented a single-step process for cellulose conversion into 5-HMF by using an ionic liquid solvent system with a pair metal chlorides ($CuCl_2$–$CrCl_2$) catalyst, and obtained a 5-HMF yield of 55 % under relatively mild conditions of 120 °C in 8 h reaction time. After the product 5-HMF was separated from the solvent, the catalytic performance of recovered [EMIM][Cl] and the catalysts were used in repeated experiments. Under these conditions, cellulose depolymerizes at a rate that is about one order of magnitude higher than when using a homogeneous acid catalysis. Single metal chlorides at the same total loading showed considerably less activity under similar conditions [59]. Binder et al. [44] studied the conversion of cellulose, corn stover, and pine dust in DMA-LiCl-EMIM][Cl] mixture using chromium(II) or chromium(III) chlorides as catalysts and hydrochloric acid as a co-catalyst. The mixed system converted cellulose into 5-HMF with a yield of 54 %. When [EMIM][Cl] was used as solvent and $CrCl_2$–HCl as the catalytic system, results were about the same. These examples show the presence of a catalyst is essential for the reaction system and a 5-HMF yield of only 4 % could be obtained in the absence of the catalyst. For lignocellulosic feedstocks, 5-HMF yields varied from 16 to 47 %. A 5-HMF yield of 16 % was obtained for corn stover substrate regardless of whether it was subjected to pretreatment of ammonia fiber expansion. Lignin and proteins did not seem to affect the results.

Li et al. [47] studied the conversion of cellulose and raw lignocellulosic biomass with microwave irradiation using $CrCl_3$ as a catalyst in [BMIM][Cl] solvent. Initially the investigation was conducted on transformation of cellulose to 5-HMF using cellulose of various degrees of polymerization (Avicel, Spruce, Sigamcell, R-cellulose). The 5-HMF yields varied from 53 to 62 % for all substrates. The authors proposed that the coordination of cellulose with $[CrCl_3 + n]^{n-}$ species is responsible for the partial weakening of the 1,4-glucosidic bonds, thus making cellulose more susceptible to attack by water in the hydrolysis step. The authors observed that the use of microwave heating has significant effect in reducing the reaction time, this was also observed by Qi et al., who examined the conversion of cellulose into 5-HMF in [BMIM][Cl] under microwave irradiation, and a 5-HMF yield of 54 % in 10 min reaction time at 150 °C was obtained [19].

A reaction system that could convert microcrystalline cellulose into 5-HMF under mild conditions was reported by Zhang et al. [60] They designed a green process for the conversion of cellulose into water-soluble reducing carbohydrates with a total yield as high as 97 % in the absence of added acid catalysts with the main point of originality of the process being added water to the ionic liquid-cellulose system. The formed carbohydrates could be transformed into 5-HMF in 89 % yield when $CrCl_2$ was added. Such a high 5-HMF yield of nearly 90 % means that not only glucose, but also other water-soluble reducing sugars were converted to product. The total reducing sugars and 5-HMF yields depend not only on reaction temperature and time but also on the amount of added water. By using ab initio calculations, the authors demonstrated that the favorable results obtained in a catalyst system was probably due to the dissociation constant (K_w) of water that increased with the addition of purified ILs. The increased K_w of water by ILs makes the IL-water mixture exhibits higher concentrations of both $[H^+]$ and $[OH^-]$ than pure water, thus enabling acid- or base-catalyzed reactions to occur. For example, water in the presence of 15 mol% of the [EMIM][Cl] at 120 °C exhibits K_w values up to three orders of magnitude higher than those of pure water under ambient conditions [60]. This intrinsic property of the IL-water mixtures can not only be used in biomass processing and conversion, but also in organic catalysis, electrochemistry, or other relative research fields.

Due to the inherent environmental risk of chromium, finding of chromium-free processes for conversion of cellulose into 5-HMF is of great significance for biomass utilization. The transformation of cellulose into 5-HMF involves three chemical processes, that is, hydrolysis of cellulose to glucose catalyzed by acid catalyst, isomerization of glucose into fructose by base catalyst, and dehydration of fructose into 5-HMF [22, 24]. Peng and co-workers synthesized acid–base bifunctionalized mesoporous silica nanoparticles with large pore sizes of around 30 nm (LPMSNs) for cooperative catalysis of one-pot cellulose to 5-HMF conversion [61]. They used a grafting method to functionalize LPMSN with an acid (SO_3H) or/and a base (NH_2) group in order to prepare SO_3H, NH_2, and both SO_3H and NH_2 functionalized LPMSN (denoted as LPMSN-SO_3H, LPMSN-NH_2, and LPMSN-both, respectively). Results show that LPMSN-SO_3H and LPMSN-NH_2 could increase the yield of 5-HMF converted from the reactions that need acid and base catalysts, respectively. The LPMSN-SO_3H catalysts were found to promote for one-pot conversion of cellulose to 5-HMF and gave enhanced 5-HMF yields.

9.2.1.5 Inulin

Inulin, which is also called fructan, is a carbohydrate consisting of fructose units that vary in the degree of polymerization (DP) from 2 to 60, or higher. The fructosyl units in inulin are linked by $\beta(2 \rightarrow 1)$ linkages with the polymer chains terminating with a glucose unit [62]. Inulin is indigestible. The production of 5-HMF from inulin in ionic liquids has not been studied as extensively as cellulose, since inulin is not so abundant in nature. However, the carbohydrate provides a different feedstock that might give insight into mechanism of its transformation into 5-HMF. The hydrolysis of inulin to fructose followed by the dehydration of fructose to produce 5-HMF is a possible two-step reaction pathway. Because both reaction steps are catalyzed by acid catalysts, it is interesting to consider the production of 5-HMF from inulin as a one pot reaction, since this would avoid the separation of the fructose in the intermediate step.

Hu et al. [63] developed a process for the direct conversion of inulin to 5-HMF in choline chloride (ChoCl)/oxalic acid and ChoCl/citric acid deep eutectic solvents (DES), for which a 5-HMF yield of 56 % was obtained at relatively low temperatures (80 °C). When a biphasic system with IL and ethyl acetate was used for the in-situ extraction of 5-HMF, the 5-HMF yield was enhanced to 64 %, since the product 5-HMF is soluble in ethyl acetate, while the reactant inulin and fructose are insoluble in the DES, and ethyl acetate is only slightly soluble in the DES, which is favorable for the reaction in the biphasic system [63]. Although $SnCl_4$/[EMIM][BF_4] system is efficient for glucose dehydration, there is no advantage in using the system for inulin conversion to 5-HMF since only moderate 5-HMF yields of 40 % are obtained [50].

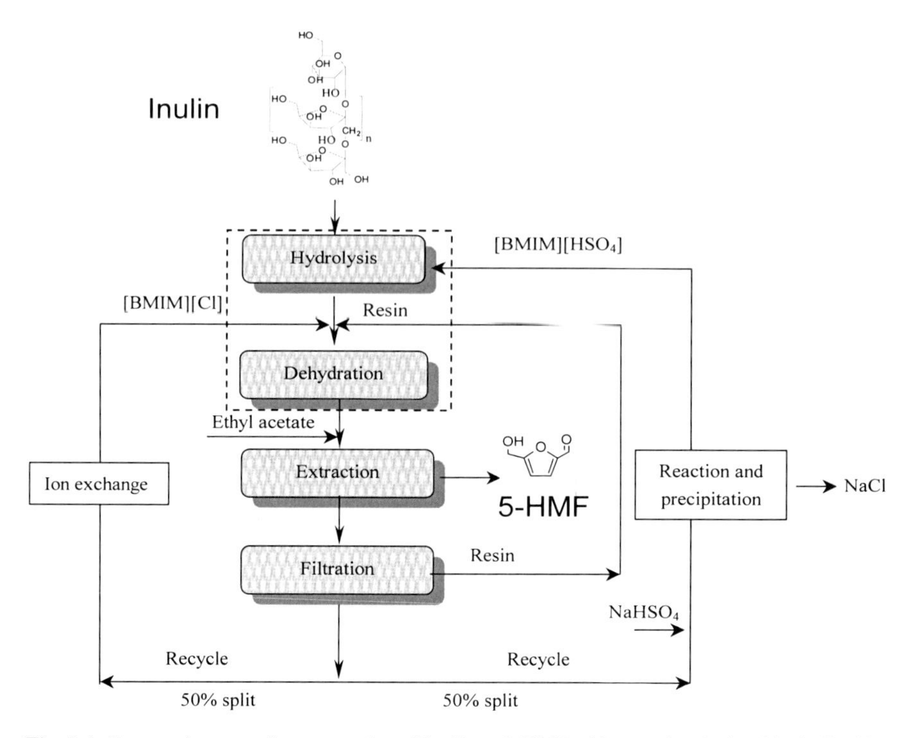

Fig. 9.4 Proposed process for conversion of inulin to 5-HMF with recycle of mixed ionic liquids. The 1-pot, 2-step reaction is enclosed in the *dashed box* (Reproduced with permission from [64]. Copyright © 2010 Royal Society of Chemistry)

Qi et al. [64] used the characteristics of two kinds of ionic liquids to develop an efficient process for the direct conversion of inulin to 5-HMF in one pot with two reaction steps under mild conditions. In the first step, the ionic liquid 1-butyl-3-methyl imidazolium hydrogen sulfate ([BMIM][HSO_4]) was employed as both solvent and catalyst for the rapid hydrolysis of inulin into fructose with 84 % yield in 5 min reaction time at 80 °C. In the second step, 1-butyl-3-methyl imidazolium chloride ([BMIM][Cl]) and a strong acidic cation exchange resin were added to the mixture to selectively convert the generated fructose into 5-HMF, achieving a 5-HMF yield of 82 % in 65 min, which is the highest 5-HMF yield reported to date for an inulin feedstock. These authors proposed a conceptual process that the ionic liquid, resin, catalyst, ethyl acetate, and the supplied chemicals are internally recycled and represent an efficient method for producing 5-HMF from inulin (Fig. 9.4).

Xie et al. [65] studied the catalytic conversion of inulin into 5-HMF in ionic liquids by lignosulfonic acid catalyst, and found that that consecutive hydrolysis of inulin was very fast, with a maximum yield of 47.0 % achieved in only 5 min, and the yield could be maintained even with prolonging the reaction time to 90 min. The effect of water addition was examined and it was demonstrated that the addition of

H_2O does not affect the reaction, with comparative yields obtained since the dehydration of one molecule of fructose produces three molecules of H_2O, which is sufficient for the previous hydrolysis of inulin into fructose, therefore, additional water does not seem to be necessary for this reaction system. Bennoit et al. [66] tried to partially substitute ionic liquids with glycerol or glycerol carbonate as cheap, safe and renewable sourced co-solvents for the acid-dehydration of inulin to 5-HMF. They found that a system that used [BMIM][Cl]/glycerol carbonate in a 10:90 ratio as solvent and a wet Amberlyst 70 acidic resin as catalyst resulted in a 5-HMF yield of 60 % at 110 °C. A major part of the ionic liquid was substituted with glycerol carbonate which allows the cost and environmental impact of the process to be potentially be lower than methods that petroleum derived co-solvents.

Although ionic liquids have been investigated as solvent for the synthesis of 5-HMF, they still have some shortcomings such as price and toxicity that hamper their industrial applications. Liu et al. [38] developed an interesting choline chloride (ChCl)/CO_2 system, where the addition of CO_2 could decrease the pH of the system through formation of carbonic acid and played as an efficient catalyst for the hydrolysis of inulin and the dehydration of fructose into 5-HMF. The inulin conversion with or without water addition in ChCl/CO_2 system was studied. The yield of 5-HMF was rather disappointing (12 %) in the absence of water (120 °C, 4 mPa CO_2, 90 min). The authors attributed the low 5-HMF yield to the low content of water at the initial stage of the reaction that does not favor the hydrolysis of inulin to fructose. Hence, water was initially added (16 wt%), and the yield of 5-HMF improved to 38 % when the reaction time was prolonged to 6 h. The 5-HMF was recovered with 41 % yield after 15 h of reaction, demonstrating the stability of 5-HMF in ChCl as compared to other solvents. This process has many advantages: (1) it is not necessary to remove the trace amounts of catalyst contained in 5-HMF after extraction; (2) the use of cheap and safe CO_2 as a source of acid catalyst; (3) the low ecological risk of the ChCl/CO_2 system that is renewable; and (4) the system is still efficient at high loadings of fructose (up to 100 wt%).

9.2.2 Catalyst

Although Ranoux et al. reported that the synthesis of 5-HMF from carbohydrates is autocatalytic that does not require any additional catalyst [67], the proper use of catalyst not only promotes the reaction rate, but also can improve product selectivity. Catalysts play a crucial role in the conversion of carbohydrates into 5-HMF. The production of 5-HMF from carbohydrates in ionic liquids has been broadly studied in the presence of a variety of catalysts such as mineral and organic acids [36, 44, 68–70], acidic ionic liquids [29, 30, 71–73], transition metal ions [41, 43–45, 50, 51, 53, 74–79] or solid acid catalysts such as ion-exchange resins [20, 28, 34, 80], molecular sieves [31], zeolite [81], or carbonaceous materials [65, 82, 83], as summarized in Table 9.1.

Table 9.1 Catalytic production of 5-HMF from mono- and polysaccharides in different ionic liquids systems

Subs. (S)	Conc. (g/g IL)	Solvent system	Catalyst (C),	Dose (g C/g S)	T (°C)	Reaction time (min)	Conv. (%)	5-HMF yield (%)	References
Fructose	0.1	[BMIM][BF_4]/DMSO	Amberlyst 15 resin	1	80	1,920	–	87	[28]
	0.1	[EMIM][Cl]	$AlCl_3$	0.045	80	180	98	75	[41]
	0.1	[EMIM][Cl]	$FeCl_3$	0.054	80	180	99	75	[41]
	0.1	[EMIM][Cl]	CuCl	0.033	80	180	90	76	[41]
	0.1	[EMIM][Cl]	CuBr	0.074	80	180	99	75	[41]
	0.1	[EMIM][Cl]	$PtCl_2$	0.089	80	180	98	80	[41]
	0.06	Choline chloride/Citric acid	–	–	80	60	93.2	78	[36]
	0.1	[BMIM][Cl]	NHC/$CrCl_2$	0.061	100	360	–	96	[45]
	0.11	DMA-[EMIM][Cl] (80:20)	H_2SO_4,	0.033	100	120	–	84	[44]
	0.2	[BMIM][Cl]	WCl_6	0.22	50	240	–	63	[84]
	0.2	[BMIM][Cl]	WCl_6	0.22	50	240	–	61	[84]
	0.67	Choline chloride	$FeCl_3$	0.09	100	30	–	59	[37]
	0.036	[EMIM][HSO_4]/IBMK	–	–	100	30	100	88	[53]
	0.05	[BMIM][Cl]	Amberlyst 15 resin	1	80	10	99	83	[20]
	0.05	[BMIM][Cl]	Amberlyst 15 resin	1	100	3	98	82	[20]
	0.05	[BMIM][Cl]/Acetone	Amberlyst 15 resin	1	25	360	97	78	[34]
	0.05	[BMIM][Cl]	$CrCl_3$	0.3	MW, 100	1	–	78	[19]
	0.1	[BMIM][Cl]	No catalyst	–	120	50	93	63	[85]
	0.05	[BMIM][Cl]	SO_4^{2-}/ZrO_2	0.4	100	30	96	88	[86]
	0.05	[BMIM][Cl]	$GeCl_4$	0.12	100	5	–	92	[77]
	0.05	[BMIM][Cl]	$HfCl_4$	0.18	100	5	–	92	[77]
	0.1	DMSO-[BMIM][Cl] mixtures	LCC	0.5	110	10	92	82	[83]

	0.1	Tetraethylammonium chloride	$CrCl_3$	0.088	130	10	–	74	[87]
	0.18	[BMIM][Br]	No catalyst	–	100	60	99	92	[71]
	0.1	[EMIM][Cl]	$CuCl_2$	0.06	80	180	90	84	[88]
	1	Choline chloride	CO_2, 4 mPa	–	120	90	–	74	[38]
	0.1	[EMIM][Cl]	LPMSN-SO_3H	0.27	120	180	–	70	[61]
	0.01	[BMIM][Cl]	CSS	5	80	10	–	83	[82]
	0.1	[BMIM][Cl]	Lignosulfonic acid	0.25	100	10	99	93	[65]
Glucose	0.1	[EMIM][Cl]	$CrCl_2$	0.041	80	180	94	68	[41]
	0.1	[EMIM][Cl]	$CrCl_3$	0.053	80	180	72	46	[41]
	0.1	[BMIM][Cl]	NHC/$CrCl_2$	0.061	100	360	–	81	[45]
	0.11	DMA-[EMIM][Cl] (80:20)	$CrCl_2$	0.041	100	360	–	67	[44]
	0.2	[EMIM][BF_4]	$SnCl_4$	0.2	100	180	99	60	[50]
	0.67	Choline chloride	$CrCl_2$	0.068	110	30	–	45	[37]
	0.1	[BMIM][Cl]	$CrCl_3$	0.036	MW	1	–	91	[47]
	0.036	[BMIM][Cl]/toluene	$CrCl_3$	0.18	100	240	91	91	[53]
	0.05	[BMIM][Cl]	$CrCl_3$	0.089	MW, 140	0.5	99	71	[19]
	0.1	[BMIM][Cl]	$CeCl_3$	0.14	140	360	–	3	[51]
	0.1	[BMIM][Cl]	$Yb(OTf)_3$	0.35	140	360	–	24	[51]
	0.1	[EMIM][Cl]	Boric acid	0.28	120	180	95	41	[68]
	0.1	[BMIM][Cl]	$GeCl_4$	0.12	100	75	–	38	[77]
	0.1	Tetraethylammonium chloride	$CrCl_3$	0.089	130	10	–	71	[87]
	0.1	[EMIM][Cl]	$CrCl_2$	0.06	100	180	100	68	[88]
	0.1	[EMIM][Cl]	LPMSN-SO_3H-NH_2	0.27	120	180	–	13	[61]
	0.1	[EMIM][Cl]	$Cr(CO)_6$	0.12	120	1,440	–	49	[49]

(continued)

Table 9.1 (continued)

Subs. (S)	Conc. (g/g IL)	Solvent system	Catalyst (C),	Dose (g C/g S)	T (°C)	Reaction time (min)	Conv. (%)	5-HMF yield (%)	References
Inulin	0.055	Choline chloride/oxalic acid	–	–	90	120	100	55	[63]
	1	Choline chloride	$FeCl_3$	0.09	90	60	–	55	[37]
	0.1	[OMIM][Cl] + ethyl acetate	HCl + $CrCl_2$	0.4	120	60	–	54	[70]
	0.2	Choline chloride	CO_2, 4 mPa	–	120	900	–	41	[38]
	0.1	[BMIM][Cl]	Lignosulfonic acid	0.25	120	10	–	47	[65]
Cellulose	0.05	DMA-LiCl-[EMIM][Cl]	$CrCl_2$-HCl	0.17	140	120	–	54	[44]
	0.05	[BMIM][Cl]	$CrCl_3$	0.1	–	2	–	61	[47]
	0.05	[BMIM][Cl]	$CrCl_3$	0.3	MW, 150	10	–	54	[19]
	0.41	[EMIM][Cl]-H_2O	$CrCl_2$	0.068	120	360	–	89	[60]
	0.1	[EMIM][Cl]	$CrCl_2/RuCl_3$	0.068	120	120	–	60	[74]
	0.1	[EMIM][Cl]	LPMSN-SO_3H	0.27	120	210	–	19	[61]
	0.1	[EMIM][Cl]	MZNs	0.27	120	180	–	29	[89]
	0.05	[BMIM][Cl]	$MnCl_2$	0.033	120	60	84	63	[90]

MW Microwave heating, *LCC* Lignin-derived carbonaceous catalyst, *CSS* Carbonaceous sulfonated solid catalyst, *MZNs* Mesoporous zirconia nanoparticles

Fig. 9.5 Typical SEM images of carbon materials obtained by hydrothermal treatment of cellulose: (**a**) product with H_2SO_4 post-treatment, carbonaceous sulfonated solid (CSS); (**b**) product with KOH and H_2SO_4 post-treatment, activated carbonaceous sulfonated solid (a-CSS) (Reproduced with permission from [82]. Copyright © 2012 Wiley-VCH Verlag GmbH & Co. KGaA)

The dehydration of fructose into 5-HMF is an acid-promoted reaction so that homogeneous acid catalysts might be thought of as being efficient for the process. However, homogenous acids have serious practical problems related to product separation, solvent recycle and equipment corrosion. Therefore, solid acid catalysts are better candidates for 5-HMF preparation from fructose than homogeneous catalysts. Qi et al. proposed an efficient process for the 5-HMF production from fructose in [BMIM][Cl] ionic liquid by using a sulfonic ion-exchange resin as catalyst. These types of catalysts are limited to temperatures below 130 °C due to the thermal stability of the resin [20]. To overcome the temperature limitations of the polymeric resins, Qi et al. [82] synthesized a novel carbonaceous solid catalysts with $-SO_3H$, –COOH, and phenolic –OH groups by incomplete carbonization of cellulose followed by either sulfonation with H_2SO_4 to give carbonaceous sulfonated solid (CSS) material or by both chemical activation with KOH and sulfonation to give activated carbonaceous sulfonated solid (*a*-CSS) material (Fig. 9.5). The catalysts were shown to be effective for the catalytic conversion of fructose into 5-HMF, and a 5-HMF yield of 83 % could be obtained in [BMIM][Cl] with CSS catalyst at 80 °C for 10 min. Catalyst *a*-CSS exhibited a somewhat lower activity than that of CSS, even though *a*-CSS had a much larger surface area than that of CSS (0.5 VS. 514 m^2/g). The authors ascribed the lower activity of the *a*-CSS catalyst compared with that of the CSS catalyst to the lower concentration of $-SO_3H$ groups of the *a*-CSS catalyst (0.172 vs. 0.953 mmol/g).

Many Lewis acids such as metal chlorides have been used for the transformation of 5-HMF from carbohydrates in ionic liquids. The most frequently used Lewis acid catalyst, chromium (II or III) chloride, has been extensively employed for the production of 5-HMF from different carbohydrates such as fructose, glucose, sucrose, cellobiose, and cellulose [41, 45, 47, 53, 87, 91]. However, there are other metal salts that have been used for efficient conversion of sugars into 5-HMF in ionic liquids. Zhang et al. [77] established a new catalytic system based on germanium(IV) chloride ($GeCl_4$) for the conversion of carbohydrate into 5-HMF in [BMIM][Cl], and this system exhibited excellent catalytic activity for fructose and moderate activity for other carbohydrates such as glucose, cellobiose, sucrose, and cellulose, in terms of 5-HMF yield. Chan et al. [84] developed an efficient tungsten salt catalytic system for the conversion of fructose into 5-HMF with a yield of 63 % at low temperatures (RT to 50 °C) in [BMIM][Cl] ionic liquid. When the THF–[BMIM][Cl] biphasic system was applied as a continuous reaction process, a 5-HMF yield of above 80 % was obtained, indicating that the system could be suitable for the large-scale synthesis of 5-HMF from fructose.

Liu et al. [38] reported some exciting results for fructose dehydrated to 5-HMF that was achieved with a ChCl/CO_2 system. Yields of up to 72 % could be obtained with the environmental strategy and 5-HMF was found to be highly stable in the presence of ChCl, presumably through the formation of a hydrogen-bonding structure with ChCl to form a DES. The catalytic process allowed 5-HMF yields higher than 60 % with a high fructose initial content (up to 100 %), whereas such high 5-HMF yields could be obtained only from a fructose initial content lower than 20 % with usual methods. In usual processes, after extraction, 5-HMF is unavoidably contaminated with Lewis and Brønsted acids, however, in this process, a decrease of pH upon addition of CO_2 allows circumventing this problem because carbonic acid is readily converted to CO_2 and water when the CO_2 pressure is released.

9.2.3 Reaction Temperature

Compared with organic and aqueous solvents, the application of ILs for the dehydration of carbohydrates can significantly reduce the reaction temperature [20, 32]. Traditional synthetic methods for 5-HMF production from carbohydrates are normally carried out at temperatures ranging from 150 to 300 °C [4, 8, 15, 18], but the reaction temperatures required for the process in ionic liquids could be reduced to 80–150 °C [19, 41, 45, 47, 85], or even room temperature in some cases [34]. Normally the conversion of fructose needs a relative low temperature ranging from 80–120 °C since it is readily converted into 5-HMF through elimination of three molecules of water. Glucose is more difficult than fructose to be transformed into 5-HMF since it tends to form a stable six-membered pyranoside structure that has a low enolization rate, and its dehydration is generally carried out at 100–140 °C [19, 41, 45, 47]. The conversion of cellulose is more difficult for efficient conversion to 5-HMF than monosaccharides and requires high temperatures of about 150 °C in the presence of catalysts [19, 47, 53]. In general, lower temperatures lead to low 5-HMF yield (ca. 10–20 %), whereas higher temperatures promote formation of side-products and affects the 5-HMF yield so that an optimum temperature exists.

The reaction temperature is an important parameter for carbohydrate conversions since lower temperatures allow one to reduce the energy requirements. Chan et al. [84] was able to lower the dehydration reaction temperature to below 50 °C by using a system containing [BMIM][Cl] and metal salts. Chloride salts of zirconium (IV), titanium (IV), ruthenium (III), and tungsten (IV) or tungsten (VI), the most efficient chloride salt was that of tungsten (VI) that gave a 5-HMF yield of 63 % at 50 °C.

Reactions that can be promoted at ambient conditions are considered as one of the key goals among the 12 principles of green chemistry [92, 93]. Remarkably, it was found that efficient dehydration of fructose to 5-HMF could be carried out at room temperature in ionic liquids provided that the ionic liquid was brought about its melting point and the substrate was pre-dissolved in the ionic liquid before cooling to room temperature. Qi et al. [34] developed a green catalytic system for the production of 5-HMF from fructose catalyzed by a strong cation exchange resin, by the addition of different cosolvents such as DMSO, acetone, methanol, ethanol, ethyl acetate, and supercritical carbon dioxide to the ionic liquid 1-butyl-3-methylimidazolium chloride ([BMIM][Cl]). In a typical reaction, fructose was first dissolved in [BMIM][Cl] in a water bath at 80 °C for 20 min. After the mixture was cooled down to room temperature, the solution appeared gel-like with a very high viscosity. Subsequently, the catalyst (Amberlyst-15 sulfonic ion-exchange resin) and some amount of co-solvent were added for viscosity reduction and the reaction proceeded smoothly at 25 °C. Through addition of a co-solvent, the viscosities could be greatly reduced from an estimated value of 6,800 mPa s to values of around 2,000 mPa s, with the best results being obtained for acetone (1,850 mPa s) and ethyl acetate (1,930 mPa s). Interestingly, a gaseous co-solvent, such as CO_2 or supercritical CO_2 (>31 °C) was tried and found to provide comparable results to the organic solvents. Thus, use of CO_2 can possibly provide viscosity reduction and make it simple to regenerate the solvent system. Reductions in viscosity allowed the transformations to be carried out at close to room temperature. For this case [34], 5-HMF was obtained at yields of 78–82 %. The time for reaction was longer than in the previous work (6 h *vs.* 10 min) [20], but this method has the advantage of being performed at room temperature.

9.2.4 Reaction Time

In the production of 5-HMF from carbohydrates, increasing reaction time leads to an increase in the 5-HMF yield for cases in which 5-HMF decomposition does not occur. However, the yield of 5-HMF generally exhibits a maximum, indicating that decomposition of 5-HMF occurs [18, 20, 34]. The reaction mixture changing in color from yellow to deep brown is evidence for 5-HMF decomposition [5]. Generally, there are three pathways for the decomposition of 5-HMF in acid catalyzed dehydration of carbohydrates [3, 21], as depicted in Fig. 9.6. The first pathway is the rehydration of 5-HMF into levulinic acid and formic acid; the second one is the self-

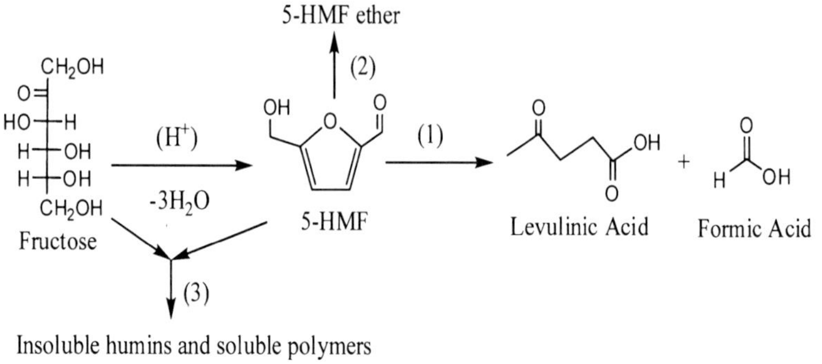

Fig. 9.6 5-hydroxymethylfurfural formation from sugars and 5-HMF decomposition pathways (fructose as an example) (Reproduced with permission from [20]. Copyright © 2009 Royal Society of Chemistry)

polymerization between 5-HMF molecules; and the third one is the cross-polymerization between 5-HMF and monosaccharides [21, 25, 40]. In non-aqueous ionic liquid systems, 5-HMF rehydration can be suppressed since the water present is limited to that of the dehydration of the carbohydrate. Control experiments without the carbohydrates exclude the possibility of self-polymerization of 5-HMF [20, 34]. Thus, it is thought that the decomposition of 5-HMF and the formation of humins are mainly due to the polymerization between 5-HMF and carbohydrates, which consume the initial carbohydrates and the formed 5-HMF, and hence reduce the 5-HMF selectivity [20].

The catalytic system and heating method have a large effect on the optimal reaction time for the production of 5-HMF. For example, when the dehydration of fructose was performed in [EMIM][Cl]-$CrCl_2$ system, an optimal 5-HMF yield of 83 % was obtained at 80 °C in 3 h reaction time [41], but a comparable 5-HMF yield was achieved in [BMIM][Cl]-Amberlyst 15 resin system at 80 °C in only 10 min reaction time [20]. Microwave heating was found to be able to accelerate the transformation of monosaccharides and di-/polysaccharides into 5-HMF and to shorten the reaction time. Qi et al. [19] investigated the catalytic conversion of glucose in [BMIM][Cl]-$CrCl_3$ system, and a 5-HMF yield of 71 % was achieved in 30 s for 96 % glucose conversion with microwave heating at 140 °C. In comparison, the reaction was performed with oil-bath heating and microwave heating at identical conditions. Microwave heating generally gives higher yields than with oil-bath heating, which was also reported by Li et al. who obtained a 91 % 5-HMF yield with microwave irradiation at 400 W in a reaction time of 1 min [47].

9.2.5 *Substrate Loading*

For practical applications, the concentration of feedstock in the reaction mixture should be as concentrated as possible. However, 5-HMF yield and selectivity

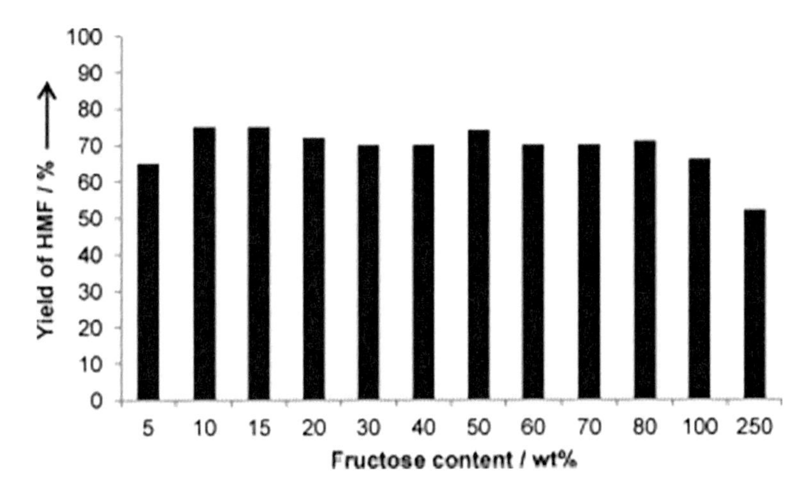

Fig. 9.7 Effect of the fructose content on the 5-HMF yield in ChCl/CO_2 system. Conditions: T = 120 °C, P_{CO2} 4 MPa, t = 90 min (Reproduced with permission from [38]. Copyright © 2012 Wiley-VCH Verlag GmbH & Co. KGaA)

normally decrease with increasing initial carbohydrate concentration since the formed 5-HMF can combine with monosaccharides, and cross-polymerize to form humins [25]. In aqueous systems including aqueous mixture systems, losses due to humin formation can be as high as 35 % for 18 wt% fructose solution, although this value decreases to 20 % for 4.5 wt% fructose solutions [25]. It was reported that in aqueous mixtures, the formation of humins for fructose conversions in ionic liquids can be largely inhibited [20, 45, 65]. Qi et al. [20] examined the effect of initial fructose concentration on fructose conversion and 5-HMF selectivity in [BMIM][Cl]. When the fructose/[BMIM][Cl] weight ratio was increased from 0.01 to 0.05, the 5-HMF selectivity changed slightly from 86 to 85.2 %. As the fructose/[BMIM][Cl] weight ratio was increased from 0.05 to 0.1, the 5-HMF selectivity decreased by about 5 %. Additional increases in the fructose/[BMIM][Cl] weight ratio did not lead to significant lowering of the 5-HMF selectivity, which is similar to the trends noted by Yong et al. for the conversion of fructose in [BMIM][Cl] with NHC/$CrCl_2$ (NHC=N-heterocyclic carbene) complexes as catalysts [45]. Xie et al. [65] studied the dehydration of fructose in [BMIM][Cl] catalyzed by lignosulfonic acid, and the 5-HMF yields of 92.7, 88.0, 84.1, and 73 % were obtained when 5, 10, 25, and 50 wt% of fructose were used, respectively. Increasing amounts of humins (0.007, 0.019, and 0.046 g) was observed for the increasing fructose concentrations, supporting the hypothesis that high concentrations of sugars and 5-HMF in ILs promote degradation and polymerization of 5-HMF.

Liu et al. [38] presented a promising choline chloride/CO_2 (ChCl/CO_2) system that could tolerate high fructose concentration with stable 5-HMF yield. The research group showed that as high as 100 wt% of fructose could be dehydrated into 5-HMF in ChCl/CO_2 system without substantial loss of yield (66 %) (Fig. 9.7).

This process provides a practical and eco-efficient synthesis method for 5-HMF. The authors ascribed the surprising tolerance of the ChCl/CO_2 system to a high content of fructose to a change of the physico-chemical properties of the ChCl/fructose mixture. Even in the presence of water, ChCl is capable of forming various deep eutectic solvents (DESs) with many hydrogen-bond donors such as polyols, urea, or carboxylic acid. After fructose is dehydrated into 5-HMF, ChCl and 5-HMF tend to form a new DES, which stabilizes the 5-HMF allowing its reactivity to be drastically reduced thus inhibiting its decomposition.

9.2.6 Water Content

The influence of water content on the production of 5-HMF from biomass has many aspects. Firstly, in the dehydration of fructose in ionic liquids, a small amount of water (5 wt%) present in the ionic liquid mixture has a negligible effect on the fructose conversion and 5-HMF yield, and it reduces the viscosity of the ionic liquid system and is beneficial to mass transfer [20, 36]. However, when the water content increases above about 5 wt%, the fructose and 5-HMF yield decrease significantly, which can be ascribed to the loss of catalytic activity due to the lowering of the dielectric constant of the reaction media by the addition of water [41]. Further, the presence of water in the reaction mixture favors the rehydration of 5-HMF to generate undesired products such as levulinic and formic acids. From one point of view, water should be completely avoided and removed from the production process of 5-HMF. On the other hand, water serves as a necessary reactant in the hydrolysis step in the conversion of oligosaccharides, thus an appropriate amount of water is required in the reaction system, although water has all the disadvantages connected to the dehydration reaction, and polysaccharides are no longer soluble in ionic liquids and precipitate from the ionic liquid solution when the water content exceeds a certain level [94, 95]. Therefore, water content should be carefully controlled in the conversion of polysaccharides in ionic liquids.

Considering that the addition of water can reduce the catalytic activity of the reaction system but increase the stability of glucose, Qi et al. developed an efficient two-step process for converting microcrystalline cellulose into 5-HMF with ionic liquids under mild conditions [95]. In the first step, cellulose was efficiently hydrolyzed by a strong acidic cation exchange resin in 1-ethyl-3-methyl imidazolium chloride ([EMIM][Cl]) with gradual addition of water. The addition of water probably allowed a balance to be achieved between glucose stability and cellulose solubility, since too much water will result in the precipitation of the cellulose substrate. Through the water addition technique, Qi et al. reported glucose yields above 80 % for converting microcrystalline cellulose under mild conditions [95]. Based on the high glucose yield, a second step was applied to produce 5-hydroxymethylfurfural by separating the resin from the reaction mixture and adding $CrCl_3$ as catalyst, which lead to a 5-HMF yield of 73 % based on cellulose substrate. The strategy described should be useful for efficient conversion of cellulose into 5-HMF as well as into other biomass-derived chemicals.

9.2.7 Reaction Kinetics

Reaction kinetic studies are not only important for uncovering the mechanism of 5-HMF formation on a molecular level, but also useful for process development studies to optimize process conditions and reactor configurations to obtain high 5-HMF yields. A number of kinetic studies have been carried out on the transformation of various mono- or poly-saccharides into 5-HMF in different reaction medias, but herein we will mainly overview the kinetics studies reported in the ionic liquids system.

Moreau et al. [29] investigated the reaction kinetic of the conversion of fructose into 5-HMF in ionic liquid, 3-methylimidazolium chloride acting as both solvent and catalyst. The reactions were carried out at the temperature range of 90–120 °C with the initial fructose concentration of 0.01–2.5 M. A reaction order of one was applied for the kinetic analysis of the dehydration of fructose, obtaining an activation energy of 143 kJ/mol for the reaction of fructose to 5-HMF. Wei et al. [96] reported the conversion of fructose to 5-HMF in 1-butyl-3-methyl imidazolium chloride using $IrCl_3$ as the catalyst. The reactions were performed at temperatures ranging from 80 to 120 °C and a fructose concentration of 10 wt%. A kinetic network involving fructose conversion to 5-HMF and byproducts was proposed to model the experimental data using first order reaction kinetics. The activation energies for fructose conversion were estimated to be 165 and 124 kJ/mol for the formation of 5-HMF and formation of byproducts, respectively. Based on a reaction order of one, Qi et al. [20] reported an activation energy of 65 kJ/mol for the formation of 5-HMF from fructose, which was performed in 1-butyl-3-methyl imidazolium chloride in the presence of a strong cation exchange resin Amberlyst 15A. The authors contributed the lower activation energy to higher acidity of the strong sulfonic ion-exchange resin over that of the other catalysts.

Compared with fructose, fewer studies have been carried out for the kinetic study on the dehydration of glucose to 5-HMF, and these studies reported so far have been performed in aqueous systems, except for one work [19], which was performed in the ionic liquid 1-butyl-3-methyl imidazolium chloride using $CrCl_3$ as catalyst under microwave irradiation. In that reaction system, an activation energy of 115 kJ/mol was reported using first-order reaction approach to model the experimental data [19]. The value of the activation energy of this work was comparable with the values (108–137 kJ/mol) reported for the dehydration of glucose in water with sulfuric acid as catalyst. However, the pre-exponential factor determined in that work was 3.5×10^{14} min^{-1}, which was 3–8 orders of magnitude larger than those of previous works. The difference in the pre-exponential factors should be a result of the enhanced effective collision among reactant molecules in the homogeneous phase [19, 97]. It has been found that the activation energy for the side reactions that glucose decomposed to undesired humins is higher than for the desired reaction of glucose to 5-HMF, indicating that lower temperatures favor 5-HMF formation [98]. Therefore, the reaction temperature should be optimized for 5-HMF production.

In many kinetic studies of carbohydrates, 5-HMF is involved as an intermediate or decomposition product in reaction pathway to obtain sugars or target compounds and so the kinetics of 5-HMF is not well known. The decomposition of cellulose in aqueous systems is often modeled with first-order approaches that give activation energies for the decomposition of cellulose in water to be between 140 and 190 kJ/mol [2]. Vanoye et al. [99] conducted a kinetic study on the acid-catalyzed hydrolysis of cellobiose in [EMIM][Cl] with 3.5 mM methanesulfonic acid catalyst in the presence of small amounts of water (3.5 mM) as the cosolvent. Activation energies of 111 and 102 kJ/mol were obtained for cellobiose hydrolysis and glucose degradation, respectively. Dee and Bell [100] performed kinetic studies on the cellulose hydrolysis in a batch set up in ionic liquids ([EMIM][Cl] and [BMIM][Cl]) with mineral acid catalysts, and glucose, 5-HMF and cellobiose were the primary reaction products. The reaction kinetics for glucose formation were fit with first-order reaction rate equation and an activation energy of 96 kJ/mol was determined.

For 5-HMF production, the undesired decomposition of 5-HMF to levulinic acid, formic acid, and humins should be suppressed as much as possible. To gain insight in the reactivity of 5-HMF, kinetic studies using 5-HMF as the starting material should be investigated. However, kinetic studies on the decomposition of 5-HMF in ionic liquids system thus far are not available. Instead, there are many works that focus on the reaction kinetic of 5-HMF decomposition with 5-HMF as starting material in aqueous systems and organic-aqueous mixtures. For the rehydration of 5-HMF to levulinic acid, activation energies vary from 47 to 210 kJ/mol, and the activation energies range between 100 and 125 kJ/mol for the decomposition of 5-HMF to humins. Asghari et al. [21] reported a higher activation energy for the formation of humins from 5-HMF (122 kJ/mol) than that for the decomposition of 5-HMF to levulinic acid (94 kJ/mol) catalyzed by HCl in subcritical water. On the contrary, Girisuta et al. [101] found similar activation energies of about 111 ± 2 kJ/mol for both the rehydration of 5-HMF to levulinic acid and the side reaction of decomposition of 5-HMF to humins.

9.3 Separation of 5-HMF from Ionic Liquids System

In most of the reported studies for the production of 5-HMF from biomass in ionic liquids, the 5-HMF was obtained in solution, and the yield of 5-HMF was determined with HPLC or GC without isolation. However, it is of great importance to develop an efficient isolation method for the synthesis of 5-HMF. Wei et al. [102] developed a novel entrainer-intensified vacuum reactive distillation process for the separation of 5-HMF from ionic liquids 1-octyl-3-methylimidazolium chloride

([OMIM][Cl]) that involves heating reaction mixtures under vacuum (ca.300 Pa) to 150–180 °C under a flow of an entrainer (nitrogen), and 95 % of the 5-HMF was recovered from the reaction mixture after 10 min at 180 °C. However, distillation is not a favorable method for the separation of 5-HMF from ionic liquids system, not only because both 5-HMF and ionic liquids have high boiling point, but also because 5-HMF is very reactive and tends to decompose at high temperatures.

Monosaccharides such as fructose and glucose have polarities similar to those of ionic liquids, while 5-HMF has different properties from ionic liquids, which allows separation of 5-HMF from ionic liquids by liquid-liquid extraction. However, an efficient separation solvent is needed to reduce the amount of solvent (low solvent/feed ratio) required for extraction of 5-HMF from ionic liquid solutions. An ideal extraction solvent should be immiscible with ionic liquids and have a large affinity for 5-HMF. Further, it should be separable from the product, with a method that does not require a large amount of energy. Therefore, it should have a relatively low boiling point to avoid thermal decomposition or polymerization of the 5-HMF and to reduce the energy costs. Many organic solvents such as MIBK (methyl isobutyl ketone) [53], ethyl acetate [20, 63], diethyl ether [29, 45], THF [84], and acetone [60] have been reported to be efficient extraction solvents. Qi et al. [20] tested the separation of 5-HMF from [BMIM][Cl] after reaction by extracting five times with 8 ml of ethyl acetate after 0.5 g of water was added to reduce the viscosity of the ionic liquid. [BMIM][Cl] and fructose were found to be insoluble in ethyl acetate and 5-HMF was the sole product in ethyl acetate phase, and above 95 % of 5-HMF was separated from the reaction mixture after five times extraction. Hu et al. [36] presented 5-HMF synthesis from fructose in a biphasic system of ethyl acetate and choline chloride (ChoCl) with citric acid, and the continuous in-situ extraction of 5-HMF with ethyl acetate led to a 15 % enhancement in the 5-HMF yield in comparison with a neat ionic liquid. For larger-scale production, more efforts have to be made on efficient separation techniques.

9.4 Conclusions and Perspectives

Growing concerns on global warming and the depletion of traditional resources have driven us to look for green and sustainable energy sources. As a biomass-derived platform chemical, the production of 5-HMF has been one of the hottest topics in biomass utilization in these few years, and a large number of publications have appeared. Many works focus on improving 5-HMF yield by applying different substrates, different catalysts, different solvents, different reaction conditions and extraction methods. Application of ionic liquids has been shown to be efficient for the preparation of 5-HMF, and the rapid growing interest in 5-HMF production with ionic liquids from biomass holds great promise for the future. However, this field is still in its infancy and some issues concerning the chemistry and process engineering should be addressed as follows for practical large-scale processes:

1. The high viscosity of ionic liquids leads to some disadvantages in the process, and most of the present used ionic liquids are prepared with petroleum-derived chemicals as starting material. The synthesis and use of ionic liquids in which cation and/or anion are originated from natural resources should be developed in the 5-HMF production. Choline based ionic liquids are good examples that have been reported [38, 63]. The other possible alternatives include glycerol, diethylethanolamine, or amino acid based ionic liquids. The use of these biorenewable ionic liquids permits the processes to be even more green and sustainable. Chapters 2 and 4 provide some exciting new avenues for ionic liquids that dissolve cellulose and that have low viscosity.
2. In the production of 5-HMF from carbohydrates in various solvents, an important side reaction is the formation of soluble or insoluble polymers known as humins that greatly affect 5-HMF yield. These humins are formed from different intermediates in the reaction and their formation increases with increasing substrate concentration. The formation of humins remains a troublesome problem for 5-HMF synthesis even in ionic liquids. More investigations are required to make clear the detailed formation mechanism of humins and restrain its formation. Interesting results have been reported for the addition of acetonitrile as co-solvent to the glucose – [BMIM][Cl] or [EMIM][Cl] reaction system. Addition of acetonitrile to the reaction system inhibited the formation of humins and enhanced the glucose conversion up to 99 %, with a 98 % 5-HMF selectivity. Moderate amounts of humins of up to 20 % were formed in the absence of acetonitrile as a co-solvent [103]. These results could help to reveal the formation mechanism of humins and find a way to inhibit their formation and to provide a practical method to produce 5-HMF from carbohydrates.
3. To date, efficient production of 5-HMF is mainly from fructose, and the only proposed efficient catalytic system for production of 5-HMF from glucose and cellulose is chromium or tin chlorides that are hazardous to environment. Therefore, more efforts are needed for more environmentally-friendly catalysts that are both efficient and renewable. Functionalized carbonaceous materials possibly could provide new avenues for research in this area [104–106].
4. The starting materials used in most of studies on 5-HMF production are model chemicals such as sugars (glucose, fructose and sucrose) and microcrystalline cellulose, but the use of crude lignocellulosic biomass is relatively lack. However, the composition complexity of the crude biomass will require that the process of 5-HMF production in ionic liquids be greatly different from that when pure chemicals are used. Consequently, more work is required for practical applications in the treatment of actual biomass for 5-HMF production.
5. More works are needed for the separation of 5-HMF product from ionic liquid systems, especially for continuous operation. The extraction of 5-HMF from ionic liquids by supercritical carbon dioxide with co-solvent addition may be a possibility for 5-HMF separation and the creative use of CO_2 in forming biphasic systems for separations could provide both an environmental and efficient separation method.

6. Since 5-HMF is unstable, it might be further reacted to produce other chemicals such as furan and esters, and these derivatives would probably be more soluble in CO_2 and thus it might be possible to extract them with supercritical CO_2.

References

1. Roman-Leshkov Y, Chheda JN, Dumesic JA. Phase modifiers promote efficient production of hydroxymethylfurfural from fructose. Science. 2006;312(5782):1933–7.
2. van Putten RJ, van der Waal JC, de Jong E, Rasrendra CB, Heeres HJ, de Vries JG. Hydroxymethylfurfural, a versatile platform chemical made from renewable resources. Chem Rev. 2013;113(3):1499–597.
3. Lewkowski J. Synthesis, chemistry and applications of 5-hydroxymethyl-furfural and its derivatives. Arkivoc. 2001;2:17–54.
4. Bicker M, Kaiser D, Ott L, Vogel H. Dehydration of D-fructose to hydroxymethylfurfural in sub- and supercritical fluids. J Supercrit Fluids. 2005;36(2):118–26.
5. Gandini A. Furans as offspring of sugars and polysaccharides and progenitors of a family of remarkable polymers: a review of recent progress. Polym Chem. 2010;1(3):245–51.
6. Ishida H, Seri K. Catalytic activity of lanthanoide(III) ions for dehydration of D-glucose to 5-(hydroxymethyl)furfural. J Mol Catal A Chem. 1996;112(2):L163–5.
7. Carlini C, Patrono P, Galletti AMR, Sbrana G. Heterogeneous catalysts based on vanadyl phosphate for fructose dehydration to 5-hydroxymethyl-2-furaldehyde. Appl Catal Gen. 2004;275(1–2):111–8.
8. Asghari FS, Yoshida H. Acid-catalyzed production of 5-hydroxymethyl furfural from D-fructose in subcritical water. Ind Eng Chem Res. 2006;45(7):2163–73.
9. Seri K, Inoue Y, Ishida H. Catalytic activity of Lanthanide(III) ions for the dehydration of hexose to 5-hydroxymethyl-2-furaldehyde in water. Bull Chem Soc Jpn. 2001;74(6):1145–50.
10. Carlini C, Giuttari M, Galletti AMR, Sbrana G, Armaroli T, Busca G. Selective saccharides dehydration to 5-hydroxymethyl-2-furaldehyde by heterogeneous niobium catalysts. Appl Catal Gen. 1999;183(2):295–302.
11. Moreau C, Durand R, Razigade S, Duhamet J, Faugeras P, Rivalier P, Ros P, Avignon G. Dehydration of fructose to 5-hydroxymethylfurfural over H-mordenites. Appl Catal Gen. 1996;145(1–2):211–24.
12. Musau RM, Munavu RM. The preparation of 5-hydroxymethyl-2-furaldehyde (Hmf) from D-fructose in the presence of DMSO. Biomass. 1987;13(1):67–74.
13. Nakamura Y, Morikawa S. The dehydration of D-fructose to 5-hydroxymethyl-2-furaldehyde. Bull Chem Soc Jpn. 1980;53(12):3705–6.
14. Seri K, Inoue Y, Ishida H. Highly efficient catalytic activity of lanthanide(III) ions for conversion of saccharides to 5-hydroxymethyl-2-furfural in organic solvents. Chem Lett. 2000;1:22–3.
15. Qi XH, Watanabe M, Aida TM, Smith RL. Selective conversion of D-fructose to 5-hydroxymethylfurfural by ion-exchange resin in acetone/dimethyl sulfoxide solvent mixtures. Ind Eng Chem Res. 2008;47(23):9234–9.
16. Bicker M, Hirth J, Vogel H. Dehydration of fructose to 5-hydroxymethylfurfural in sub- and supercritical acetone. Green Chem. 2003;5(2):280–4.
17. Huber GW, Chheda JN, Barrett CJ, Dumesic JA. Production of liquid alkanes by aqueous-phase processing of biomass-derived carbohydrates. Science. 2005;308(5727):1446–50.
18. Qi XH, Watanabe M, Aida TM, Smith RL. Catalytic dehydration of fructose into 5-hydroxymethylfurfural by ion-exchange resin in mixed-aqueous system by microwave heating. Green Chem. 2008;10(7):799–805.

19. Qi XH, Watanabe M, Aida TM, Smith RL. Fast transformation of glucose and di-/polysaccharides into 5-hydroxymethylfurfural by microwave heating in an ionic liquid/catalyst system. ChemSusChem. 2010;3(9):1071–7.
20. Qi XH, Watanabe M, Aida TM, Smith RL. Efficient process for conversion of fructose to 5-hydroxymethylfurfural with ionic liquids. Green Chem. 2009;11(9):1327–31.
21. Asghari FS, Yoshida H. Kinetics of the decomposition of fructose catalyzed by hydrochloric acid in subcritical water: formation of 5-hydroxymethylfurfural, levulinic, and formic acids. Ind Eng Chem Res. 2007;46(23):7703–10.
22. Qi XH, Watanabe M, Aida TM, Smith RL. Catalytical conversion of fructose and glucose into 5-hydroxymethylfurfural in hot compressed water by microwave heating. Catal Commun. 2008;9(13):2244–9.
23. Asghari FS, Yoshida H. Dehydration of fructose to 5-hydroxymethylfurfural in sub-critical water over heterogeneous zirconium phosphate catalysts. Carbohydr Res. 2006;341(14):2379–87.
24. Watanabe M, Aizawa Y, Iida T, Aida TM, Levy C, Sue K, Inomata H. Glucose reactions with acid and base catalysts in hot compressed water at 473 K. Carbohydr Res. 2005;340(12):1925–30.
25. Kuster BFM. 5-Hydroxymethylfurfural (HMF). A review focussing on its manufacture. Starch. 1990;42:314–21.
26. Kroon MC, Toussaint VA, Shariati A, Florusse LJ, van Spronsen J, Witkamp GJ, Peters CJ. Crystallization of an organic compound from an ionic liquid using carbon dioxide as anti-solvent. Green Chem. 2008;10(3):341–4.
27. Fayet C, Gelas J. Nouvelle méthode de préparation du 5-hydroxyméthyl-2-furaldéhyde par action de sels d'ammonium ou d'immonium sur les mono-, oligo- et poly-saccharides. Accès direct aux 5-halogénométhyl-2-furaldéhydes. Carbohydr Res. 1983;122(1):59–68.
28. Lansalot-Matras C, Moreau C. Dehydration of fructose into 5-hydroxymethylfurfural in the presence of ionic liquids. Catal Commun. 2003;4(10):517–20.
29. Moreau C, Finiels A, Vanoye L. Dehydration of fructose and sucrose into 5-hydroxymethylfurfural in the presence of 1-H-3-methyl imidazolium chloride acting both as solvent and catalyst. J Mol Catal A Chem. 2006;253(1–2):165–9.
30. Jadhav AH, Kim H, Hwang IT. Efficient selective dehydration of fructose and sucrose into 5-hydroxymethylfurfural (HMF) using dicationic room temperature ionic liquids as a catalyst. Catal Commun. 2012;21:96–103.
31. Guo XC, Cao Q, Jiang YJ, Guan J, Wang XY, Mu XD. Selective dehydration of fructose to 5-hydroxymethylfurfural catalyzed by mesoporous SBA-15-SO3H in ionic liquid BmimCl. Carbohydr Res. 2012;351:35–41.
32. Zakrzewska ME, Bogel-Lukasik E, Bogel-Lukasik R. Ionic liquid-mediated formation of 5-hydroxymethylfurfural – a promising biomass-derived building block. Chem Rev. 2011;111(2):397–417.
33. Lai LK, Zhang YG. The effect of imidazolium ionic liquid on the dehydration of fructose to 5-hydroxymethylfurfural, and a room temperature catalytic system. ChemSusChem. 2010;3(11):1257–9.
34. Qi XH, Watanabe M, Aida TM, Smith RL. Efficient catalytic conversion of fructose into 5-hydroxymethylfurfural in ionic liquids at room temperature. ChemSusChem. 2009;2(10):944–6.
35. Shi CY, Zhao YL, Xin JY, Wang JQ, Lu XM, Zhang XP, Zhang SJ. Effects of cations and anions of ionic liquids on the production of 5-hydroxymethylfurfural from fructose. Chem Commun. 2012;48(34):4103–5.
36. Hu SQ, Zhang ZF, Zhou YX, Han BX, Fan HL, Li WJ, Song JL, Xie Y. Conversion of fructose to 5-hydroxymethylfurfural using ionic liquids prepared from renewable materials. Green Chem. 2008;10(12):1280–3.

37. Ilgen F, Ott D, Kralisch D, Reil C, Palmberger A, Konig B. Conversion of carbohydrates into 5-hydroxymethylfurfural in highly concentrated low melting mixtures. Green Chem. 2009;11 (12):1948–54.
38. Liu F, Barrault J, Vigier KD, Jerome F. Dehydration of highly concentrated solutions of fructose to 5-hydroxymethylfurfural in a cheap and sustainable choline chloride/carbon dioxide system. ChemSusChem. 2012;5(7):1223–6.
39. Román-Leshkov Y, Barrett CJ, Liu ZY, Dumesic JA. Production of dimethylfuran for liquid fuels from biomass-derived carbohydrates. Nature. 2007;447(7147):982–5.
40. Chheda JN, Roman-Leshkov Y, Dumesic JA. Production of 5-hydroxymethylfurfural and furfural by dehydration of biomass-derived mono- and poly-saccharides. Green Chem. 2007;9(4):342–50.
41. Zhao HB, Holladay JE, Brown H, Zhang ZC. Metal chlorides in ionic liquid solvents convert sugars to 5-hydroxymethylfurfural. Science. 2007;316(5831):1597–600.
42. Yang FL, Liu QS, Bai XF, Du YG. Conversion of biomass into 5-hydroxymethylfurfural using solid acid catalyst. Bioresour Technol. 2011;102(3):3424–9.
43. Zhang ZH, Zhao ZBK. Microwave-assisted conversion of lignocellulosic biomass into furans in ionic liquid. Bioresour Technol. 2010;101(3):1111–4.
44. Binder JB, Raines RT. Simple chemical transformation of lignocellulosic biomass into furans for fuels and chemicals. J Am Chem Soc. 2009;131(5):1979–85.
45. Yong G, Zhang YG, Ying JY. Efficient catalytic system for the selective production of 5-hydroxymethylfurfural from glucose and fructose. Angew Chem Int Ed. 2008;47 (48):9345–8.
46. Zhang ZH, Zhao ZB. Production of 5-hydroxymethylfurfural from glucose catalyzed by hydroxyapatite supported chromium chloride. Bioresour Technol. 2011;102(4):3970–2.
47. Li CZ, Zhang ZH, Zhao ZBK. Direct conversion of glucose and cellulose to 5-hydroxymethylfurfural in ionic liquid under microwave irradiation. Tetrahedron Lett. 2009;50(38):5403–5.
48. Florence TM. The speciation of trace-elements in waters. Talanta. 1982;29(5):345–64.
49. He JH, Zhang YT, Chen EYX. Chromium(0) nanoparticles as effective catalyst for the conversion of glucose into 5-hydroxymethylfurfural. ChemSusChem. 2013;6(1):61–4.
50. Hu SQ, Zhang ZF, Song JL, Zhou YX, Han BX. Efficient conversion of glucose into 5-hydroxymethylfurfural catalyzed by a common Lewis acid SnCl4 in an ionic liquid. Green Chem. 2009;11(11):1746–9.
51. Stahlberg T, Sorensen MG, Riisager A. Direct conversion of glucose to 5-(hydroxymethyl) furfural in ionic liquids with lanthanide catalysts. Green Chem. 2010;12(2):321–5.
52. Qi XH, Watanabe M, Aida TM, Smith RL. Synergistic conversion of glucose into 5-hydroxymethylfurfural in ionic liquid-water mixtures. Bioresour Technol. 2012;109:224–8.
53. Lima S, Neves P, Antunes MM, Pillinger M, Ignatyev K, Valente AA. Conversion of mono/di/polysaccharides into furan compounds using 1-alkyl-3-methylimidazolium ionic liquids. Appl Catal Gen. 2009;363(1–2):93–9.
54. Rinaldi R, Schuth F. Acid hydrolysis of cellulose as the entry point into biorefinery schemes. ChemSusChem. 2009;2(12):1096–107.
55. Saka S, Ueno T. Chemical conversion of various celluloses to glucose and its derivatives in supercritical water. Cellulose. 1999;6(3):177–91.
56. Ishikawa Y, Saka S. Chemical conversion of cellulose as treated in supercritical methanol. Cellulose. 2001;8(3):189–95.
57. Li CZ, Wang Q, Zhao ZK. Acid in ionic liquid: an efficient system for hydrolysis of lignocellulose. Green Chem. 2008;10(2):177–82.
58. Li CZ, Zhao ZK. Efficient acid-catalyzed hydrolysis of cellulose in ionic liquid. Adv Synth Catal. 2007;349:1847–50.

59. Yu S, Brown HM, Huang XW, Zhou XD, Amonette JE, Zhang ZC. Single-step conversion of cellulose to 5-hydroxymethylfurfural (HMF), a versatile platform chemical. Appl Catal Gen. 2009;361(1–2):117–22.
60. Zhang YT, Du HB, Qian XH, Chen EYX. Ionic liquid-water mixtures: enhanced K-w for efficient cellulosic biomass conversion. Energy Fuel. 2010;24:2410–7.
61. Peng WH, Lee YY, Wu CN, Wu KCW. Acid–base bi-functionalized, large-pored mesoporous silica nanoparticles for cooperative catalysis of one-pot cellulose-to-HMF conversion. J Mater Chem. 2012;22(43):23181–5.
62. Sirisansaneeyakul S, Worawuthiyanan N, Vanichsriratana W, Srinophakun P, Chisti Y. Production of fructose from inulin using mixed inulinases from Aspergillus niger and Candida guilliermondii. World J Microbiol Biotechnol. 2007;23(4):543–52.
63. Hu SQ, Zhang ZF, Zhou YX, Song JL, Fan HL, Han BX. Direct conversion of inulin to 5-hydroxymethylfurfural in biorenewable ionic liquids. Green Chem. 2009;11(6):873–7.
64. Qi XH, Watanabe M, Aida TM, Smith RL. Efficient one-pot production of 5-hydroxymethylfurfural from inulin in ionic liquids. Green Chem. 2010;12(10):1855–60.
65. Xie HB, Zhao ZK, Wang Q. Catalytic conversion of inulin and fructose into 5-hydroxymethylfurfural by lignosulfonic acid in ionic liquids. ChemSusChem. 2012;5 (5):901–5.
66. Benoit M, Brissonnet Y, Guelou E, Vigier KD, Barrault J, Jerome F. Acid-catalyzed dehydration of fructose and inulin with glycerol or glycerol carbonate as renewably sourced co-solvent. ChemSusChem. 2010;3(11):1304–9.
67. Ranoux A, Djanashvili K, Arends IWCE, Hanefeld U. 5-Hydroxymethylfurfural synthesis from hexoses is autocatalytic. ACS Catal. 2013;3(4):760–3.
68. Stahlberg T, Rodriguez-Rodriguez S, Fristrup P, Riisager A. Metal-free dehydration of glucose to 5-(hydroxymethyl)furfural in ionic liquids with boric acid as a promoter. Chemistry. 2011;17(5):1456–64.
69. Sievers C, Musin I, Marzialetti T, Olarte MBV, Agrawal PK, Jones CW. Acid-catalyzed conversion of sugars and furfurals in an ionic-liquid phase. ChemSusChem. 2009;2 (7):665–71.
70. Yi YB, Lee JW, Hong SS, Choi YH, Chung CH. Acid-mediated production of hydroxymethylfurfural from raw plant biomass with high inulin in an ionic liquid. J Ind Eng Chem. 2011;17(1):6–9.
71. Li YN, Wang JQ, He LN, Yang ZZ, Liu AH, Yu B, Luan CR. Experimental and theoretical studies on imidazolium ionic liquid-promoted conversion of fructose to 5-hydroxymethylfurfural. Green Chem. 2012;14(10):2752–8.
72. Tao FR, Song HL, Chou LJ. Efficient conversion of cellulose into furans catalyzed by metal ions in ionic liquids. J Mol Catal A Chem. 2012;357:11–8.
73. Liu W, Holladay J. Catalytic conversion of sugar into hydroxymethylfurfural in ionic liquids. Catal Today. 2013;200:106–16.
74. Kim B, Jeong J, Lee D, Kim S, Yoon HJ, Lee YS, Cho JK. Direct transformation of cellulose into 5-hydroxymethyl-2-furfural using a combination of metal chlorides in imidazolium ionic liquid. Green Chem. 2011;13(6):1503–6.
75. Binder JB, Cefali AV, Blank JJ, Raines RT. Mechanistic insights on the conversion of sugars into 5-hydroxymethylfurfural. Energy Environ Sci. 2010;3(6):765–71.
76. Su Y, Brown HM, Huang XW, Zhou XD, Amonette JE, Zhang ZC. Single-step conversion of cellulose to 5-hydroxymethylfurfural (HMF), a versatile platform chemical. Appl Catal Gen. 2009;361(1–2):117–22.
77. Zhang ZH, Wang QA, Xie HB, Liu WJ, Zhao ZB. Catalytic conversion of carbohydrates into 5-hydroxymethylfurfural by germanium(IV) chloride in ionic liquids. ChemSusChem. 2011;4(1):131–8.
78. Pidko EA, Degirmenci V, Hensen EJM. On the mechanism of Lewis acid catalyzed glucose transformations in ionic liquids. ChemCatChem. 2012;4(9):1263–71.

79. Zhang ZH, Liu B, Zhao ZB. Conversion of fructose into 5-HMF catalyzed by GeCl4 in DMSO and [Bmim]Cl system at room temperature. Carbohydr Polym. 2012;88(3):891–5.
80. Takagaki A, Ohara M, Nishimura S, Ebitani K. A one-pot reaction for biorefinery: combination of solid acid and base catalysts for direct production of 5-hydroxymethylfurfural from saccharides. Chem Commun 2009;45(41):6276–8.
81. Jadhav H, Taarning E, Pedersen CM, Bols M. Conversion of D-glucose into 5-hydroxymethylfurfural (HMF) using zeolite in [Bmim]Cl or tetrabutylammonium chloride (TBAC)/CrCl2. Tetrahedron Lett. 2012;53(8):983–5.
82. Qi XH, Guo HX, Li LY, Smith RL. Acid-catalyzed dehydration of fructose into 5-hydroxymethylfurfural by cellulose-derived amorphous carbon. ChemSusChem. 2012;5 (11):2215–20.
83. Guo F, Fang Z, Zhou TJ. Conversion of fructose and glucose into 5-hydroxymethylfurfural with lignin-derived carbonaceous catalyst under microwave irradiation in dimethyl sulfoxide-ionic liquid mixtures. Bioresour Technol. 2012;112:313–8.
84. Chan JYG, Zhang YG. Selective conversion of fructose to 5-hydroxymethylfurfural catalyzed by tungsten salts at low temperatures. ChemSusChem. 2009;2(8):731–4.
85. Cao Q, Guo XC, Yao SX, Guan J, Wang XY, Mu XD, Zhang DK. Conversion of hexose into 5-hydroxymethylfurfural in imidazolium ionic liquids with and without a catalyst. Carbohydr Res. 2011;346(7):956–9.
86. Qi XH, Guo HX, Li LY, Smith RL. Efficient conversion of fructose to 5-hydroxymethylfurfural catalyzed by sulfated zirconia in ionic liquids. Ind Eng Chem Res. 2011;50(13):7985–9.
87. Hu L, Sun Y, Lin L. Efficient conversion of glucose into 5-hydroxymethylfurfural by chromium(III) chloride in inexpensive ionic liquid. Ind Eng Chem Res. 2012;51 (3):1099–104.
88. Liu DJ, Chen EYX. Ubiquitous aluminum alkyls and alkoxides as effective catalysts for glucose to HMF conversion in ionic liquids. Appl Catal Gen. 2012;435:78–85.
89. Kuo IJ, Suzuki N, Yamauchi Y, Wu KCW. Cellulose-to-HMF conversion using crystalline mesoporous titania and zirconia nanocatalysts in ionic liquid systems. RSC Adv. 2013;3 (6):2028–34.
90. Shi JC, Gao HY, Xia YM, Li W, Wang HJ, Zheng CG. Efficient process for the direct transformation of cellulose and carbohydrates to 5-(hydroxymenthyl)furfural with dual-core sulfonic acid ionic liquids and co-catalysts. RSC Adv. 2013;3(21):7782–90.
91. Zhang YM, Pidko EA, Hensen EJM. Molecular aspects of glucose dehydration by chromium chlorides in ionic liquids. Chemistry. 2011;17(19):5281–8.
92. Anastas PT, Zimmerman JB. Design through the 12 principles of green engineering. Environ Sci Tech. 2003;37(5):94A–101.
93. Tang SY, Bourne RA, Poliakoff M, Smith RL. The 24 principles of green engineering and green chemistry: "IMPROVEMENTS PRODUCTIVELY". Green Chem. 2008;10(3):276–7.
94. Swatloski RP, Spear SK, Holbrey JD, Rogers RD. Dissolution of cellose with ionic liquids. J Am Chem Soc. 2002;124(18):4974–5.
95. Qi XH, Watanabe M, Aida TM, Smith RL. Catalytic conversion of cellulose into 5-hydroxymethylfurfural in high yields via a two-step process. Cellulose. 2011;18 (5):1327–33.
96. Wei ZJ, Li Y, Thushara D, Liu YX, Ren QL. Novel dehydration of carbohydrates to 5-hydroxymethylfurfural catalyzed by Ir and Au chlorides in ionic liquids. J Taiwan Inst Chem Eng. 2011;42(2):363–70.
97. Hosseini M, Stiasni N, Barbieri V, Kappe CO. Microwave-assisted asymmetric organocatalysis. A probe for nonthermal microwave effects and the concept of simultaneous cooling. J Org Chem. 2007;72(4):1417–24.
98. Girisuta B, Janssen L, Heeres HJ. A kinetic study on the conversion of glucose to levulinic acid. Chem Eng Res Des. 2006;84(A5):339–49.

99. Vanoye L, Fanselow M, Holbrey JD, Atkins MP, Seddon KR. Kinetic model for the hydrolysis of lignocellulosic biomass in the ionic liquid, 1-ethyl-3-methyl-imidazolium chloride. Green Chem. 2009;11(3):390–6.
100. Dee SJ, Bell AT. A study of the acid-catalyzed hydrolysis of cellulose dissolved in ionic liquids and the factors influencing the dehydration of glucose and the formation of humins. ChemSusChem. 2011;4(8):1166–73.
101. Girisuta B, Janssen LPBM, Heeres HJ. A kinetic study on the decomposition of 5-hydroxymethylfurfural into levulinic acid. Green Chem. 2006;8(8):701–9.
102. Wei Z, Liu Y, Thushara D, Ren Q. Entrainer-intensified vacuum reactive distillation process for the separation of 5-hydroxylmethylfurfural from the dehydration of carbohydrates catalyzed by a metal salt-ionic liquid. Green Chem. 2012;14(4):1220–6.
103. Chidambaram M, Bell AT. A two-step approach for the catalytic conversion of glucose to 2,5-dimethylfuran in ionic liquids. Green Chem. 2010;12(7):1253–62.
104. Titirici MM, White RJ, Falco C, Sevilla M. Black perspectives for a green future: hydrothermal carbons for environment protection and energy storage. Energy Environ Sci. 2012;5(5):6796–822.
105. Qi XH, Li LY, Tan TF, Chen WT, Smith RL. Adsorption of 1-butyl-3-methylimidazolium chloride ionic liquid by functional carbon microspheres from hydrothermal carbonization of cellulose. Environ Sci Tech. 2013;47(6):2792–8.
106. Guo HX, Lian YF, Yan LL, Qi XH, Smith RL. Cellulose-derived superparamagnetic carbonaceous solid acid catalyst for cellulose hydrolysis in an ionic liquid or aqueous reaction system. Green Chem. 2013;15(8):2167–74.

Part IV
Compatibility of Ionic Liquids with Enzyme in Biomass Treatment

Chapter 10
Compatibility of Ionic Liquids with Enzymes

Ngoc Lan Mai and Yoon-Mo Koo

Abstract The potential of ionic liquids as a green alternative to environmentally harmful volatile organic solvents has been well recognized. Being considered as "designer solvents", ionic liquids have been used extensively in a wide range of applications including biotransformations. As compared to those in traditional organic solvents, enzyme performance in ionic liquids is showed enhance in their activity, enantioselectivity, stability, as well as their recoverability and recyclability. This chapter will cover the biocompatibility issue of ionic liquids with enzymes. The effects of ionic liquid properties on the enzymatic reactions and conformation of enzyme as well as methods for activation and stabilization of enzymes in ionic liquids will be described. In addition, the current attempts for rational design of biocompatible ionic liquids will be also discussed.

Keywords Enzyme • Biocompatible • Biotransformation • Ionic liquids • Molecular simulation • Rational design • QSAR

10.1 Introduction

Ionic liquids, which are composed entirely of ions and are liquid at room temperature, have been extensively used as a potential alternative to toxic, hazardous, flammable and highly volatile organic solvents due to their unusual and useful properties. Unlike traditional organic solvents, ionic liquids have many favorable properties such as negligible vapor pressure, wide liquid temperature range, non-flammability, high thermal and excellent chemical stability, high ionic conductivity, large electrochemical window, and ability to dissolve a variety of solutes. In addition, the physicochemical properties of ionic liquids such as melting point,

N.L. Mai • Y.-M. Koo (✉)
Department of Marine Science and Biological Engineering, Inha University, Incheon 402-751, Republic of Korea
e-mail: ymkoo@inha.ac.kr

Z. Fang et al. (eds.), *Production of Biofuels and Chemicals with Ionic Liquids*, Biofuels and Biorefineries 1, DOI 10.1007/978-94-007-7711-8_10,

viscosity, density, hydrophobicity, polarity, and solubility can be finely tuned by simply selecting appropriate combination of cations and anion as well as attached substituents to customize ionic liquids for many specific demands, leading to the use of the terms "designer" and "task specific" ionic liquids [1]. In fact, these unique and tunable properties of ionic liquids make them as promising solvents, co-solvents, and reagents in wide range of applications including electrochemistry [2], analytical chemistry [3], organic and inorganic synthesis [4, 5], nanomaterial synthesis [6], polymerization [7], separation [8, 9], and biotechnology [10] for more than a decade and their applications continue to expand.

In fields of biotechnology, ionic liquids have been widely used as solvents for biomolecule purification [11, 12], pretreatment of cellulosic biomass for biofuel production [13], and enzymatic reactions [14]. The extraction of biomolecules such as protein, enzymes and amino acids using ionic liquids as media or as ionic liquid-based aqueous two-phase systems (APTS) is gaining increasing attention in recent years which showed high extraction efficiency and recyclability. In addition, ionic liquids have been effectively used to pretreat cellulosic biomass for biofuel production. The pretreated cellulose showed an improvement in enzymatic hydrolysis compared to untreated cellulose. However, the most tremendous uses of ionic liquids in biotechnology are in biotransformation. Many excellent reviews have summarized the variety of enzymes used in ionic liquids [14–16]. However, this is not of focus in this chapter, but rather the issues regarding the compatibility of enzyme in ionic liquids are discussed.

10.2 Enzymes in Ionic Liquids

Many enzymatic reactions in ionic liquids have been reported over the last decades. The performance of biocatalyst in ionic liquids reveals that ionic liquids are not only environmentally friendly alternatives to organic solvents but also good solvents for many enzymes and whole cell catalysts. The first example of biotransformation ionic liquids was reported by Lye et al. in 2000 [17]. It involved a whole cell biotransformation of 1,3-dicyanobenzene to 3-cyanobenzamide, with a *Rhodococcus* sp. in a biphasic [Bmim][PF_6]/H_2O medium. In this example, ionic liquids essentially act as a reservoir for the substrate and product, thereby decreasing the substrate and product inhibition observed in water. In principle, an organic solvent could be used for the same purpose but it was found that ionic liquids caused less damage to the microbial cell than organic solvents, for example, toluene [18].

The first use of isolated enzymes in ionic liquids was commonly recognized by the report of Erbeldinger et al. in 2000 [19] although Magnuson et al. earlier demonstrated the activity and stability of alkaline phosphatase in aqueous mixtures of [$EtNH_3$][NO_3] in 1984. However, the finding of Magnuson et al. did not attract much attention due to the lack of knowledge of ionic liquids at that time. In the work of Erbeldinger et al., thermolysin-catalyzed synthesis of Z-aspartame in [Bmim][PF_6]/H_2O (95/5, v/v) showed comparable reaction rate to those observed

Table 10.1 Examples of using enzymes in ionic liquids

Biocatalysts	Reactions	Ionic liquids	Refs.
Lipase	Transesterification	[Bmim][PF_6]	[21]
	Alcoholysis, ammonialysis, perhydrolysis	[Bmim][PF_6], [Bmim][BF_4]	[20]
	Kinetic resolution of chiral alcohols	[Bmim][Tf_2N]	[22]
	Resolution of amino acid ester	[Epy][BF_4], [Emim][BF_4]	[23]
	Esterification of carbohydrates	[MOEmim][BF_4]	[24]
	Synthesis of polyesters	[Bmim][PF_6]	[21]
Alcohol dehydrogenase	Enantioselective reduction of 2-octanone	[Bmim][Tf_2N]-Buffer	[25]
Thermolysin	Synthesis of *Z*-aspartame	[Bmim][PF_6]-H_2O	[19]
α-Chymotrypsin	Transesterification	[Bmim][PF_6]	[26]
Esterase	Transesterification	[Bmim][PF_6]	[27]
Subtilisin	Resolution of amino acid ester	[EPy][TFA]-H_2O	[28]
β-Galactosidase	Synthesis of *N*-acetyllactosamine	[Mmim][$MeSO_4$]-H_2O	[29]
Peroxidase	Oxidation of guaiacol	[Bmim][PF_6]	[30]
Laccase	Oxidation of anthracene	[Bmim][PF_6]	[31]
Formate dehydrogenase	Regeneration of NADH	[Mmim][$MeSO_4$]-H_2O	[29]
Baker's yeast	Enantioselective reduction of ketones	[Bmim][PF_6]-H_2O	[32]

in ethylacetate/H_2O. In addition, the enzyme exhibited a higher stability in the ionic liquids/water medium although the small amount ($3.2\ mg \cdot mL^{-1}$) of enzyme that dissolved in ionic liquids was catalytically inactive. At the same time, Sheldon et al. [20] showed that *Candida antarctica* lipase B(CALB), either free enzyme (SP525) or an immobilized form (Novozym 435) is able to catalyze a variety of biotransformation in [Bmim][BF_4] or [Bmim][PF_6]. Since then, a wide number of enzymes have been examined in ionic liquids such as lipase, protease, oxidoreductase, peroxidase, cellulase, whole cells, and so forth (Table 10.1). Among them, lipase is the major enzyme reported so far in ionic liquids since it is considered as "work horses" of biocatalyst in various potential applications from fine chemical, chiral compounds, biopharmaceuticals to bulk chemicals such as surfactants, and biodiesel.

The uses of ionic liquids in biocatalysis can be classified into anhydrous ionic liquid system, aqueous/ionic liquids system using ionic liquids as co-solvent or additives, aqueous/ionic liquid two phase system, organic solvent/ionic liquids system, and supercritical CO_2/ionic liquid systems. A large number of enzymes (e.g. lipases, proteases, peroxidases, dehydrogenases, glycosidases) and reactions (e.g. esterification, kinetic resolution, reductions, oxidations hydrolysis, etc.) have been tested in monophasic system based on ionic liquid [14, 15, 33]. While most water-miscible ionic liquids have been shown to act as enzyme -deactivating agents at low water content, water-immiscible ionic liquids (e.g. [Bmim][Tf_2N], [Bmim][PF_6], etc.) act as suitable reaction media for enzymatic catalysis in the same conditions. The hygroscopic character of water-immiscible ionic liquids could be regarded as an additional advantage of these solvents, because enzymes require a certain degree of hydration to become active [34]. In addition, biphasic systems based on ionic liquid–water or ionic liquid–organic solvent have been assayed for biocatalytic processes [35, 36]. For example, the lipase catalyzed kinetic resolution of *rac*-ibuprofen in a water –ionic liquids biphasic medium [37], lipase catalyzed production of isoamyl acetate in an ionic liquid–alcohol biphasic system [38], and lipase catalyzed kinetic resolution of *rac*-1-phenylethanol in both ionic liquid/ hexane [39]. In these biphasic reaction systems, the products can be easily recovered by liquid-liquid extraction. However, in the case of hydrophobic compounds, liquid–liquid extraction with organic solvents might breakdown the greenness of the process. However, product recovery from ionic liquid–enzyme reaction media by another non-aqueous green solvent, such as supercritical CO_2, is nowadays considered the most interesting strategy for developing integral clean chemical processes [34].

Many works and excellent reviews have been published during the last decade regarding the use of enzymes in ionic liquids [14–16, 40–47]. The results showed that enzymes in ionic liquids have many advantages compared to those in organic solvents such as high activity, high selectivity including substrate, region- and enantioselectivity, better enzyme stability as well as better recoverability and recyclability. In addition, ionic liquids can be used for carrying out biotransformation with polar or hydrophilic substrates (e.g. amino acid, carbohydrates) which are not soluble or sparingly soluble in most organic solvents.

Advantages of using enzymes in ionic liquids are summarized as follows:

- Ionic liquids can be designed for particular bioprocesses
- Enzymes show excellent operational and thermal stability in ionic liquids. Therefore, bioprocesses can be conducted at high temperature due to high thermal stability of enzymes in ionic liquids
- Enzymes can be easily recovered and recycled simply by filtration or centrifugation
- Products and substrates can be recovered by evaporation, extraction with organic solvents or with supercritical CO_2

10.3 Effect of Physicochemical Properties of Ionic Liquids on Enzyme Activity and Stability

It is well known that the performance of enzymes in organic solvents as well as in ionic liquids is affected by common factors such as water activity, pH as well as the existence of excipients and impurities. Moreover, the physicochemical properties of ionic liquids such as viscosity, polarity, hydrophobicity, kosmotropicity, *etc.* also have strong effect on the activity and stability of enzymes in ionic liquids. The effect of these physicochemical properties of ionic liquids on enzyme activity and stability are summarized in Table 10.2.

In general, enzymes are optimized in nature to perform best in aqueous environment, at neutral pH, temperature below 40 °C and at low osmotic pressure. When enzymes are used in either pure organic solvents or ionic liquids, the

Table 10.2 Effect of ionic liquid's physicochemical properties on enzyme activity and stability

Properties	Enzyme activity and stability
Water content	Minimum water content is required for maintaining enzyme activity
	Might cause the hydrolysis of fluorine based ionic liquids resulting the inactivation of enzymes
Impurities	Affect the physicochemical properties of ionic liquids and therefore have impact on enzymes
	Halide present in ionic liquids might inactivate enzymes
Viscosity	High activity in less viscous ionic liquids
	High viscous ionic liquids lower reaction rate due to mass transfer limitation
Polarity	Activities and stabilities of enzyme in ionic liquids are strongly depend on ionic liquids polarity and increase as increasing ionic liquids polarity
Hydrophobicity	Hydrophilic anions might denature enzyme through hydrogen bonding interaction with protein
	Ionic liquids with long alkyl chain cations may behave as surfactants in aqueous solution and have strong stabilizing impact on enzyme
Ion kosmotropicity	The kosmotropic anions and chaotropic cations stabilize enzymes, while chaotropic cations and kosmotropic cations destabilize them (Hofmeister series)
	The influence of Hofmeister series is complicated when enzymes present in nearly anhydrous ionic liquids

minimum amount of water, which is best described by the water activity, is of crucial importance to maintain the enzyme activity. For ionic liquids, the same methods can be used to maintain a constant water activity as those established for organic solvents [33]. However, in some ionic liquids (e.g. fluoride contained ionic liquids), the water present in the reaction system may cause hydrolysis of ionic liquids and result in reduced enzyme activity and stability. This is attributed by the change in pH of system and the inhibition effect of the hydrolyzed products. The pH strongly effects the activity and stability of enzyme in ionic liquids media as shown in work of Tavares et al. [48]. In this study, the activity of laccase was well maintained at pH 9.0 for 7 days (with activity loss about 10 %) in aqueous solution of 1-ethyl-3-methylimidazolium 2-(2-methoxyethoxy) ethylsulfate, [Emim][$MDEGSO_4$], while significantly reduced at pH 5.0. Moreover, the impurities present in ionic liquids are also important factor that need to be taken into account when carrying out enzymatic reactions in ionic liquids. The impurities may influence the physicochemical properties of ionic liquids and hence, enzymatic reaction. For example, the activity of immobilized CALB (Novozym 435) in [Omim][Tf_2N] decrease linearly with chloride content while the activity of immobilized lipase from *Rhizomer miehei* (Lipozyme IM) drastically decreases in the presence of [Omim][Cl] [49].

Since ionic liquids are composed only of ions, the effect of ions on the enzyme activity and stability is also an important factor. Ions can affect the stability of proteins through the interactions between the ions and charged amino acid groups in the proteins [50]. Some enzymes require metal ions such as cobalt, manganese, zinc, *etc.* for their activity. If these ions are removed or interfered by interaction

← Stabilizing Destabilizing →

Anion $F^{-} > PO_4^{3-} > SO_4^{2-} > CH_3COO^{-} > Cl^{-} > Br^{-} > I^{-} > SCN^{-}$

Cation $(CH_3)_4N^{+} > (CH_3)NH^{2+} > NH_4^{+} > K^{+} > Na^{+} > Cs^{+} > Li^{+} > Mg^{2+} > Ca^{2+} > Ba^{2+}$

Fig. 10.1 The Hofmeister series as an order of the ion effect on protein stability (Reprinted with permission from Ref. [43]. Copyright © 2013, Elsevier)

with ionic liquids, enzyme might be inactivated. Several researchers have proposed Hofmeister series to explain the behavior of enzymes in aqueous solution of ionic liquids. Hofmeister series (Fig. 10.1) which indicates the kosmotropicity of individual cations and anions of ionic liquids may be a good guide for choosing ionic liquids for enzyme activity and stability in aqueous solutions [43, 44, 46]. The anion such as PO_4^{3-}, SO_4^{2-}, CO_3^{2-} (kosmotropic anions), and cations such as NH_4^{+}, K^{+}, Cs^{+} (chaotropic cations) stabilize enzymes, while chaotropic anions and kosmotropic cations destabilize them. However, the influence of Hofmeister series is complicated when enzymes are present in nearly anhydrous ionic liquids.

Properties of ionic liquids such as viscosity, polarity, and hydrophobicity also affect the enzymes in ionic liquids. Ionic liquids are well recognized to have higher viscosity than conventional organic solvents. As a general trend, enzymes are more active in less viscous ionic liquids that can be attributed to the mass transfer limitation in high viscous media. For example, α-chymotrypsin maintained higher activity in less viscous [Emim][Tf_2N] (34 mPa · s) than that in high viscous [MTOA][Tf_2N] (574 mPa · s) for the synthesis of N-acetyl-L-tyrosine propyl ester [26]. Ionic liquids are considered as highly polar account on their ionic nature. The polarity of ionic liquids has been empirically determined by means of a variety of solvatochromic probe dyes [51]. For instance, betaine dye no. 30 or Reichardt's dye has been used to characterize ionic liquid polarity by the solvent polarity parameter $E_T(30)$ and the corresponding normalized polarity scale E_T^N. In addition, the Kamlet-Taft parameters (α, β and π*, which quantify hydrogen-bond donating ability (acidity), hydrogen-bond accepting ability (basicity) and polarity/polarizability, respectively), determined by set of dyes, have also been used to quantify the polarity of ionic liquids [52, 53]. Based on solvatochromic probes studies, ionic liquids have polarity close to low chain alcohol or formamide [54]. Several studies have demonstrated that activities and stabilities of enzyme in ionic liquids were increased with increasing ionic liquids polarity [24, 26] although there were some reports showing no clear effect of polarity on enzyme [21, 45]. Hydrophobicity may be considered as narrow concept of polarity. However, it is practically important to separate hydrophobicity to polarity since the former is often related to the miscibility with water [55]. Depending on the structure of cation and anions, ionic liquids can be hydrophilic or hydrophobic. As a general rule, the structure of anions has strong influence on the hydrophobicity of ionic liquids than the contribution of cations. Hydrophilic anions such as halides, carboxyl groups, having high hydrogen bonding capability, strongly interact with enzyme resulting in the conformational change and deactivation of enzymes. On the other hand, the hydrophobicity of ionic liquids increases as the alkyl chain length in cations increases. In some reports, ionic liquids with long alkyl chain in cations may behave as surfactants in aqueous solution and have strong stabilizing impact on enzyme [56, 57]. However, in some cases, the long alkyl chain might have negative effect on the activity and stability of enzyme due to the resulted higher viscosity [58]. In practice, the log *P* (logarithm of partition coefficient between octanol and water) scale can be used to quantify the hydrophobicity of ionic liquids, and in some reports this scale can be used to optimize enzyme activity and stability in ionic liquids [21, 46, 59, 60].

10.4 Methods to Enhance Enzyme Activity and Stability in Ionic Liquids

Although enzymes in ionic liquids have shown enhance in activity and stability, the relatively low solubility and therefore low activity and stability of lyophilized or free enzymes in ionic liquids are still the major drawback of enzymatic reactions in ionic liquids. As with organic solvents, proteins are not soluble in most of the pure ionic liquids. Although some ionic liquids can dissolve enzymes through the weak hydrogen bonding interactions, they often induce enzyme conformational change resulting in enzyme inactivation [41, 61]. That might limit their potential applications in biotechnology. Various attempts have been made to improve the activity and stability of enzyme in ionic liquids and in general these attempts can be divided into two categories: (1) modification of enzymes and (2) modification of solvent environment (Table 10.3).

Among these methods for modification of enzymes, immobilization is the most common technique. Immobilization of enzymes on the solid carriers or supports is a routine method for improving the enzyme stability in organic solvents as well as in ionic liquids. The supports can be resins [14] (e.g., Novozym 435, lipase B from

Table 10.3 Methods for activation and stabilization of enzymes

Modification of enzymes	Modification of solvent environment
Immobilization of enzyme	Water-in-ionic liquid microemulsions
Immobilization of support materials	Using surfactants to form aqueous micelles in ionic liquids
Sol-gel encapsulation	
Protein cross-linking	
PEG modification	Coating enzymes with ionic liquids
Physical adsorption on PEG	
Covalent conjugating	
Pretreatment with polar organic solvents	Manipulation and design of ionic liquids structures
Precipitate and rinse with n-propanol	Having large molecule structure in order to minimize the H-bond basicity and nucleophilicity of anion
	Containing multiple ether and/or hydroxyl groups to optimize the solvent properties for enzymes

Candida antartica immobilized on acrylic resin, the most commercially available well-known immobilized enzyme), carbon nanotube [62–64], agarose hydrogel film [65–67], and magnetic silica particles [68]. Furthermore, enzymes can be immobilized by sol-gel encapsulation which is a technique for entrapping biomolecules in polymer matrix via non-covalent interactions between the polymer network and biomolecules [69, 70]. To overcome the limitations of sol-gel process such as gel shrinkage, pore collapse and inhibition effect of released alcohol during the preparation of sol-gel materials, different additives such as sugars, amino acids, carbon nanotubes, and ionic liquids have been added to reduce the gel shrinkage, adjust the protein hydration and to increase the activity and stability of enzyme [71, 72]. Interestingly, the polymerized ionic liquid microparticles could be also used to encapsulate enzymes, *e.g.* horseradish peroxidase, with activity more than two times higher than that encapsulated in polyacrylamide microparticles [73]. Moreover, enzymes can be self-immobilized by using cross linking agent such as glutaraldehyde. Several different techniques have been tried for self-immobilization of enzyme using glutaraldehyde such as cross-linked enzymes (CLEs) [74, 75], cross-linked enzyme crystals (CLECs or CLCs) [76–78] and cross-linked enzyme aggregates (CLEAs) [79–81]. Among these techniques, CLEAs are shown to have more advantages such as ease of preparation and recycling, enhanced activity and stability. CLEAs can be used for variety of enzymes, and higher enzymes stability of CLEAs in ionic liquid than free enzyme have been reported [82–84]. In addition, the modification of enzymes with PEG (having both hydrophilic and hydrophobic characteristic) either by physical adsorption or covalent attachment in order to make the modified enzymes soluble in organic solvents and ionic liquids [85–92] as well as pretreatment of enzyme and supports/carriers with organic solvents such as n-propanol before immobilization also result in enhanced enzyme activity and stability [84, 93–96].

The activity and stability of enzymes can be improved by the modification of solvent environments. In this approach, the micro-environment surrounding the enzymes or ionic liquids media are modified to become more compatible with enzymes. Several approaches have been done in recent years such as using surfactant to form aqueous micelles in ionic liquids [97–100], coating enzyme with ionic liquids [101–104] and manipulating or designing enzyme compatible with ionic liquids [105–107]. Regarding the activation and stabilization of enzymes in ionic liquids, several excellent reviews are available in literature [16, 55, 108].

10.5 Molecular Dynamics of Enzyme Structure and Conformation in Ionic Liquids

Understanding the mechanism of how ionic liquids stabilize and activate enzymes is crucial for researchers and engineers to optimize enzymatic reactions as well as synthesize the enzyme compatible ionic liquids. Many efforts with different techniques such as spectroscopy, molecular dynamics simulation have been paid to explore the dynamic structure and conformation of protein in ionic liquid in recent years [109–121]. In general, the ionic liquids that have strong interaction with protein such as halide containing ionic liquids tend to change the conformation of proteins and therefore inactivate enzymes while other ionic liquids such as $[Tf_2N]^-$ based ionic liquids strengthen the protein conformation resulting in enhanced enzyme stability. For example, Sasmal et al. by using fluorescence correlation spectroscopy studied the conformation dynamics of human serum albumin (HSA) protein in [Pmim][Br] and observed the denaturant effect of ionic liquids [121]. De Diego et al. used fluorescence and circular dichroism (CD) spectroscopy to analysis the α-chymotrypsin stabilization of [Emim][Tf_2N] and found out that this ionic liquids shows ability to compact the native conformation of protein by great enhancement of the β-strand of protein. In our recent studies, molecular dynamics (MD) simulation was employed to investigate the structure of CALB enzyme in different ionic liquids and organic solvents and their corresponding enzyme activities. The MD simulations indicate that the structure and dynamics of the cavity that holds the catalytic triad are solvent dependent: the cavity can be opened or closed in water; the cavity is open in [Bmim][TfO] and tert-butanol; the cavity is closed in [Bmim][Cl]. The closed or narrow cavity conformation observed in our simulations obstructs passage for substrates, thus lowering their probability of reaching the catalytic triad (Fig. 10.2). In addition, we observed that two isoleucines, ILE-189 and ILE-285, play a pivotal role in the open-close dynamics of cavity. Specifically, ILE-285 situated on a helix (α-10) that can significantly change conformation in different solvents. This change is acutely evident in [Bmim][Cl] where interactions of LYS-290 with chlorine anions induces a conformational switch from an α-helix into a turn (Fig. 10.3). Disruption of the α-10 helix structure results in a narrower entrance to the catalytic triad site and this change is responsible for reduced activity of CALB in [Bmim][Cl]. Moreover, the cavity profile's size is well agreed with the enzyme activity for the synthesis of butyl acetate. The activity of the enzyme decreases with the size of the cavity in the following order: [Bmim][TfO] > tert-butanol > [Bmim][Cl].

10.6 Future Perspective on Rational Design of Ionic Liquids for Biotransformation

Ionic liquids are considered as "designer solvents" because the physical, chemical and biological properties of ionic liquids can be tuned by altering the combination of their ionic constituents. However, finding the proper combination of anions and cations and their mixtures among 10^{18} possibilities to yield required properties is a major challenge. Many attempts have been tried to design and synthesize "task-specific" ionic liquids in the last decade. For example, Abe et al. [122] based on their experimental observations that hydrophilic imidazolium ionic liquids having alkyl ether functionalizing sulfate salts were appropriate for lipase-catalyzed reaction have anticipated that phosphonium

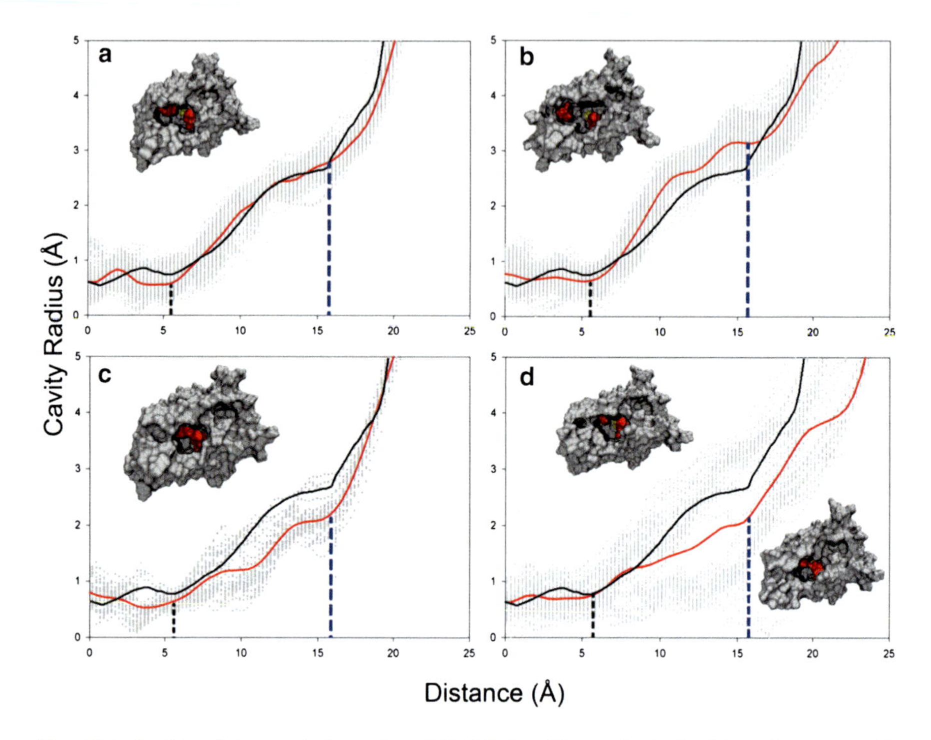

Fig. 10.2 Profile of the catalytic cavity of CALB in different ionic liquids and solvents. (**a**) [Bmim][TfO], (**b**) tert-butanol, (**c**) [Bmim][Cl], (**d**) 0.3 M NaCl. Catalytic cavity profile (*grey points*). Average value of all points is shown in *red line*. *Black line* indicates cavity profile of crystalized CALB. Illustrations *above* and *below* the graph indicate the surface rendering of CALB model with *red* labeled molecules representing ILE-189 and ILE-285. The approximate position of cavity entrance is shown by *blue dotted line* whereas the black dotted line approximates the catalytic triad position

salt which alkyl ether group might also be appropriate for lipase. Their anticipation relied on the fact that phosphonium salt commonly exists in living creatures. Several types of phosphonium ionic liquids have been prepared and tested for the activity of ionic liquid coating lipase. A novel ionic liquids, 2-methoxyethoxymethyl (tri-n-butyl)-phosphonium bis (trifluoromethanesulfonyl) amide ([P_{444}MEM][Tf_2N]) was successfully found to enhance the activity two times while perfectly maintaining enantioselectivity of ionic liquid –coated lipase. However, this approach is mainly based on the experimental trial-error means. More recently, many methods including *ab initio* calculation, molecular simulation, quantum chemistry, and correlation have been successfully applied to calculate and/or predict the physical properties of ionic liquids [123, 124]. Not only physical, but also chemical and biological properties of ionic liquids can be predicted by these methods. This opens a new path for designing task specific ionic liquids which depends less on the experimental trial-and-errors. For instance, in our studies, quantitative structure-activity relationship (QSAR) model based on the information from the structure of ionic liquids (structural molecular

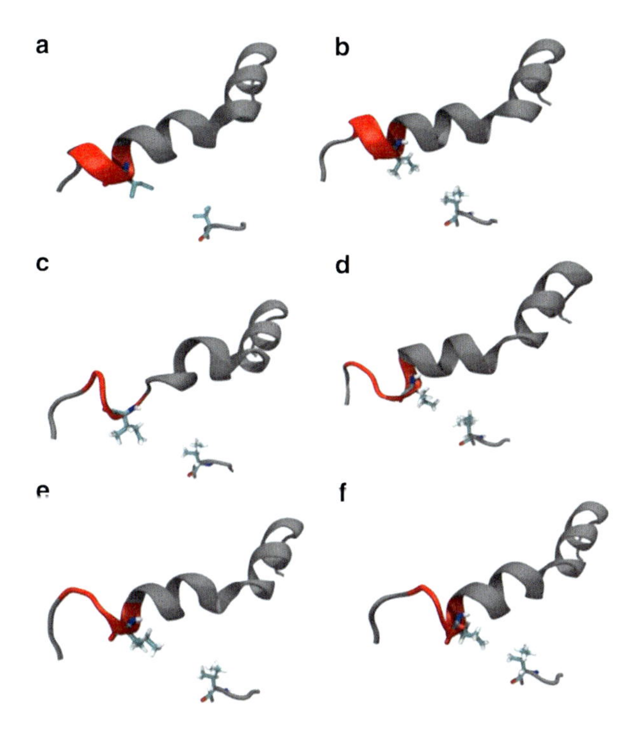

Fig. 10.3 Conformational changes in the α-10 helix region of CALB in various solvents. (**a**) Initial X-crystal structure, (**b**) [Bmim][TfO], (**c**) *tert*-butanol, (**d**) [Bmim][Cl], (**e**) 0.3 M NaCl (*Opened*), (**f**) 0.3 M NaCl (*Closed*). The region with significantly conformational changes is shown in *red*. ILE-189 and ILE-285 are shown as stick molecular model

descriptors derived from CODESSA program) were used to predict the activity of *Candida antarctica* lipase B (CALB) in the kinetic resolution of sec-phenylethanol in ionic liquids. An optimal QSAR model with 5 structural molecular descriptors was established with the correlation efficient (R^2) of 0.9481 and 0.9208 for 18 training and 5 testing ionic liquids set, respectively. This indicates that the performance of enzymatic reaction in new ionic liquids could be predicted by QSAR model based on the structural molecular descriptors of ionic liquids.

10.7 Conclusions and Remarks

In this chapter, a summary of issues regarding enzymes in ionic liquids has been provided. In general, most of enzymes can be used in adequate ionic liquids. In most cases, it is found that ionic liquids with hydrophobic nature, less viscosity, kosmotropic anion and chaotropic cation usually enhance the activity and stability of enzymes. The activity and stability of enzyme in ionic liquids can be improved by immobilization, modification with activated stabilizing agents, pretreatment with polar organic solvents or by reaction media engineering such as mircroemulsion, using co-solvent, and design of biocompatible ionic liquids, *etc.* Furthermore, the information regarding structure and conformation dynamics of protein in ionic liquids could be helpful for engineering and scientific communities to understand how ionic liquids enhance the stability and activity of enzymes, and thereafter be useful for choosing or designing ionic liquids for specific enzymatic reactions with the help of quantum chemistry.

References

1. Freemantle M. Designer solvents. Chem Eng News. 1998;76:32–7.
2. Lewandowski A, Świderska-Mocek A. Ionic liquids as electrolytes for Li-ion batteries—an overview of electrochemical studies. J Power Sour. 2009;194:601–9.
3. Sun P, Armstrong DW. Ionic liquids in analytical chemistry. Anal Chim Acta. 2010;661:1–16.
4. Yue C, Fang D, Liu L, Yi T-F. Synthesis and application of task-specific ionic liquids used as catalysts and/or solvents in organic unit reactions. J Mol Liq. 2011;163:99–121.
5. Olivier-Bourbigou H, Magna L, Morvan D. Ionic liquids and catalysis: recent progress from knowledge to applications. Appl Catal A. 2010;373:1–56.
6. Li Z, Jia Z, Luan Y, Mu T. Ionic liquids for synthesis of inorganic nanomaterials. Curr Opin Solid State Mater Sci. 2008;12:1–8.
7. Kubisa P. Ionic liquids as solvents for polymerization processes—progress and challenges. Prog Polym Sci. 2009;34:1333–47.
8. Berthod A, Ruiz-Ángel MJ, Carda-Broch S. Ionic liquids in separation techniques. J Chromatogr A. 2008;1184:6–18.
9. Vidal L, Riekkola M-L, Canals A. Ionic liquid-modified materials for solid-phase extraction and separation: a review. Anal Chim Acta. 2012;715:19–41.
10. Quijano G, Couvert A, Amrane A. Ionic liquids: applications and future trends in bioreactor technology. Bioresour Technol. 2010;101:8923–30.
11. Pei Y, Wang J, Wu K, Xuan X, Lu X. Ionic liquid-based aqueous two-phase extraction of selected proteins. Sep Purif Technol. 2009;64:88–95.
12. Tonova K. Separation of poly- and disaccharides by biphasic systems based on ionic liquids. Sep Purif Technol. 2012;89:57–65.
13. Vancov T, Alston A-S, Brown T, McIntosh S. Use of ionic liquids in converting lignocellulosic material to biofuels. Renew Energy. 2012;45:1–6.
14. van Rantwijk F, Sheldon RA. Biocatalysis in ionic liquids. Chem Rev. 2007;107:2757–85.
15. Park S, Kazlauskas RJ. Biocatalysis in ionic liquids – advantages beyond green technology. Curr Opin Biotechnol. 2003;14:432–7.
16. Moniruzzaman M, Nakashima K, Kamiya N, Goto M. Recent advances of enzymatic reactions in ionic liquids. Biochem Eng J. 2010;48:295–314.
17. SG C, JD H, V-M V, KR S, GJ L. Room-temperature ionic liquids as replacements for organic solvents in multiphase bioprocess operations. Biotechnol Bioeng. 2000;69:227–33.
18. Sheldon RA, Rantwijk F v, Lau RM. Biotransformations in ionic liquids: an overview. ACS Symp Ser. 2003;856:192–205.
19. Erbeldinger M, Mesiano AJ, Russell AJ. Enzymatic catalysis of formation of Z-Aspartame in ionic liquid – an alternative to enzymatic catalysis in organic solvents. Biotechnol Progr. 2001;16:1129–31.
20. Lau RM, Rantwijk F v, Seddon KR, Sheldon RA. Lipase-catalyzed reactions in ionic liquids. Org Lett. 2000;2:4189–91.
21. Kaar JL, Jesionowski AM, Berberich JA, Moulton R, Russell AJ. Impact of ionic liquid physical properties on lipase activity and stability. JACS. 2003;125:4125–31.
22. Schöfer SH, Kaftzik N, Wasserscheid P, Kragl U. Enzyme catalysis in ionic liquids: lipase-catalysed kinetic resolution of 1-phenyl ethanol with improved enantioselectivity. Chem Commun. 2001;5:425–6.

23. Zhao H, Luo RG, Malhotra SV. Kinetic study on the enzymatic resolution of homophenylalanine ester using ionic liquids. Biotechnol Progr. 2003;19:1016–18.
24. Park S, Kazlauskas RJ. Improved preparation and use of room-temperature ionic liquids in lipase-catalyzed enantio- and regioselective acylations. J Org Chem. 2001;66:8395–401.
25. Eckstein M, Villela FM, Liese A, Kragl U. Use of an ionic liquid in a two-phase system to improve an alcohol dehydrogenase catalysed reduction. Chem Commun. 2004;10:1084–5.
26. Lozano P, Diego T g, Guegan J-P, Vaultier M, Iborra JL. Stabilization of α-chymotrypsin by ionic liquids in transesterification reactions. Biotechnol Bioeng. 2001;75:563–9.
27. Persson M, Bornscheuer UT. Increased stability of an esterase from *Bacillus stearothermophilus* in ionic liquids as compared to organic solvents. J Mol Catal B: Enzym. 2003;22:21–7.
28. Zhao H, Malhotra SV. Enzymatic resolution of amino acid esters using ionic liquid N-ethyl pyridinium trifluoroacetate. Biotechnol Lett. 2002;24:1257–9.
29. Kaftzik N, Wasserscheid P, Kragl U. Use of ionic liquids to increase the yield and enzyme stability in the β-galactosidase catalysed synthesis of N-acetyllactosamine. Org Process Res Dev. 2002;6:553–7.
30. Laszlo JA, Compton DL. Comparison of peroxidase activities of hemin, cytochrome c and microperoxidase-11 in molecular solvents and imidazolium-based ionic liquids. J Mol Catal B: Enzym. 2002;18:109–20.
31. Hinckley G, Mozhaev VV, Budde C, Khmelnitsky YL. Oxidative enzymes possess catalytic activity in systems with ionic liquids. Biotechnol Lett. 2002;24:2083–7.
32. Howarth J, James P, Dai J. Immobilized baker's yeast reduction of ketones in an ionic liquid, [bmim]PF6 and water mix. Tetrahedron Lett. 2001;42:7517–19.
33. Klembt S, Dreyer S, Eckstein M, Kragl U. Biocatalytic reactions in ionic liquids. In: Wasserscheid P, Welton T, (eds). Ionic liquids in synthesis. Wiley-VCH Verlag GmbH & Co. KGaA. 2008;641–61.
34. Lozano P. Enzymes in neoteric solvents: from one-phase to multiphase systems. Green Chem. 2010;12:555–69.
35. Eckstein M, Villela Filho M, Liese A, Kragl U. Use of an ionic liquid in a two-phase system to improve an alcohol dehydrogenase catalysed reduction. Chem Commun. 2004:1084–5.
36. Freire MG, Claudio AFM, Araujo JMM, Coutinho JAP, Marrucho IM, Lopes JNC, Rebelo LPN. Aqueous biphasic systems: a boost brought about by using ionic liquids. Chem Soc Rev. 2012;41:4966–95.
37. Miyako E, Maruyama T, Kamiya N, Goto M. Enzyme-facilitated enantioselective transport of (S)-ibuprofen through a supported liquid membrane based on ionic liquids. Chem Commun. 2003;0:2926–7.
38. Fehér E, Illeová V, Kelemen-Horváth I, Bélafi-Bakó K, Polakovič M, Gubicza L. Enzymatic production of isoamyl acetate in an ionic liquid–alcohol biphasic system. J Mol Catal B: Enzym. 2008;50:28–32.
39. Lozano P, De Diego T, Sauer T, Vaultier M, Gmouh S, Iborra JL. On the importance of the supporting material for activity of immobilized *Candida antarctica* lipase B in ionic liquid/hexane and ionic liquid/supercritical carbon dioxide biphasic media. J Supercrit Fluids. 2007;40:93–100.
40. Kragl U, Eckstein M, Kaftzik N. Enzyme catalysis in ionic liquids. Curr Opin Biotechnol. 2002;13:565–71.
41. Sheldon RA, Lau RM, Sorgedrager MJ, van Rantwijk F, Seddon KR. Biocatalysis in ionic liquids. Green Chem. 2002;4:147–51.
42. van Rantwijk F, Madeira Lau R, Sheldon RA. Biocatalytic transformations in ionic liquids. Trends Biotechnol. 2003;21:131–8.
43. Zhao H. Effect of ions and other compatible solutes on enzyme activity, and its implication for biocatalysis using ionic liquids. J Mol Catal B: Enzym. 2005;37:16–25.
44. Zhao H, Campbell SM, Jackson L, Song Z, Olubajo O. Hofmeister series of ionic liquids: kosmotropic effect of ionic liquids on the enzymatic hydrolysis of enantiomeric phenylalanine methyl ester. Tetrahedron: Asymmetry. 2006;17:377–83.

45. Zhao H, Olubajo O, Song Z, Sims AL, Person TE, Lawal RA, Holley LA. Effect of kosmotropicity of ionic liquids on the enzyme stability in aqueous solutions. Bioorg Chem. 2006;34:15–25.
46. Yang Z. Hofmeister effects: an explanation for the impact of ionic liquids on biocatalysis. J Biotechnol. 2009;144:12–22.
47. Naushad M, Alothman ZA, Khan AB, Ali M. Effect of ionic liquid on activity, stability, and structure of enzymes: a review. Int J Biol Macromol. 2012;51:555–60.
48. Tavares APM, Rodriguez O, Macedo EA. Ionic liquids as alternative co-solvents for laccase: study of enzyme activity and stability. Biotechnol Bioeng. 2008;101:201–7.
49. Lee SH, Ha SH, Lee SB, Koo Y-M. Adverse effect of chloride impurities on lipase-catalyzed transesterifications in ionic liquids. Biotechnol Lett. 2006;28:1335–9.
50. Vrbka L, Jungwirth P, Bauduin P, Touraud D, Kunz W. Specific ion effects at protein surfaces: a molecular dynamics study of bovine pancreatic trypsin inhibitor and horseradish peroxidase in selected salt solutions. J Phys Chem B. 2006;110:7036–43.
51. Reichardt C. Polarity of ionic liquids determined empirically by means of solvatochromic pyridinium N-phenolate betaine dyes. Green Chem. 2005;7:339–51.
52. Ab Rani MA, Brant A, Crowhurst L, Dolan A, Lui M, Hassan NH, Hallett JP, Hunt PA, Niedermeyer H, Perez-Arlandis JM, Schrems M, Welton T, Wilding R. Understanding the polarity of ionic liquids. Phys Chem Chem Phys. 2011;13:16831–40.
53. Jessop PG, Jessop DA, Fu D, Phan L. Solvatochromic parameters for solvents of interest in green chemistry. Green Chem. 2012;14:1245–59.
54. Carmichael AJ, Seddon KR. Polarity study of some 1-alkyl-3-methylimidazolium ambient-temperature ionic liquids with the solvatochromic dye, Nile Red. J Phys Org Chem. 2000;13:591–5.
55. Zhao H. Methods for stabilizing and activating enzymes in ionic liquids—a review. J Chem Technol Biotechnol. 2010;85:891–907.
56. Greaves TL, Drummond CJ. Ionic liquids as amphiphile self-assembly media. Chem Soc Rev. 2008;37:1709–26.
57. Tariq M, Freire MG, Saramago B, Coutinho JAP, Lopes JNC, Rebelo LPN. Surface tension of ionic liquids and ionic liquid solutions. Chem Soc Rev. 2012;41:829–68.
58. Lou W-Y, Zong M-H. Efficient kinetic resolution of (R, S)-1-trimethylsilylethanol via lipase-mediated enantioselective acylation in ionic liquids. Chirality. 2006;18:814–21.
59. Irimescu R, Kato K. Investigation of ionic liquids as reaction media for enzymatic enantioselective acylation of amines. J Mol Catal B: Enzym. 2004;30:189–94.
60. Zhao H, Baker GA, Song Z, Olubajo O, Zanders L, Campbell SM. Effect of ionic liquid properties on lipase stabilization under microwave irradiation. J Mol Catal B: Enzym. 2009;57:149–57.
61. Lau RM, Sorgedrager MJ, Carrea G, Van Rantwijk F, Secundo F, Sheldon RA. Dissolution of Candida antarctica lipase B in ionic liquids: effects on structure and activity. Green Chem. 2004;6:483–7.
62. Karajanagi SS, Vertegel AA, Kane RS, Dordick JS. Structure and function of enzymes adsorbed onto single-walled carbon nanotubes. Langmuir. 2004;20:11594–9.
63. Eker B, Asuri P, Murugesan S, Linhardt R, Dordick J. Enzyme–carbon nanotube conjugates in room-temperature ionic liquids. Appl Biochem Biotechnol. 2007;143:153–63.
64. Shah S, Solanki K, Gupta M. Enhancement of lipase activity in non-aqueous media upon immobilization on multi-walled carbon nanotubes. Chem Cent J. 2007;1:30.
65. Wang S-F, Chen T, Zhang Z-L, Shen X-C, Lu Z-X, Pang D-W, Wong K-Y. Direct electrochemistry and electrocatalysis of heme proteins entrapped in agarose hydrogel films in room-temperature ionic liquids. Langmuir. 2005;21:9260–6.
66. Wang S-F, Chen T, Zhang Z-L, Pang D-W, Wong K-Y. Effects of hydrophilic room-temperature ionic liquid 1-butyl-3-methylimidazolium tetrafluoroborate on direct electrochemistry and bioelectrocatalysis of heme proteins entrapped in agarose hydrogel films. Electrochem Commun. 2007;9:1709–14.

67. Brusova Z, Gorton L, Magner E. Comment on "direct electrochemistry and electrocatalysis of heme proteins entrapped in agarose hydrogel films in room-temperature ionic liquids". Langmuir. 2006;22:11453–5.
68. Jiang Y, Guo C, Xia H, Mahmood I, Liu C, Liu H. Magnetic nanoparticles supported ionic liquids for lipase immobilization: enzyme activity in catalyzing esterification. J Mol Catal B: Enzym. 2009;58:103–9.
69. Reetz M. Practical protocols for lipase immobilization via sol-gel techniques. In: Guisan J (ed) Immobilization of Enzymes and Cells. Methods in Biotechnology. Humana Press. 2006;22:65–76.
70. Campás M, Marty J-L. Encapsulation of enzymes using polymers and sol-gel techniques. In: Guisan J (ed) Immobilization of Enzymes and Cells. Methods in Biotechnology. Humana Press; 2006;22:77–85.
71. Lee SH, Doan TTN, Ha SH, Koo Y-M. Using ionic liquids to stabilize lipase within sol–gel derived silica. J Mol Catal B: Enzym. 2007;45:57–61.
72. Lee SH, Doan TTN, Ha SH, Chang W-J, Koo Y-M. Influence of ionic liquids as additives on sol–gel immobilized lipase. J Mol Catal B: Enzym. 2007;47:129–34.
73. Nakashima K, Kamiya N, Koda D, Maruyama T, Goto M. Enzyme encapsulation in microparticles composed of polymerized ionic liquids for highly active and reusable biocatalysts. Org Biomol Chem. 2009;7:2353–8.
74. Habeeb AFSA. Preparation of enzymically active, water-insoluble derivatives of trypsin. Arch Biochem Biophys. 1967;119:264–8.
75. Jansen EF, Olson AC. Properties and enzymatic activities of papain insolubilized with glutaraldehyde. Arch Biochem Biophys. 1969;129:221–7.
76. Quiocho FA, Richards FM. Intermolecular cross linking of a protein in the crystalline state: carboxypeptidase-A. PNAS. 1964;52:833–9.
77. St. Clair NL, Navia MA. Cross-linked enzyme crystals as robust biocatalysts. JACS. 1992;114:7314–16.
78. Margolin AL, Navia MA. Protein crystals as novel catalytic materials. Angew Chem Int Ed. 2001;40:2204–22.
79. Cao L, Langen L v, Sheldon RA. Immobilised enzymes: carrier-bound or carrier-free? Curr Opin Biotechnol. 2003;14:387–94.
80. Sheldon R, Schoevaart R, Langen L. Cross-linked enzyme aggregates. In: Guisan J (ed) Immobilization of Enzymes and Cells. Methods in Biotechnology. Humana Press; 2006;22:31–45.
81. Sheldon RA, Schoevaart R, Van Langen LM. Cross-linked enzyme aggregates (CLEAs): a novel and versatile method for enzyme immobilization (a review). Biocatal Biotransform. 2005;23:141–7.
82. Cao L, van Langen LM, van Rantwijk F, Sheldon RA. Cross-linked aggregates of penicillin acylase: robust catalysts for the synthesis of β-lactam antibiotics. J Mol Catal B: Enzym. 2001;11:665–70.
83. Toral AR, de los Ríos AP, Hernández FJ, Janssen MHA, Schoevaart R, van Rantwijk F, Sheldon RA. Cross-linked *Candida antarctica* lipase B is active in denaturing ionic liquids. Enzyme Microb Technol. 2007;40:1095–9.
84. Shah S, Gupta MN. Kinetic resolution of (±)-1-phenylethanol in [Bmim][PF6] using high activity preparations of lipases. Bioorg Med Chem Lett. 2007;17:921–4.
85. Nakashima K, Maruyama T, Kamiya N, Goto M. Homogeneous enzymatic reactions in ionic liquids with poly(ethylene glycol)-modified subtilisin. Org Biomol Chem. 2006;4:3462–7.
86. Maruyama T, Nagasawa S, Goto M. Poly(ethylene glycol)-lipase complex that is catalytically active for alcoholysis reactions in ionic liquids. Biotechnol Lett. 2002;24:1341–5.
87. Laszlo JA, Compton DL. α-Chymotrypsin catalysis in imidazolium-based ionic liquids. Biotechnol Bioeng. 2001;75:181–6.

88. Woodward CA, Kaufman EN. Enzymatic catalysis in organic solvents: polyethylene glycol modified hydrogenase retains sulfhydrogenase activity in toluene. Biotechnol Bioeng. 1996;52:423–8.
89. Ohno H, Suzuki C, Fukumoto K, Yoshizawa M, Fujita K. Electron transfer process of poly (ethylene oxide)-modified cytochrome c in imidazolium type ionic liquid. Chem Lett. 2003;32:450–1.
90. Nakashima K, Maruyama T, Kamiya N, Goto M. Comb-shaped poly(ethylene glycol)-modified subtilisin Carlsberg is soluble and highly active in ionic liquids. Chem Commun. 2005;0:4297–9.
91. Kodera Y, Tanaka H, Matsushima A, Inada Y. Chemical modification of L-asparaginase with a comb-shaped copolymer of polyethylene glycol derivative and maleic anhydride. Biochem Biophys Res Commun. 1992;184:144–8.
92. Kazunori N, Jun O, Tatsuo M, Noriho K, Masahiro G. Activation of lipase in ionic liquids by modification with comb-shaped poly(ethylene glycol). Sci Technol Adv Mater. 2006;7:692.
93. Partridge J, Halling PJ, Moore DB. Practical route to high activity enzyme preparations for synthesis in organic media. Chem Commun. 1998;0:841–2.
94. Theppakorn T, Kanasawud P, Halling P. Activity of immobilized papain dehydrated by n-propanol in low-water media. Biotechnol Lett. 2004;26:133–6.
95. Roy I, Gupta MN. Preparation of highly active α-chymotrypsin for catalysis in organic media. Bioorg Med Chem Lett. 2004;14:2191–3.
96. Shah S, Gupta MN. Obtaining high transesterification activity for subtilisin in ionic liquids. Biochim Biophys Acta. 2007;1770:94–8.
97. Qiu Z, Texter J. Ionic liquids in microemulsions. Curr Opin Colloid Interface Sci. 2008;13:252–62.
98. Moniruzzaman M, Kamiya N, Nakashima K, Goto M. Water-in-ionic liquid microemulsions as a new medium for enzymatic reactions. Green Chem. 2008;10:497–500.
99. Zhou G-P, Zhang Y, Huang X-R, Shi C-H, Liu W-F, Li Y-Z, Qu Y-B, Gao P-J. Catalytic activities of fungal oxidases in hydrophobic ionic liquid 1-butyl-3-methylimidazolium hexafluorophosphate-based microemulsion. Colloids Surf B. 2008;66:146–9.
100. Pavlidis IV, Gournis D, Papadopoulos GK, Stamatis H. Lipases in water-in-ionic liquid microemulsions: structural and activity studies. J Mol Catal B: Enzym. 2009;60:50–6.
101. Lee JK, Kim M-J. Ionic liquid-coated enzyme for biocatalysis in organic solvent. J Org Chem. 2002;67:6845–7.
102. Itoh T, Han S, Matsushita Y, Hayase S. Enhanced enantioselectivity and remarkable acceleration on the lipase-catalyzed transesterification using novel ionic liquids. Green Chem. 2004;6:437–9.
103. Itoh T, Matsushita Y, Abe Y, Han S-h, Wada S, Hayase S, Kawatsura M, Takai S, Morimoto M, Hirose Y. Increased enantioselectivity and remarkable acceleration of lipase-catalyzed transesterification by using an imidazolium PEG–alkyl sulfate ionic liquid. Chem Eur J. 2006;12:9228–37.
104. Dang DT, Ha SH, Lee S-M, Chang W-J, Koo Y-M. Enhanced activity and stability of ionic liquid-pretreated lipase. J Mol Catal B: Enzym. 2007;45:118–21.
105. Guo Z, Chen B, Lopez Murillo R, Tan T, Xu X. Functional dependency of structures of ionic liquids: do substituents govern the selectivity of enzymatic glycerolysis? Org Biomol Chem. 2006;4:2772–6.
106. Das D, Dasgupta A, Das PK. Improved activity of horseradish peroxidase (HRP) in 'specifically designed' ionic liquid. Tetrahedron Lett. 2007;48:5635–9.
107. Vafiadi C, Topakas E, Nahmias VR, Faulds CB, Christakopoulos P. Feruloyl esterase-catalysed synthesis of glycerol sinapate using ionic liquids mixtures. J Biotechnol. 2009;139:124–9.
108. Moniruzzaman M, Kamiya N, Goto M. Activation and stabilization of enzymes in ionic liquids. Org Biomol Chem. 2010;8:2887–99.
109. Bihari M, Russell TP, Hoagland DA. Dissolution and dissolved state of cytochrome c in a neat, hydrophilic ionic liquid. Biomacromolecules. 2010;11:2944–8.

110. Geng F, Zheng L, Yu L, Li G, Tung C. Interaction of bovine serum albumin and long-chain imidazolium ionic liquid measured by fluorescence spectra and surface tension. Process Biochem. 2010;45:306–11.
111. Yan H, Wu J, Dai G, Zhong A, Chen H, Yang J, Han D. Interaction mechanisms of ionic liquids [$C_n$mim]Br (n = 4, 6, 8, 10) with bovine serum albumin. J Lumin. 2012;132:622–8.
112. Akdogan Y, Hinderberger D. Solvent-induced protein refolding at low temperatures. J Phys Chem B. 2011;115:15422–9.
113. Bekhouche M, Blum LJ, Doumèche B. Contribution of dynamic and static quenchers for the study of protein conformation in ionic liquids by steady-state fluorescence spectroscopy. J Phys Chem B. 2011;116:413–23.
114. McCarty TA, Page PM, Baker GA, Bright FV. Behavior of acrylodan-labeled human serum albumin dissolved in ionic liquids. Ind Eng Chem Res. 2007;47:560–9.
115. De Diego T, Lozano P, Gmouh S, Vaultier M, Iborra JL. Understanding structure – stability relationships of Candida antartica lipase B in ionic liquids. Biomacromolecules. 2005;6:1457–64.
116. De Diego T, Lozano P, Gmouh S, Vaultier M, Iborra JL. Fluorescence and CD spectroscopic analysis of the α-chymotrypsin stabilization by the ionic liquid, 1-ethyl-3-methylimidazolium bis[(trifluoromethyl)sulfonyl]amide. Biotechnol Bioeng. 2004;88:916–24.
117. Lin Huang J, Noss ME, Schmidt KM, Murray L, Bunagan MR. The effect of neat ionic liquid on the folding of short peptides. Chem Commun. 2011;47:8007–9.
118. Baker SN, McCleskey TM, Pandey S, Baker GA. Fluorescence studies of protein thermostability in ionic liquids. Chem Commun. 2004;0:940–1.
119. Page TA, Kraut ND, Page PM, Baker GA, Bright FV. Dynamics of loop 1 of domain I in human serum albumin when dissolved in ionic liquids. J Phys Chem B. 2009;113:12825–30.
120. Shu Y, Liu M, Chen S, Chen X, Wang J. New insight into molecular interactions of imidazolium ionic liquids with bovine serum albumin. J Phys Chem B. 2011;115:12306–14.
121. Sasmal DK, Mondal T, Sen Mojumdar S, Choudhury A, Banerjee R, Bhattacharyya K. An FCS study of unfolding and refolding of CPM-labeled human serum albumin: role of ionic liquid. J Phys Chem B. 2011;115:13075–83.
122. Abe Y, Yoshiyama K, Yagi Y, Hayase S, Kawatsura M, Itoh T. A rational design of phosphonium salt type ionic liquids for ionic liquid coated-lipase catalyzed reaction. Green Chem. 2010;12:1976–80.
123. Turner EA, Pye CC, Singer RD. Use of ab initio calculations toward the rational design of room temperature ionic liquids. Russ J Phys Chem A. 2003;107:2277–88.
124. Coutinho JAP, Carvalho PJ, Oliveira NMC. Predictive methods for the estimation of thermophysical properties of ionic liquids. RSC Adv. 2012;2:7322–46.

Chapter 11
Biocompatibility of Ionic Liquids with Enzymes for Biofuel Production

Teresa de Diego, Arturo Manjón, and José Luis Iborra

Abstract This chapter focuses on the application of enzyme technology in non-aqueous green solvents as ionic liquids (ILs) to transform biomass, mainly non-edible biomass (e.g. cellulose, lignocellulose, wood, forest residues, etc.), into fermentable monomeric compounds, and low cost vegetable oils or animal fats in biodiesel. This review aims to identify the key parameters that determine the biocompatibility of ionic liquids with enzymes for the rational design of ionic liquid-based formulations in biocatalysis for biofuel production.

Keywords Biofuels • Ionic liquids • Enzymatic-saccharification • Enzymatic-transesterification • Biodiesel • Bioethanol

Abbreviations

Anions

[Cl]	Chloride
[ClO_4]	Perchlorate
[Br]	Bromide
[H_2PO_4]	Phosphate
[BF_4]	Tetrafluoroborate
[PF_6]	Hexafluorophosphate
[BPh4]	Tetraphenylborate
[NO_2]	Nitrite
[NO_3]	Nitrate

T. de Diego • A. Manjón • J.L. Iborra (✉)
Department of Biochemistry and Molecular Biology B and Immunology, University of Murcia, P.O. Box 4021, E-30100 Murcia, Spain
e-mail: jliborra@um.es

Z. Fang et al. (eds.), *Production of Biofuels and Chemicals with Ionic Liquids*, Biofuels and Biorefineries 1, DOI 10.1007/978-94-007-7711-8_11,

$[(MeO)_2PO_2]$	Dimethylphosphate
[Ac]	Acetate
[TFA]	Trifluoroacetate
[DMP]	Dimethylphosphate
$[MeSO_4]$	Methylsulfate
[TfO]	Trifluoromethylsulfonate
$[NTf_2]$	Bis[(trifluoromethyl)sulfonyl] amide
[SCN]	Thiocyanate
$[SbF_6]$	Hexafluoroantimonate

Cations

[MMIM]	1-Methyl-3-methylimidazolium
[EMIM]	1-Ethyl-3-methylimidazolium
[BMIM]	1-Butyl-3-methylimidazolium
[OMIM]	1-Octyl-3-methylimidazolium
[BMpy]	1-Butyl-1-methylpyrrolidinium
$[Et_3MeN]$	Triethyl-3-methylimidazolium
$[C_{16}MIM]$	1-Hexadecyl-3-methylimidazolium
$[C_{18}MIM]$	1-Octadecyl-methylimidazolium
[MTOA]	Methyl trioctylammonium
[BTMA]	Butyl trimethylammonium
AMMOENG 100	Ethyloctadecanoyl oligoethyleneglycol ammonium
AMMOENG 102	Ethyloctadecanoyl oligoethyleneglycol ammonium
Choline	2-Hydroxy-N,N,N-trimethylethanammonium

11.1 Introduction

Increasing energy demands inevitably lead to an increase in crude oil prices, directly affecting global economic activity [1]. The progressive depletion of conventional fossil fuels with increasing energy consumption and gas emissions have led to a move towards alternative, renewable, sustainable, efficient, and cost-effective energy sources with lower emissions. Biomass appears to be the most feasible feedstock for current routes towards the production of biofuels since it is renewable, cheap, has low sulphur content and involves no net release of carbon dioxide, meaning that it has a high potential to become economically feasible at the present time [2].

Bioethanol is a major biofuel on the market worldwide. In 2011, total fuel bioethanol production worldwide was 28.94 billion gallons (109.4 billion litres) [3]. It is estimated that bioethanol production could reach more than 227.4 billion litres of bioethanol thereby displacing a substantial portion of the fossil fuel currently consumed by the transportation sector [4]. Bioethanol is used to partially replace gasoline to make gasoline-ethanol mixtures, E15 (15 % ethanol and 85 % gasoline) and E85 (85 % ethanol and 15 % gasoline). The current commercial fuel ethanol is produced mainly from sugarcane and corn, depending on the climatic conditions of the producers' locations. The feedstock used for fuel ethanol production is mainly sugarcane in tropical areas such as Brazil and Colombia, while it is predominantly corn in other areas such as the United States, the European Union and China [2].

However, the production of these raw materials is competing for the limited arable land available for food and feed production. Therefore, it is critical to investigate advanced or second generation biofuel production technologies. Bioethanol can also be produced from lignocellulosic materials, which is commonly called second generation bioethanol [5]. The feedstocks for the second generation bioethanol include agricultural residues, grasses, and forestry and wood residues [6].

Biodiesel which is produced using vegetable oil, plant oil and animal fat is another major biofuel. Nigam and Singh [7] consider biodiesel as a "carbon neutral" fuel, as any carbon dioxide released from its burning was previously captured from the atmosphere during the growth of the vegetative crop that was used for its production. Obviously, biodiesel is an alternative fuel for diesel, and most diesel engines can use 100 % biodiesel. The main feedstock currently used for biodiesel production includes soy bean, canola seed or rapeseed, sunflower and palm oil. There are going research activities into using alternative oils such as waste oils from kitchens and restaurants and microalgal oils for biodiesel production. However, these biofuels represent a tiny portion (<4 %) of the total fuels consumed because it is not feasible to greatly increase biofuel production using the current technologies available [7].

11.2 Biotransformation in Ionic Liquids

In recent years, the application of ionic liquids (ILs) as (co)solvents and/or reagents in enzymatic catalysis for the production of biofuels is an emerging research area (Fig. 11.1).

ILs are composed entirely of ions and are liquids at ambient or far below ambient temperature, and have been extensively used as a potential alternative to toxic, hazardous, flammable and highly volatile organic solvents. ILs, particularly, have been shown to be exceptionally interesting non-aqueous reaction media for enzymatic transformations [8, 9]. Typical ILs are based on organic cations, e.g. 1,3-dialkylimidazolium, tetraalkylammonium, etc., paired with a variety of anions that have a strongly delocalized negative charge (e.g. $[BF_4]$, $[PF_6]$, etc.), which results in colourless and easily handled materials with very interesting properties as solvents [10]. Their interest as green solvents resides in their negligible vapour pressure, excellent thermal stability, high ability to dissolve a wide range of organic and inorganic compounds, and their non-flammable nature, which avoids the problem of the emission of volatile organic solvents to the atmosphere. Moreover, their solvent properties, such as miscibility or immiscibility with water or some

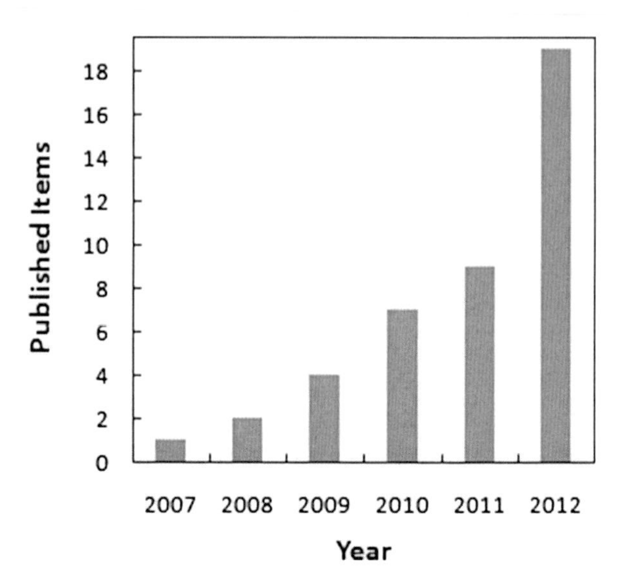

Fig. 11.1 Published items in each year (Source: Web of Science http://apps.webofknowledge.com (keywords for the search: ionic liquids, biofuels and enzyme))

organic solvents (e.g. hexane), can be tuned by selecting the appropriate cation and anion, which increases their usefulness for recovering products from the reaction mixture [8]. So, an important aspect for an IL is that it able to partially or completely solubilize the reactants, as well as low solubility for the reaction products. Solubility with reactants improves the reaction by allowing reactants to come into contact with each other, while the product can be separated by simple decantation as it is insoluble with ionic liquids, which can therefore be recycled.

The use of ILs as a reaction medium for enzymes, the application of such green compounds as cosolvents and/or reagents for biotransformation is well recognized. Studies on enzymatic reactions in ILs over the last 8–9 years have revealed not only that ILs are environmentally friendly alternatives to volatile organic solvents, but also that they have excellent selectivity, including substrate, regio- and enantioselectivity. Besides, many enzymes maintain very high thermal and operational stability in ILs. In this strategy, enzymes are simply suspended in the ILs, and then the resulting mixture can be used for biocatalytic reactions. It has been reported that lipase from *Candida antarctica* immobilized with IL is active at very high temperatures (95 °C) in hexane- and solvent-free conditions [11].

The most interesting biotransformations in ILs were observed at low water contents or in nearly anhydrous conditions because of the ability of hydrolases to carry out synthetic reactions. Furthermore, it is possible to design two-phase reaction systems that easily permit product recovery [12–14]. In anhydrous conditions, most of the ILs miscible with water clearly act as enzyme deactivating agents (e.g. [BMIM][Cl] or [BMIM][NO_3]), with a few exceptions (e.g. [BMIM][BF_4]) [8, 15, 16]. However, all the water-immiscible ILs assayed (e.g. [BMIM][PF_6] or [BMIM][NTf_2]) were shown to be suitable reaction media for biocatalytic reactions at low water content (<2 %, v/v) or in anhydrous conditions. In this context, lipases are enzymes that have been most widely studied because of the high level of activity and stereoselectivity displayed in synthesizing many different compounds, e.g. aliphatic and aminoacid esters [17], chiral esters by kinetic resolution of sec-alcohols [12–14, 18, 19], flavonoid derivatives [20], polymers [15, 21], etc. Numerous other enzymatic reactions have also been reported in ILs [22]. Furthermore, water-immiscible ILs also have an important stabilizing effect on hydrolases (lipases, esterases, proteases, etc.) in nearly anhydrous conditions [23–26].

We will focus on the use of enzymes in ILs for biofuels production.

11.2.1 Biodiesel Production in Ionic Liquids

Due to their unique proprieties, ILs have been used for biodiesel production processes through lipase catalyzed transesterification (alcoholysis) of vegetable oils (or animal fats) (Fig. 11.2) by several research groups [27–34]. Biodiesel is a renewable diesel fuel that is also known as FAME (fatty-acid methyl esters).

The enzymatic transesterification method offers many advantages over the chemical methods, such as mild reaction conditions, low energy demand, low waste treatment, the reusability of enzymes (lipases in most cases), flexibility in choosing different enzymes for different substrates, and the fact that it allows a small amount of water to be present in substrates. Besides, in chemical processes, some oil or fats may need a pretreatment for deacidification, depending on the composition of the materials, to remove free fatty acids (FFAs), which form soap with alcohols. Lipase can convert both triglycerides and FFAs into biodiesel [2]. Some authors have even successfully used waste cooking oil to obtain enzymatic biodiesel which may be a promising alternative for reducing the cost of biodiesel [27, 35–38].

The production yield was improved markedly when immobilized *Candida antarctica* lipase B (CALB) on an acrylic resin was used as a biocatalyst compared with other microbial lipases [27, 29, 30, 39, 40]. Ha et al. [30] screened 23 ionic liquids as solvents in the production of biodiesel from soybean oil using *Candida antarctica* lipase as catalyst. [EMIM][TfO] produced the highest biodiesel yield (80 % in 12 h of reaction), a yield that was better than for the solvent-free system and other commonly used solvents (tert-butanol). Nineteen ILs were studied to determine their effectiveness as solvents in the transesterification process using *Burkholderia cepacia* lipase (BSL) as catalyst [32]. The ionic liquid used have combinations of cations and anions, being the cations based mainly on imidazolium, while the anions were [NTf_2] and [PF_6] to get a suitable reaction media (Table 11.1).

Lipase-catalyzed methanolysis when conducted in a solvent-free medium led to the deactivation of lipase with increased molar ratio of methanol to sunflower oil >3 [45]. A similar deactivation of lipases was also observed during lipase catalyzed methanolysis in a biphasic oil-aqueous system for FAME production [46]. Methanolysis was conducted at different molar ratios of methanol to oil

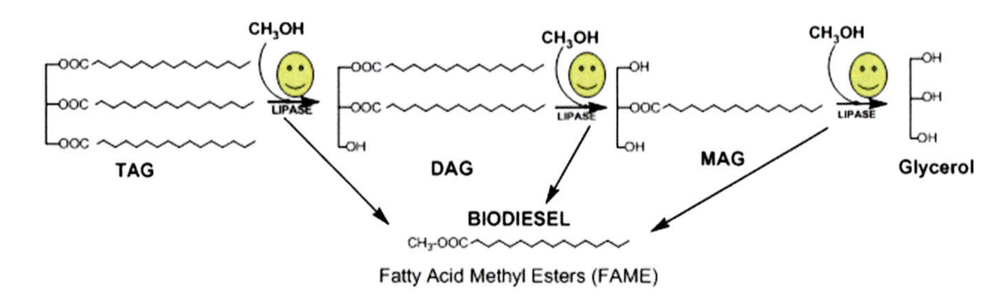

Fig. 11.2 Biocatalytic production of biodiesel

Table 11.1 Ionic liquids used in biodiesel synthesis with enzyme as catalyst

Ionic liquid	Enzyme	Oil source	Reaction conditions	Biodiesel % yield	References
Ammoeng 100	Lipase (*Candida antárctica*)	Triolein	60 °C, 10 h	99	[40]
[EMIM][TfO]	Lipase (*Candida antárctica*)	Soybean	50 °C, 12 h	80	[30]
[EMIM][PF_6]	Lipase (*Candida antárctica*)	Sunflower	60 °C, 4 h	99	[41]
[BMIM][NTf_2]	Lipase (*Pseudomonas cepacia*)	Soybean	25 °C, 30 h	96.3	[29]
[BMIM][PF_6],	Lipase (*Penicillium expansum*)	Corn oil	40 °C, 25 h	69.7	[9]
[BMIM][PF_6],	Lipase (*Candida antarctica*)	Triolein	48–55 °C, 6 h	80	[42]
[BMIM][PF_6]	Lipase (*Penicillium expansum*)	Corn oil	40 °C, 25 h	86	[43]
[OmPy][BF_4]	Lipase (*Burkholderia cepacia*)	Soybean	40 °C, 16 h	82.2	[32]
[BMIM][NTf_2]	Lipase (*Candida antárctica*)	Miglyols® oil 812	50 °C, 3 h	93–97	[44]
[C_{16}MIM][NTf_2]	Lipase (*Candida antarctica*)	Triolein	60 °C, 6 h	98	[27]
[BMIM][PF_6]	Lipase (*Penicillium expansum*) (CLEAs)	Microalgal oil	60 °C	85.7	[31]

(4:1 to 10:1) and similar results were obtained. Thus the concentration of methanol did not have a great effect on product formation in the presence of IL, which protects the lipase from methanol-induced deactivation.

Among the various types of ILs used, hydrophobic ILs were found to be the most effective for the production of biodiesel, the biodiesel yield increasing with both cation chain length and IL hydrophobicity, and decreasing when ILs with strong water miscible properties were used ([BMIM][NTf_2], [EMIM][PF_6], AMOENG 100, AMOENG 102, [C_{16}MIM][NTf_2] or [C_{18}MIM][NTf_2] etc). Hydrophilic ILs were not suitable as solvent in enzyme-catalyzed transesterification as only 10 %

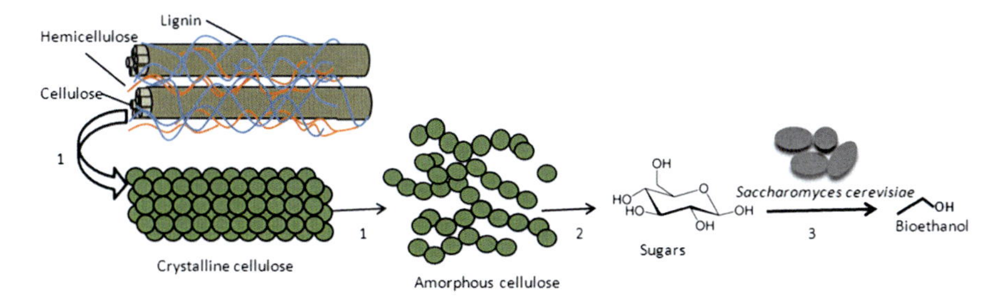

Fig. 11.3 Bioethanol production process. Steps: (1) Pretreatment of substrates, (2) Saccharification process, (3) Fermentation and Distillation

FAME yield was obtained for [HMIM][BF_4], while no FAME was observed when [BMIM][BF_4] was used as the solvent [33].

This trend can be explained in terms of methanol-induced enzyme deactivation. Hydrophobic ILs protects the enzyme from such deactivation because lipase is entrapped in the IL matrix. The most notable advantages of the use of ILs in such bioconversions are that the biodiesel can be separated by simple decantation and the recovered ionic liquid/enzyme catalytic system can be re-used several times without loss of catalytic activity and selectivity. More recently, our group [27] used [C_{16}MIM][NTf_2] as a homogeneous reaction mixture and, when the reaction was complete, a triphasic system was created through the appearance of a FAMEs phase (upper layer), a glycerol phase (middle layer) and a lower layer with the ionic liquid containing the enzyme, which could be solidified by decreasing the reaction temperature of the media (the melting point for [C_{16}MIM][NTf_2] is 42.6 °C), in this way facilitating extraction of the biodiesel product. Furthermore, ILs provides the ideal medium for removal of the by-product glycerol, thus accounting for the increase in biodiesel yield. A promising strategy employed by Lai et al. [47] was to use cross-linked enzyme aggregates (CLEAs) of lipase from *Penicillium expansum* as catalyst for biodiesel production in [BMIM][PF_6] from microalgal oil, with a conversion of 85.7 % in 48 h.

11.2.2 Fermentable Sugar Production in Ionic Liquids

Bioethanol can be produced from lignocellulosic biomass feedstocks (wood, grasses, agricultural residues, and waste materials). The general steps for producing ethanol include pretreatment of substrates, saccharification to release the fermentable sugars from polysaccharides, fermentation of the released sugars and, finally, a distillation step to separate the ethanol (Fig. 11.3).

Hydrolysis is usually catalyzed by cellulase enzymes and the fermentation is carried out by yeast or bacteria. Pretreatment of lignocellulosic materials is a prerequisite to facilitate the separation of cellulose, hemicellulose and lignin, so that complex carbohydrate molecules constituting the cellulose and hemicellulose can be broken down by enzyme-catalysed hydrolysis into their constituent simple sugars. Lignin consists of phenols and, for practical purposes, is not fermentable, while hemicellulose consists of 5-carbon sugars, and, although they are easily broken down into their constituent sugars such as xylose and pentose, the fermentation process is much more difficult, and requires efficient microorganisms that are able to ferment 5-carbon sugars to ethanol [2, 7, 48, 49]. Besides, in their natural state, cellulose fibers are highly crystalline and tightly packed, so pretreatment is necessary to increase the porosity and the accessible surface area for hydrolytic enzymes. Various biomass pretreatment methods have been used for the production of bioethanol, either simple such as steam explosion alone, or combined treatments such as steam explosion/ammonia, explosion/acid or alkali, fiber explosion/CO_2, chemical hydrolysis/enzymatic processes [50–53]. The advantages of biological pretreatment include low energy requirement and mild environmental conditions, but the hydrolysis rate is very low.

In a pioneering study by Swatloski et al. [41], several ILs, in particular [BMIM][Cl], were found to be capable of dissolving up to 25 % cellulose (by weight), forming highly viscous solutions. This prompted several other groups to test a variety of other ILs for their ability to dissolve cellulose [34, 43, 44, 54–61]. To optimize the use of lignocellulosic materials any pretreatment method should extract lignin and decrease the cellulose crystal structure. Lignocellulosic biomass can also be dissolved in ILs such as [AMIM][Cl], [BMIM][Cl] and [EMIM][Ac] [62–64] and, for ionic liquids containing the same cation [BMIM], the ability to dissolve the residual lignin was dependent on the anion, as follows (in order): [MSO_4] > [Cl] >> [Br] >>> [PF_6] [65]. However, not all ILs have the capacity to dissolve cellulose. For example, it has been reported that, due to the presence of cationic hydroxy and allyl groups, alkanolammonium ILs cannot dissolve the crystal structure of cellulose [66]. In this context, ILs possessing coordinating anions (e.g., [Cl], [NO_3], [Ac], [$(MeO)_2PO_2$]) which are strong hydrogen bond acceptors, have been found to be capable of dissolving cellulose in mild conditions by forming strong hydrogen-bonds with cellulose and other carbohydrates at high temperatures [22]. A good compromise between the solubility of lignin and cellulose is achieved with [EMIM][Ac] [67].

Liu et al. [68] provided an extensive review on the mechanism of cellulose dissolution in ionic liquids, demonstrating that the key parameters in the capacity of ILs, for cellulose dissolution are the cation and anion size and the ability to form hydrogen bonds with cellulose. Besides, the presence of water is disadvantageous to the solubility of carbohydrates, but is necessary for cellulose hydrolysis. However, high concentrations of water solvate the ions of the ionic liquid and thus prevent it from interacting with carbohydrates. A compromise must be reached between the water content and the cellulose hydrolysis rate in imidazolium ionic liquids [69].

Froschauer et al. [70] reported that a mixture of the cellulose dissolving IL EMIM OAc with 15–20 wt% of water is able to selectively extract hemicelluloses when mixed with a paper-grade kraft pulp for 3 h at 60 °C. This fractionation

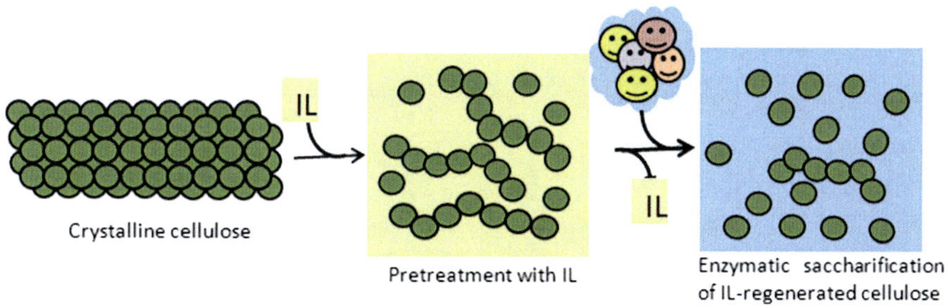

Fig. 11.4 Pre-treatment of cellulose with IL for bioethanol synthesis

method suggests the use of a cellulose solvent, which can serve repeatedly for complete dissolution of the purified cellulose fraction when applied undiluted.

It seems that although hydrophilic ILs are effective for the dissolution of cellulose, the activity of cellulases decreases significantly in their presence, which is consistent with what has been found for other enzymatic reactions. To overcome the negative effect of ILs on the enzymes, many research groups have regenerated cellulose from ILs prior to enzymatic saccharification, observing the faster hydrolysis of IL-regenerated cellulose compared to untreated cellulose [61, 67, 71]. Ionic liquid-treated cellulose was found to be essentially amorphous and more porous than native cellulose, both of which are effective parameters for enhancing the enzymatic action [72]. Figure 11.4 shows a scheme for regenerating cellulose from ILs prior to enzymatic saccharification for bioethanol synthesis.

After the pretreatment of cellulose with ionic liquids, the ILs must be removed; for this, methanol, ethanol and deionized water can be used as anti-solvents to regenerate cellulose from the cellulose/IL solutions. Enzymatic hydrolysis is affected by the anti-solvents used, the pretreatment temperature and the residual amount of ILs [73]. However, it has been shown that microwave irradiation (or sonication to a lesser degree) enhances the efficiency of dissolution compared with thermal heating, and, when used along with ionic liquid pretreatment, it increases the conversion of cellulose during the enzymatic reaction by making the external and internal surface area of cellulose more accessible. Similarly, Kamiya et al. [74] reported enzymatic *in situ* saccharification of cellulose in aqueous–ILs by adjusting the ratio of [EMIM] [DEP] to water, and Yang et al. [75] presented a new approach for enzymatic saccharification of cellulose in ionic liquids ([MMIM][DMP)-aqueous media, in which ultrasonic pretreatment is used to enhance the conversion of cellulose. Another strategy proposed to improve the stability of the cellulose in [BMIM][Cl] was to coat the immobilized enzyme particles with hydrophobic ILs. In this way, the stability of cellulase in hydrophobic IL/[BMIM][Cl] mixtures being greatly improved with respect to [BMIM] [Cl] alone [76].

Besides, it must be taken into account that cellulase activity is inhibited by cellobiose and, to a lesser extent, by glucose. Several methods have been developed to reduce such inhibition, including the use of high enzyme concentrations, supplementation with β-glucosidases during hydrolysis, and removing sugars during hydrolysis. At this point, the importance of the purity should be mentioned, since water, halides, unreacted organic salts and organics easily accumulate in ionic liquids, thus influencing the solvent properties of the IL, and/or interfering with the biocatalyst.

11.3 Biocompatibility of Ionic Liquids with Enzymes for the Production of Biofuel

Ionic liquids are biocompatible with enzymes if they allow enzymes to operate efficiently in them. These ILs must maintain an appropriate balance between the activity and stability of enzymes or even exalt their properties. It is well known that the catalytic activity of an enzyme strongly depends on this 3D structure or native conformation, which is maintained by a high number of weak internal interactions (e.g. hydrogen bonds, van der Waals, hydrophobic interactions, etc.), as well as interactions with other molecules, mainly water, as a natural solvent of living systems. Thus, water is the key component of all non-conventional media, because of the importance that enzyme-water interactions have in maintaining the active conformation of the enzyme. Few clusters of water molecules are required for the catalytic function, in which hydrophobic solvents typically permit higher enzymatic activity than hydrophilic ones due to their tendency to strip some of these essential water molecules [77].

Generally, there are three ways to use ionic liquids in a biocatalytic process:

1. As a pure solvent in nearly anhydrous conditions;
2. In aqueous solutions as a co-solvent;
3. In biphasic systems.

When an ionic liquid is used as a pure solvent, proper control of the water content, or, better, of water activity, is of crucial importance as since minimum amount of water is necessary to maintain the enzyme's activity. But, when aqueous solutions of ILs are used as reaction media, the IL and the assayed concentration are key criteria because of the high ability of water-miscible ILs to deactivate enzymes. Therefore, the water content is one of the key factors for enzymatic transesterification reactions for biodiesel production because excess water causes reverse hydrolysis reaction. So, the amount of water required to provide optimal enzyme activity differs according to the type of enzyme and composition of the reaction medium.

The most important properties of ionic liquids when they are used with biocatalysts are their polarity, hydrophilicity, viscosity and purity. The significance of these factors depends on the system used. Recently, Yang [78] has reported a good review on the effect of ionic liquids on enzymes for biotransformations.

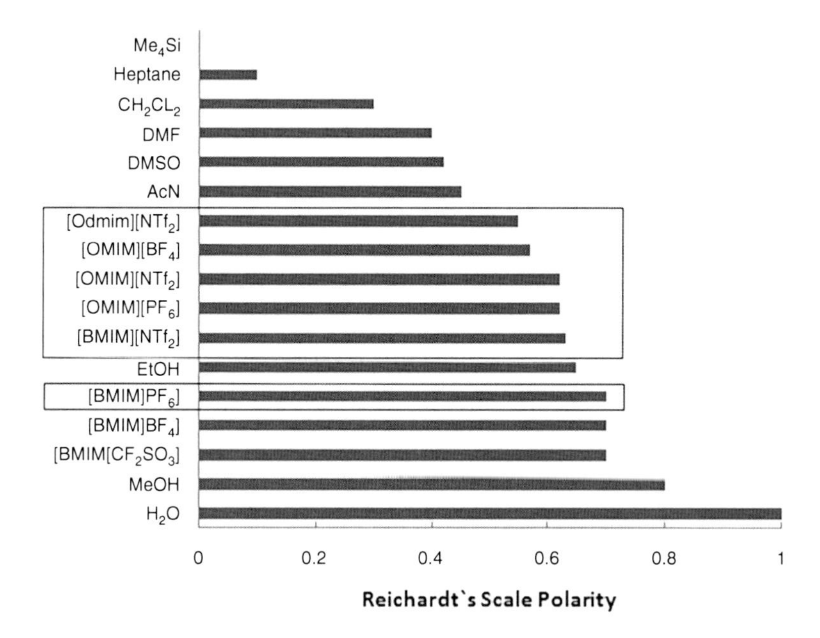

Fig. 11.5 Polarity of some ILs and organic solvents according to the Reichardt scale. Water-immiscible ILs are inside the *box*

11.3.1 Major Factors that Affect Enzyme Activity and Stability in Ionic Liquids

11.3.1.1 IL Polarity

One of the key parameters in the biocompatibility of ILs with enzymes is their polar character, which is not to be confused with the non-water miscibility of some of them (Fig. 11.5).

The subject of ionic liquid polarity has been addressed using a variety of methodologies, including measuring the absorption maximum of a solvatochromic dye, eg. Nile Red or Reichardt's dye, using a fluorescent probe, measuring the keto-enol equilibria (which are known to depend on the polarity of the medium), microwave dielectric spectroscopy measurements, etc. [79–82]. As a rule, the polar character of anions decreases as a function of the size/delocalization of the negative charge (e.g. [Cl] > $[NO_2]$ > $[NO_3]$ > $[BF_4]$ > $[NTf_2]$ > $[PF_6]$), while, in the case of cations, polarity is mainly determined by the length of the alkyl chain groups. Simple chemical reasoning predicts that a polar medium will dissolve polar compounds, such as carbohydrates. By this measure, the ionic liquid [BMIM]$[BF_4]$, which is hydrophilic, and [BMIM]$[PF_6]$, which is hydrophobic, fail the polarity test, because both dissolve less than 0.5 g/L of glucose at room temperature [68]. The concept of solvent polarity with ILs is too elusive to serve as a basis to predict, for example, solubility or reaction rates. There are, moreover, indications that solvent-solute interactions of ionic liquids obey a dual interaction model (i.e. ionic liquids behave as non-polar solvents with non-polar solutes but display a polar character with polar solutes), even to the extent that ionic liquids should be regarded as nanostructured materials [8]. However, the IL polarity-enzyme activity correlation has not been established for many enzymatic reactions performed in ILs [15, 19, 83].

Yang and Pan [9] suggested that enzyme activity may be related more to the viscosity and less to the polarity of ionic liquids. Reaction rates have usually been compared in different ionic liquids when the same amount of water is present in the reaction system (e.g. 2 % v/v of water). Therefore, the higher reaction rates in more polar ionic liquids [25, 84] can be explained by the effect of viscosity. Under such conditions, the solvent of higher polarity would leave less water associated with the enzyme and more water remaining in the solvent: the former would result in a lower reaction rate, while the latter would lead to a reduction in the viscosity of the ionic liquid and, in turn, to an improvement in the mobility of the protein molecule.

11.3.1.2 Hydrophobicity and Water Miscibility

Another property of ILs is hydrophobicity, which should not be confused with solvent polarity. The terms hydrophilic/hydrophobic ions are often used synonymously with water miscibility based on the miscibility of ILs with water, so that ionic liquids can be divided into two categories: hydrophobic (water-immiscible) and hydrophilic (water-miscible). The hydrophobicity of ILs is usually quantified by the *log P* scale, a concept derived from the partition coefficient of ILs between 1-octanol and water. Russell's group measured the *log P* values (< -2.0) of several ILs, and suggested that they are very hydrophilic in nature based on Laane's scale [31]. They also observed that free lipase (*Candida rugosa*) was only active in hydrophobic [BMIM][PF_6] (*log P* $= -2.39$), but inactive in other hydrophilic ILs, including [BMIM][Ac] (*log P* $= -2.77$), [BMIM][NO_3] (*log P* $= -2.90$) and [BMIM][TfA] [15]. Through a systematic investigation of lipase-catalyzed transesterification in over 20 ILs, it was observed that lipase activity increased with the *log P* value to reach a maximum, and then decreased as *log P* further increased (a bell-shaped dependence). These examples implied that the high hydrophobicity (high *log P*) of ILs could be beneficial for enzyme stabilization [34]. However, many exceptions to this rule have been reported, and ionic liquids have not been treated according to this *log P* concept to relate them to enzyme activity.

The water miscibility of ILs generally depends on the anions they contain, and the solubility of water an ionic liquid can be varied by changing the anion from [Cl] to [PF_6]. However, this behaviour varies widely and sometimes unpredictably; for example, [BMIM][BF_4], does not dissolve simple sugars to an appreciable degree and [BMIM][Cl], in contrast, dissolves massive amounts of cellulose. And yet, these ionic liquids are of similar polarity on the Reichardt's scale. It was demonstrated that the ability of ionic liquids to dissolve complex compounds, such as sugars and proteins, mainly depends on the H-bond accepting properties of the anion. A recent measurement of the H-bond accepting properties of such ILs revealed that $[BF_4]$ or $[MeSO_4]$ were better H-bond acceptors ($\beta = 0.61$ and 0.75, respectively) than $[PF_6]$ ($\beta = 0.50$), which can be considered a reasonable explanation for the difference in water miscibility [85]. The high hydrogen-bonding basicity and overall hydrophilic nature of water-miscible ILs enable to dissolve enzymes (to a greater or lesser degree) while enzymes are barely soluble in hydrophobic ILs [86]. The Sheldon group [87] maintains that hydrogen bonding could be the key to understanding the interactions of proteins and ionic liquids. Water is a powerful hydrogen bonding medium and an ionic liquid must mimic water in this respect to dissolve proteins, in particular as regards the hydrogen bond-accepting properties of the anion. The interaction should not be too strong, however, because, otherwise the hydrogen bonds that maintain the structural integrity of the α-helices and β-sheets will dissociate, causing the protein to unfold. So, to maintain the activity of ionic liquid-dissolved enzymes, a balance of mild hydrogen bond-accepting and donating properties is required. In contrast, in ionic liquids that do not dissolve enzymes, the enzyme preserves its native structure in this ionic liquid just as it preserves its catalytic activity. Besides, enzyme-compatible anions exhibit lower hydrogen bond basicity, which minimizes interference with the internal hydrogen bonds of an enzyme [84]. They also exhibit lower nucleophilicity and thus a lower tendency to change the enzyme's conformation by interacting with the positively charged sites in the enzyme structure [87].

The approximate ionic association strength in aprotic solvents is listed below in increasing order [88]:

$[NTf_2] < [PF_6] < [ClO_4] < [SCN] < [BF_4] < [TfO] < [Br] < [NO_3] < [TFA] < [Cl]$

This order represents the strength of an anion in its interaction with solvated cations through ionic attraction, or may even represent the strength of interactions between the anion and the changed surface of macromolecules (such as proteins). Dupont [10] suggested the strength of hydrogen-bond basicity in the similarly increasing order of:

$[BPh_4] < [PF_6] < [BF_4] < [TFA]$

These sequences confirm that enzyme activity is probably related with the hydrogen-bond acceptor strength of the anion: anions with low hydrogen-bond basicity are enzyme stabilizing.

In regards to IL cations, these are usually accepted to show a lower dominant effect than anions of the same charge density, because anions are more polarizable and hydrate more strongly [43]. So, cations seem to interact indirectly via interaction with anions, depending on the degree of coordination and the length of the alkyl chain of the cations.

Other authors suggest that the stability of the enzyme depends on the nucleophilicity of the anion. Kaar et al. [15] observed that free *Candida rugosa* lipase was only active in hydrophobic $[BMIM][PF_6]$, but inactive in all hydrophilic ILs based on $[NO_3]$, [Ac] and [TFA] during the transesterification of methylmethacrylate with

2-ethyl-1-hexanol. They indicated that the last three anions are more nucleophilic than [PF_6], and thus could interact with the enzyme causing protein conformational changes. However, in another study, a contradictory result was reported. Irimescu and Kato [89] carried out the CALB-catalyzed enantioselective acylation of 1-phenylethylamine with 4-pentenoic acid, and found that the reaction rate relied on the type of IL anions (reaction rate in a decreasing order of [TfO] > [BF_4] > [PF_6], keeping the cation unchanged). This suggests higher anion nucleophilicity, correlating with higher enzymatic activity. On the other hand, Lee et al. [90] measured the initial transesterification rates of three lipases (Novozym 435, *Rhizomucor miehei* lipase, and *Candida rugosa* lipase) in different ILs with the same water activity, and observed that the anion effect on the initial rates followed a decreasing order: [NTf_2] > [PF_6] > [TFO] > [SbF_6] ~ [BF_4]. They suggested that [TFO] and [BF_4] are more nucleophilic than [PF_6], although these results could be better explained by the anion hydrophobicity of IL.

An interesting phenomenon observed by Zhao et al. [34] during the transesterification of ethyl butyrate with 1-butanol was that when the solvents (dichloromethane or ionic liquids) and substrates were dried but the lipase was not totally dried (~3 % wt water) higher reaction rates were observed with microwaves than in a water-bath. However, when the enzyme was also dried, the differences in reaction rates became insignificant. This interesting behaviour has actually been reported in a number of papers, the authors of which maintain that in a fairly dry hydrophobic solvent and substrate environment, the enzyme particle is surrounded by (at least) one layer of water molecules. The solvent is hydrophobic so it does not strip off the water layer. In this microenvironment, the water layer near the enzyme surface has a much higher relative dielectric constant ($\varepsilon_r = 80.1$ at 20 °C) than the surrounding IL. Therefore, under microwave irradiation, the enzyme surface is likely to have a higher temperature than the bulk solvent due to the superheating of the water layer.

For bioethanol production, hydrophilic ILs are effective in dissolving cellulose, but the activity of cellulases decrease significantly in their presence. Therefore, IL residues should be entirely removed after cellulose regeneration [83].

When dealing with ILs in nearly anhydrous conditions, at low water content (<2 % v/v), all the assayed water-immiscible ILs (i.e. [BMIM][PF_6], [BTMA][NTf_2], etc.) were shown to be suitable reaction media for biocatalytic reactions, because water-immiscible ionic liquids are nevertheless hygroscopic, as noted above, and readily absorb a small percentage of water [91], but sufficient to maintain the active conformation of the enzyme. For example, in the case of the lipase B (Novozym 435 from Novo) catalyzed transesterification of vegetable oil for biodiesel production, the best enzyme activity and a biodiesel yield of 97–98 % was obtained with [EMIM][PF_6] and [BMIM][PF_6], while hydrophilic ILs were poor solvents for this methanolysis for two reasons: oil insolubility (heterogeneous system) and enzyme deactivation [22, 32]. Besides, ILs with long alkyl chains (e.g., [C_{16}MIM][NTf_2] and [C_{18}MIM][NTf_2]) have been used in a homogeneous one-phase system for lipase-catalyzed biodiesel production which this avoids direct interactions between the enzyme and pure methanol and allows for the stable reuse of lipase in the ILs. These long chain and lipophilic ILs create a non-aqueous system for oil transesterification, and at the end of the reaction, a triphasic system was formed as a result of lowering the temperature, which facilitates biodiesel extraction [27, 28].

The remarkable results obtained for enzymatic reactions in water-immiscible ILs in nearly anhydrous conditions underline the suitability of these solvents as a clear alternative for biodiesel production. In this strategy, enzymes (immobilized or not) are simply suspended in the ILs, and the resulting a mixture can be used for biocatalytic reactions. Enzymes coated with ionic liquids can also be used. For example, for citronellyl ester synthesis, immobilized lipase from *C. antarctica* coated with ionic liquids at a very high temperature (95 °C) in hexane and solvent-free conditions was used [11].

11.3.1.3 Viscosity

Another important property of ILs for use as reaction media in biotransformations is their viscosity. Indeed, high viscosity is probably inherent to ionic liquids. Strong intermolecular forces between solvent molecules cause high viscosity. For ethylene glycol, the strong forces are hydrogen bonds, whereas for ionic liquids they are charge–charge interactions. Weaker interactions such as van der Waals forces also exist between molecules [92]. So, like other solvents, the viscosity of ILs is dependent on the ion-ion interactions, such as van der Waals interactions and hydrogen bonding. Therefore, the value of the viscosity varies significantly with the chemical structure, composition, temperature, and the presence of solutes such as, water, co-solvents or impurities. The viscosity of ionic liquids is high compared with that of molecular solvents and it has been shown that such viscosity generally increases with an increase in the alkyl chain length for a fixed anionic group due to the stronger van der Waals interactions. Consequently, solvent viscosity could affect the biocatalytic reaction rate in terms of the mass transfer limitation when the reaction is rapid and the IL is relatively viscous. For example, higher enzyme activity was observed in [EMIM][NTf_2] than in [MTOA][NTf_2] due to the former lower viscosity [25]. However, this trend is not true for all biocatalytic reactions performed in ILs, particularly when reaction rates are measured in equilibrium conditions rather than by kinetic control. A recent study of the lipase-catalyzed transesterification of ethyl butyrate and 1-butanol in more than 20 ILs further confirms that IL viscosity affects the reaction rates in some cases, but is not the primary factor in controlling the enzyme's activity [34]. Different researchers explain that the high viscosity of ILs slows the conformational changes of biomolecules, allowing enzymes to maintain a more compact structure [93]. van Rantwijk and Sheldon [8] interpreted that the underlying cause of this stabilizing effect is the high viscosity of ionic liquids, which slows the migration of protein domains from the active conformation into the inactive one. The possible explanation is that IL viscosity affects the preferential hydration interactions, and that the functional groups of the enzyme absorb more water molecules, which could explain why ILs with a long alkyl chain in their cations were better stabilizers of the protein structure. Besides, ILs bearing long alkyl chains in their cations often self-assemble in aqueous solutions and behave like surfactants, creating microemulsions [94, 95] and interactions with enzymes through two main mechanisms: (1) electrostatic interactions between the surfactant head group and charged amino acid residues of the protein, and (2) hydrophobic interactions between the alkyl chains of the surfactant and hydrophobic amino acid residues. So, knowledge of surfactant–protein interactions could be useful for understanding the effect of IL on enzyme activity and stability in some systems (both aqueous and non-aqueous).

11.3.1.4 Impurities

Several groups have disagreed on whether or not a given enzyme is active in a particular ionic liquid. For example, Schöfer et al. [19] reported that CALB had no activity in [BMIM][BF_4] or [BMIM][PF_6], but other groups reported good activity during transesterification or ammoniolysis in the same ionic liquids. Such inconsistencies may be the result of impurities.

By measuring the effect of some additives on lipase catalyzed acetylation, Park and Kazlauskas [84] concluded that the most likely causes of slow or no reactions in some ionic liquids are traces of silver ion or acidic impurities. More recently, Lee et al. [90] reported that the activity of Novozym 435 in [OMIM][NTf_2] decreased linearly with the chloride content, and that 1 % (wt.) increase in [OMIM][Cl] ($\sim$1,540 ppm [Cl^-]) caused a 5 % decrease in enzyme activity.

Washing with water followed by vacuum drying can be used to purify water-immiscible ionic liquids. However, the purification of water-miscible ionic liquids, such as [BMIM][BF_4] involves filtering through silica gel followed by washing with aqueous sodium carbonate solution.

Although these purified ionic liquids worked reliably, purification still needs further research. Besides, it has to be borne in mind that the effects of impurities may vary from enzyme to enzyme in ionic liquids.

Most ionic liquids based on common cations and anions should be colourless, with minimal absorbance at wavelengths greater than 300 nm. In practice, the salts often take on a yellow hue, particularly during the quaternization step. Impurities or unwanted side reactions involving oligomerization or polymerization of small amounts of free amine are a major limitation for studying enzyme structure by UV/visible spectroscopic techniques.

Clearly, the impurity most likely to be present in high concentrations in ionic liquids is water. Other reaction solvents are generally easily removed by heating the ionic liquid under vacuum. This observation is very important when ILs (as pure solvents) in nearly anhydrous conditions are used as reaction media.

11.3.2 Structural Organization of Enzyme-IL Systems

ILs form a strong ionic matrix and the added enzyme molecules could be considered as being included rather than dissolved in the media, meaning that ILs should be regarded as liquid enzyme immobilization supports, rather than reaction media, since they enable the enzyme-IL system to be reused in consecutive operation cycles [96]. Finally, after an enzymatic transformation process in ILs, products can usually be recovered by liquid-liquid extraction, although the organic solvents used in this step represent a clear breakdown point for the integral green design of any chemical process.

To understand the biocompatibility of ionic liquids with enzymes it is necessary to consider the structure-function relationships of enzymes in water immiscible ILs and to discern how water is partitioned between the enzyme surface and the bulk IL solution. Complementary spectroscopy measurements (e.g. fluorescence, circular dichroism, FTIR) have classically been used to investigate changes in the secondary structure of enzymes in an attempt to explain the stabilization or denaturation phenomena associated with their molecular environment. Such spectroscopic methods have been used to correlate changes in the secondary structure of monellin [97], CALB [98, 99] or α-chymotrypsin [100] with enzyme stability in ILs. Iborra's group were pioneers in carrying out structural studies that revealed that the synthetic activity and stability exhibited by CALB in ILs was much higher than that observed in hexane, and was related with the associated conformational changes that take place in the native structure of CALB, as demonstrated by fluorescence and CD spectroscopic techniques [98]. The stabilization of CALB by hydrophobic ILs seems to be related with the observed evolution of α-helix to β-sheet secondary structures of the enzyme, resulting in a more compact enzyme conformation, that is able to exhibit high catalytic activity, suggesting that the stability of enzyme in this medium was improved by the formation of a compact, but flexible, native-like conformation of the enzyme. Turner et al. [16] described how the deactivation of the enzyme cellulase produced by water-miscible ILs (e.g., [BMIM][Cl]) is accompanied by a fall in the fluorescence intensity maxima of the Trp parameter with respect to the native conformation in water as a result of the enhancing exposure of Trp residues to the bulk solvent and enzyme denaturation. Fluorescence spectroscopy demonstrated that monellin in a low water content (2 % v/v) in [BMpy][NTf_2] resisted thermal unfolding. Fujita et al. [101] elucidated the power of hydrated [Choline][H_2PO_4] to maintain the activity of cytochrome c after 18 months of storage in the dissolved form at room temperature because of its ability to maintain its native secondary structure and conformation, as monitored by ATR-FTIR (attenuated total reflection Fourier transform infrared) and resonance Raman spectroscopies. A later study found that CALB aggregates can deactivate in 1-ethyl-3-methylimidazolium-based ILs in an anion-dependent manner [42]. Studies of papain in 15 % (v/v) aqueous solutions of 1-alkyl-3-methylimidazolium-based ILs using ATR-FTIR demonstrated that the choice of anion has a significant impact on the structure, specificity and stability of the enzyme [102]. Again, the β-sheet content in the secondary structure increased, while the α-helical content decreased.

Micaêlo and Soares [103] presented a molecular dynamics simulation study of the serine protease, cutinase, in two different ILs, [BMIM][PF_6] and [BMIM][NO_3]. Their work showed that the enzyme is preferentially stabilized in [BMIM][PF_6], which allows a suitable degree of hydration to be maintained at the enzyme surface and hence renders a more native-like enzyme structure, while [BMIM][NO_3] tended to be more destabilizing. These findings are in accordance with previous experimental observations [15, 86] which attributed these results to the difference in the hydrophobicity of the two ILs: [BMIM][PF_6] is more hydrophobic than [BMIM][NO_3] and hence is less likely to dissociate into ions to destabilize the enzyme.

A study of human serum albumin (HSA) and equine heart cytochrome c (cyt c) by CD spectroscopy and small-angle neutron scattering (SANS) demonstrated that the IL 1-butyl-3-methylimidazolium chloride ([BMIM][Cl]) not only caused significant unfolding of the α-helical proteins when present as a cosolvent with water, but [BMIM][Cl] also changed the aggregation state of HSA, suggesting that the interaction depends on the protein sequence [104].

The secondary structure can also be analyzed with FTIR spectroscopy since proteins absorb infrared wavelengths due to peptide bond vibrations. Liu et al. [32] reported that a significant decrease in the α-helix content of lipase from *Burkholderia cepacia* probably affects the lipase active site: the lower the α-helix, the higher the "open" conformation of the active site, allowing easier access to the substrate.

More recently, Fan et al. [105] suggested that ILs could quench the intrinsic fluorescence of papain, probably by means of a static quenching mechanism. The calculated binding constants were very small compared with that of volatile organic solvents, indicating that only very weak interaction between ILs and papain existed. The Gibbs free energy change (ΔG), enthalpy change (ΔH), and entropy change (ΔS) during the interaction of papain and ILs were estimated. The negative values of these parameters obtained, indicated that the interaction between ILs and papain was a spontaneous process, also implying that hydrogen bonding and van der Waals forces played important roles in the interaction processes.

The impact of water-miscible ILs on proteins was characterized by structural changes of green fluorescent protein (GFP) in aqueous solutions containing 25 and 50 % (v/v) of [BMIM][Cl]. The SANS and spectroscopic results indicated that GFP is a great deal less compact in 50 % (v/v) [BMIM][Cl] than in neat water, suggesting unfolding from the native structure. The oligomerization state of the protein in IL-containing aqueous solution changes from a dimer to a monomer in response to the IL, but does not change as a function of temperature of the IL solution. The SANS and spectroscopic results also demonstrate that the addition of this hydrophilic ionic liquid to the solution lowers the thermal stability of GFP, allowing the protein to unfold at lower temperatures than in aqueous solution [106].

An aqueous solution of free-enzyme molecules added to the hydrophobic IL phase could be regarded as being included, but not dissolved, in the medium, the essential water shell around the protein being preserved, and providing an adequate microenvironment for the catalytic action [14]. Usually, enzymes fold by placing the non-polar residues in a hydrophobic core, while polar residues are located on the hydrated surface. A "memory" phenomenon is observed when an enzyme is placed in a dry hydrophobic system, because the biocatalyst is trapped in the native state as a consequence of the low dielectric constant of the medium. This intensifies intramolecular electrostatic interactions and enables the catalytic activity to be maintained [103, 107, 108]. The extremely ordered supramolecular structure of ILs in solid and liquid phase has been described as an extended network of cations and anions connected by hydrogen bonds [109]. This network might be able to act as a mould, maintaining an active three-dimensional structure of the enzyme in non-aqueous environments, and avoiding classical thermal unfolding. Therefore the incorporation of molecules and macromolecules in the ionic liquid network causes changes to the physico-chemical properties of these materials and can cause, in some cases, the formation of polar and non-polar regions [109]. So, enzymes in water immiscible ILs should also be considered as being included in the hydrophilic gaps of the network, where the observed enzyme stability could be attributed to the maintenance of this strong net around the protein. ILs can clearly be considered as both solvents and liquid immobilization supports because multipoint enzyme-IL interactions (ionic, hydrogen bonds, van der Waals, etc.) may occur, resulting in a supramolecular net able to maintain an active protein conformation [98] (Fig. 11.6).

A theoretical basis for predicting the compatibility of enzymes and anhydrous ionic liquids has not yet been developed, although a number of possibly contributing factors have been discussed, such as the cation H-bond donating capability, *log P*, formation of hydrogen-bonded nanostructures, and solvent viscosity [8]. With regard to the compatibility of enzymes and hydrophobic ionic liquids, hydrogen bonding could be the key to understanding. It is well known that the thermal stability of enzymes is enhanced in both aqueous and anhydrous media containing polyols as a consequence of an increase in hydrogen bond interactions. Thus, both the solvophobic interactions essential for maintaining the native structure and the water shell around the protein molecule are preserved by the "inclusion" of the aqueous solution of free enzyme in the IL network, resulting in a clear enhancement of enzyme stability (Fig. 11.7).

Yang [110] maintains that an ion may affect enzyme performance by playing the role of substrate, cofactor, or even inhibitor. But more generally, the effect of specific ions could be better understood by considering an ion's ability to alter the bulk water structure, to affect the protein–water interaction, and to directly interact with the enzyme molecules. So, the effect of ions on enzyme activity and stability has usually been linked to the Hofmeister series (or the kosmotropicity order): kosmotropic anions and chaotropic cations stabilize enzymes, while chaotropic anions and kosmotropic cations destabilize them. The influence of hydrophilic ILs on the protein activity and stability usually follows the Hofmeister series when ILs dissociate into individual ions in water [111] but, unfortunately, there are many cases in which this series is not followed, especially when there is little or no water present in the IL media, and, furthermore, some authors associate

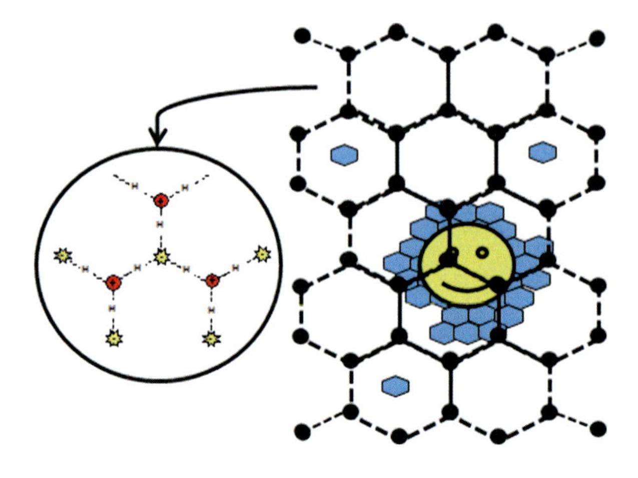

Fig. 11.6 Structural organization of enzyme hydrophobic ionic liquid systems

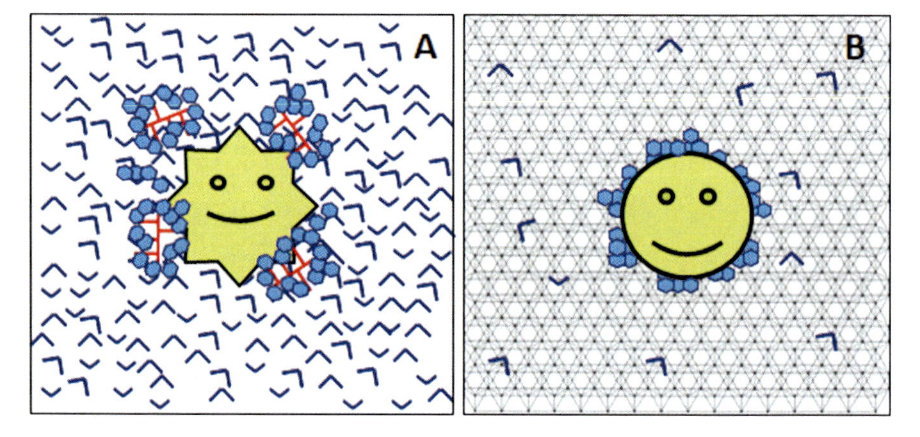

Fig. 11.7 Thermal stability of enzyme in water media containing polyols (**a**) and in water-immiscible ILs (**b**)

Hofmeister effects only with anions [112]. Micaêlo and Soares [103] presented a molecular simulation study of an enzyme in two ionic liquids, [BMIM][PF_6] and [BMIM][NO_3], observing that the enzyme structure is highly dependent on the amount of water present in the IL media and that [BMIM][PF_6] significantly increases protein thermostability at high temperatures, especially at low hydration values. These ILs "strip" most of the water from the enzyme surface in a degree similar to that found in the case of polar organic solvents, while the remaining water molecules at the enzyme surface are organized in many small clusters. [BMIM][PF_6] seems to retain similar amounts of water at the enzyme surface, as acetonitrile, and supports the evidence of the polar nature of this IL. This IL [BMIM][NO_3], in contrast, replaces almost all the water at the enzyme surface, which may be the reason for its destabilizing effect on the enzyme. A more detailed analysis of enzyme solvation by the two ILs shows that the anion species dominates the non-bonded interactions with the enzyme, as judged by the number of hydrogen bonds observed between the enzyme and the cation and anion species of each IL. The ability of ILs to dissolve molecules depends mainly on the hydrogen bond-accepting properties of the anion, as stated by Anderson et al. [113]. Moreover, Zhao et al. [83] observed that the dissolution of lipase in most hydrophilic ILs is an indication of strong interactions between the enzyme and solvent molecules. If such interactions disturb the active sites and/or are strong enough to disrupt the protein structures, the enzyme activity is lost. However, if such interactions are not too strong but allow the enzyme's structures to be maintained, these hydrophilic ILs do not inactivate the enzyme (such as $[Et_3MeN][MeSO_4]$, or AMMOENG series ILs).

Weingärtner et al. [112] observed the importance of "microheterogeneity" in ILs. The charged ionic groups and non-polar residues of cations and anions give rise to the nanoscale structural heterogeneity of ILs, which is not encountered in simple molecular solvents. The resulting hydrophilic and hydrophobic patches of the IL structure have intriguing consequences for solvation because they enable dual solvent behaviour: an IL can incorporate a non-polar solute in non-polar domains, while hydrophilic domains solvate polar solutes.

Finally, the type of reaction medium used is conditioned by the type of biotransformation; for example, water-immiscible ILs were found to be the most effective for the production of fermentable sugars from cellulose at low water content or in nearly anhydrous conditions. Moreover, immobilized CALB was the most efficient biocatalyst for the transesterification (alcoholysis) of vegetable oils (or animal fats). Both the IL and CALB can be recycled for at least four successive reactions without any loss of activity. Furthermore, aqueous solutions of hydrophilic ILs were necessary to produce bioethanol because the presence of water is necessary for cellulose hydrolysis. In this type reaction medium, the IL and the assayed IL-to-water concentration ratio are the key criteria.

Other strategies proposed to improve the efficiency of bioethanol and biodiesel transformations include biphasic systems based on IL and $scCO_2$ (supercritical carbon dioxide), the addition of cosolvents and a IL coating of immobilized enzyme particles. The enormous potential of immobilized multi-enzymatic or cross-linked enzyme aggregates for bioethanol synthesis in ionic liquid media has only just been realized.

11.4 Conclusions

There is, as yet, no theoretical basis for predicting the compatibility of ionic liquids with enzymes, although key parameters for this relationship depend on the type of reaction system.

(a) Water-immiscible ILs.

The enzyme structure is highly dependent on the amount of water present in the IL medium. So, the hydrophobicity of a water-immiscible ionic liquid may be considered as a constrainer of polarity, because hydrophobicity is related to miscibility with water, and the water shell around the protein molecule is essential for maintaining the activity/stability of the enzyme. Water-immiscible ionic liquids are nevertheless hygroscopic, as noted above, and readily absorb a low percentage of water. Besides, some local hydrophobic ion-enzyme macromolecule interactions are also important for enzyme stability. Thus, for the compatibility of enzymes and anhydrous ionic liquids, a hydrophobic effect could be the key.

(b) Water-miscible ILs.

These ILs are used as aqueous ionic liquid mixtures and the ratio of the ionic liquid-to-water used is crucial to the effect it has on the enzyme. To maintain the activity of hydrophilic ionic liquid-dissolved enzymes, a balance of mild hydrogen bond-accepting and donating properties is required, so cation and anion size and the ability to form hydrogen bonds are important for these systems because stabilization primarily results from hydrophobic forces and hydrogen-bond. The hydrogen-bond donating ability is usually a property of the cation, while the anions act as hydrogen-bond acceptors, and it has been demonstrated that the ability of ionic liquids to dissolve complex compounds, such as sugars and proteins, mainly depends on the hydrogen-bond accepting properties of the anion. So, with regard to the compatibility of enzymes and hydrophilic ionic liquids, hydrogen bonding could be the key. Ionic liquids, in particular their anions which form strong hydrogen bonds, may dissociate the hydrogen bonds that maintain the structural integrity of the α-helices and β-sheets, causing the protein to unfold wholly or partially. As discussed above, another key property for these systems is viscosity, which is strongly influenced by cation chain length.

References

1. He Y, Wang S, Lai KK. Global economic activity and crude oil prices: a cointegration analysis. Energy Econ. 2010;32:868–76.
2. Cheng JJ, Timilsina GR. Status and barriers of advanced biofuel technologies: a review. Renew Energy. 2011;36(12):3541–9.
3. RFA, Renewable Fuels Association. http://ethanolrfa.org/pages/World-Fuel-Ethanol-Production
4. Simmons BA, Loque D, Blanch HW. Next−generation biomass feedstocks for biofuel production. Genome Biol. 2008;9(12):242.1–6.
5. Naik SN, Goud VV, Rout PK, Dalai AK. Production of first and second generation biofuels: a comprehensive review. Renew Sustain Energy Rev. 2010;14(2):578–97.
6. Eisentraut E. Sustainable production of second − generation biofuels: potential and perspectives in major economies and developing countries. Paris: Int Energy Agency; 2010, p. 211.
7. Nigam PS, Singh A. Production of liquid biofuels from renewable resources. Prog Energy Combust Sci. 2011;37:52–68.
8. van Rantwijk F, Sheldon RA. Biocatalysis in ionic liquids. Chem Rev. 2007;107:2757–85.
9. Yang Z, Pan W. Ionic liquids: green solvents for nonaqueous biocatalysis. Enzyme Microb Technol. 2005;37:19–28.

10. Dupont J, de Souza RF, Suarez PAZ. Ionic liquids (molten salts) phase organometallic catalysis. Chem Rev. 2002;102:3667–91.
11. Lozano P, Piamtongkam R, Kohns K, De Diego T, Vaultier M, Iborra JL. Ionic liquids improve citronellyl ester synthesis catalyzed by immobilized *Candida antarctica* lipase B in solvent – free media. Green Chem. 2007;9:780–4.
12. Abe Y, Kude K, Hayase S, Kawatsura M, Tsunashima K, Itoh T. Design of phosphonium ionic liquids for lipase – catalyzed transesterification. J Mol Catal B: Enzym. 2008;51:81–5.
13. Itoh T, Nishimura Y, Ouchi N, Hayase S. 1 – Butyl – 2,3 – dimethylimidazolium tetrafluoroborate: the most desirable ionic liquid solvent for recycling use of enzyme in lipase – catalyzed transesterification using vinyl acetate as acyl donor. J Mol Catal B: Enzym. 2003;26:41–5.
14. Lozano P, De Diego T, Gmouh S, Vaultier M, Iborra JL. Criteria to design green enzymatic processes in ionic liquid/supercritical carbon dioxide system. Biotechnol Prog. 2004;20:661–9.
15. Kaar JL, Jesionowski AM, Berberich JA, Moulton R, Russell AJ. Impact of ionic liquid physical properties on lipase activity and stability. J Am Chem Soc. 2003;125:4125–31.
16. Turner MG, Spear SK, Huddleston JG, Holbrey JD, Rogers RD. Ionic liquid salt-induced inactivation and unfolding of cellulase from *Trichoderma reesei*. Green Chem. 2003;5:443–7.
17. Lozano P, De Diego T, Carrié D, Vaultier M, Iborra JL. Enzymatic ester synthesis in ionic liquids. J Mol Catal B: Enzym. 2003;21:9–13.
18. Itoh T, Matsushita Y, Abe Y, Han SH, Wada S, Hayase S, Kawatsura M, Takai S, Morimoto M, Hirose Y. Increased enantioselectivity and remarkable acceleration of lipase – catalyzed transesterification by using an imidazolium PEG – alkyl sulfate ionic liquid. Chem Eur J. 2006;12:9228–37.
19. Schöfer SH, Kaftzik N, Wasserscheid P, Kragl U. Enzymatic catalysis in ionic liquids: lipase catalysed kinetic resolution of 1 – phenylethanol with improved enantioselectivity. Chem Commun. 2001;30(5):425–6.
20. Katsoura MH, Polydera AC, Tsironis L, Tselepis AD, Stamatis H. Use of ionic liquids as media for the biocatalytic preparation of flavonoid derivatives with antioxidant potency. J Biotechnol. 2006;123:491–503.
21. Gorke JT, Okrasa K, Louwagie A, Kazlauskas RJ, Srienc F. Enzymatic synthesis of poly (hydroxyalkanoates) in ionic liquids. J Biotechnol. 2007;132:306–13.
22. Moniruzzaman M, Nakashimab K, Kamiyaa N, Gotoa M. Recent advances of enzymatic reactions in ionic liquids. Biochem Eng J. 2010;48:295–314.
23. Erbeldinger M, Mesiano AJ, Russell AJ. Enzymatic catalysis of formation of Z – aspartame in ionic liquid: an alternative to enzymatic catalysis in organic solvents. Biotechnol Prog. 2000;16:1129–31.
24. Feher E, Major B, Belafi – Bako K, Gubicza L. On the background of enhanced stability and reusability of enzymes in ionic liquids. Biochem Soc Trans. 2007;35:1624–7.
25. Lozano P, De Diego T, Carrié D, Vaultier M, Iborra JL. Over – stabilization of *Candida antarctica* lipase B by ionic liquids in ester synthesis. Biotechnol Lett. 2001;23:1529–33.
26. Persson M, Bornscheuer UT. Increased stability of an esterase from *Bacillus stearothermophilus* in ionic liquids as compared to organic solvents. J Mol Catal B: Enzym. 2003;22:21–7.
27. De Diego T, Manjon A, Lozano P, Iborra JL. A recyclable enzymatic biodiesel production process in ionic liquids. Biores Technol. 2010;102(10):6336–9.
28. De Diego T, Manjon A, Lozano P, Vaultier M, Iborra JL. An efficient activity ionic liquid – enzyme system for biodiesel production. Green Chem. 2010;13(2):444–51.
29. Gamba M, Lapis AAM, Dupont J. Supported ionic liquid enzymatic catalysis for the production of biodiesel. Adv Synth Catal. 2008;350:160–4.
30. Ha SH, Lan MN, Lee SH, Hwang SM, Koo Y–M. Lipase – catalyzed biodiesel production from soybean oil in ionic liquids. Enzyme Microb Technol. 2007;41:480–3.

31. Laane C, Boeren S, Vos K, Veeger C. Rules for optimization of biocatalysis in organic solvents. Biotechnol Bioeng. 1987;30:81–7.
32. Liu Y, Chen D, Yan Y, Peng C, Xu L. Biodiesel synthesis and conformation of lipase from *Burkholderia cepacia* in room temperature ionic liquids and organic solvents. Biores Technol. 2011;102:10414–18.
33. Sunitha S, Kanjilal S, Reddy PS, Prasad RBM. Ionic liquids as a reaction medium for lipase – catalyzed methanolysis of sunflower oil. Biotechnol Lett. 2007;29(12):1881–5.
34. Zhao H, Baker GA, Song Z, Olubajo O, Zanders L, Campbell SM. Effect of ionic liquid properties on lipase stabilization under microwave irradiation. J Mol Catal B: Enzym. 2009;57:149–57.
35. Chen ZF, Zong MH, Wu H. Improving enzymatic transformation of waste edible oil to biodiesel by adding organic base. Div Fuel Chem. 2005;50:2.
36. De Paola MG, Ricca E, Calabró V, Curcio S, Iorio G. Factor analysis of transesterification reaction of waste oil for biodiesel production. Biores Technol. 2009;100:5126–31.
37. Li N, Zong M–H, Zong H. Highly efficient transformation of waste oil to biodiesel by immobilized lipase from *Penicillium expansum*. Process Biochem. 2009;44:685–8.
38. Maceiras R, Vega M, Costa C, Ramos P, Marquez MC. Effect of methanol content on enzymatic production of biodiesel from waste frying oil. Fuel. 2009;88:2130–4.
39. Guimaraes DM, de Sousa JS, d'Avila E. Biotechnological methods to produce biodiesel. In: Pandey A, Larroche C, Ricke SC, Dussap CG, Gnansounou E, editors. Biofuels: alternative feedstocks and conversion processes. San Diego: Elsevier; 2011. p. 315–37.
40. Guo Z, Chen B, Murillo RL, Tan T, Xu X. Functional dependency of structures of ionic liquids: do substituents govern the selectivity of enzymatic glycerolysis? Org Biomol Chem. 2006;4:2772–6.
41. Swatloski RP, Spear SK, Holbrey JD, Rogers RD. Dissolution of cellulose with ionic liquids. J Am Chem Soc. 2002;124(18):4974–5.
42. Sate D, Janssen MHA, Stephens G, Sheldon RA, Seddon KR, Lu JR. Enzyme aggregation in ionic liquids studied by dynamic light scattering and small angle neutron scattering. Green Chem. 2007;9:859–67.
43. Zhang YHP, Ding SY, Mielenz JR, Cui JB, Elander RT, Laser M, Himmel ME, McMillan JR, Lynd LR. Fractionating recalcitrant lignocellulose at modest reaction conditions. Biotechnol Bioeng. 2007;97(2):214–23.
44. Zhao H, Baker GA, Song ZY, Olubajo O, Crittle T, Peters D. Designing enzyme – compatible ionic liquids that can dissolve carbohydrates. Green Chem. 2008;10(6):696–705.
45. Soumanou MM, Bornscheuer UT. Improvement in lipase-catalysed synthesis of fatty acid methyl esters from sunflower oil. Enzyme Microb Technol. 2003;33:97–103.
46. Oliveira AC, Rosa MF. Enzymatic transesterification of sunflower oil in an aqueous-oil biphasic system. J Am Oil Chem Soc. 2006;83:21–5.
47. Lai J-Q, Hub Z-L, Sheldon RA, Yang Z. Catalytic performance of cross-linked enzyme aggregates of *Penicillium expansum* lipase and their use as catalyst for biodiesel production. Process Biochem. 2012;47:2058–63.
48. Lozano P, De Diego T, Gmouh S, Vaultier M, Iborra JL. Enzyme catalysis in ionic liquids and supercritical carbon dioxide. In: Ionic liquid applications: pharmaceuticals, therapeutics, and biotechnology. Washington, DC: American Chemical Society; 2010. p. 181–96.
49. Mosier N, Wyman C, Dale B, Elander R, Lee YY, Holtzapple M, Ladisch M. Features of promising technologies for pretreatment of lignocellulosic biomass. Biores Technol. 2005;96 (6):673–86.
50. Ben – Ghedalia D, Miron J. The effect of combined chemical and enzyme treatment on the saccharification and in vitro digestion rate of wheat straw. Biotechnol Bioeng. 1981;23:823–31.
51. Morjanoff PJ, Gray PP. Optimization of steam explosion as method for increasing susceptibility of sugarcane bagasse to enzymatic saccharification. Biotechnol Bioeng. 1987;29:733–41.

52. Blanchette RA. Delignification by wood – decay fungi. Annu Rev Phytopathol. 1991;29:381–98.
53. Sun Y, Cheng J. Hydrolysis of lignocellulosic material for ethanol production: a review. Biores Technol. 2002;83:1–11.
54. Barthel ST, Heinze T. Acylation and carbanilation of cellulose in ionic liquids. Green Chem. 2006;8(3):01–306.
55. Heinze T, Schwikal K, Barthel S. Ionic liquids as reaction medium in cellulose functionalization. Macromol Biosci. 2005;5(6):520–5.
56. Fukaya Y, Hayashi K, Wada M, Ohno H. Cellulose dissolution with polar ionic liquids under mild conditions: required factors for anions. Green Chem. 2008;10(1):44–6.
57. Fukaya Y, Sugimoto A, Ohno H. Superior solubility of polysaccharides in low viscosity, polar, and halogen–free 1,3–dialkylimidazolium formats. Biomacromolecules. 2006;7(12):3295–7.
58. Wu J, Zhang J, Zhang H, He JS, Ren Q, Guo M. Homogeneous acetylation of cellulose in a new ionic liquid. Biomacromolecules. 2004;5(2):266–8.
59. Xie HL, Shi TJ. Wood liquefaction by ionic liquids. Holzforschung. 2006;60(5):509–12.
60. Morales – de la Rosa S, Campos – Martin JM, Fierro JLG. High glucose yields from the hydrolysis of cellulose dissolved in ionic liquids. Chem Eng J. 2012;181–182(1):538–41.
61. Zhi S, Liu Y, Yu X, Wang X, Lu X. Enzymatic hydrolysis of cellulose after pre-treatment by ionic liquids: focus on one – pot. Energy Procedia. 2012;14:1741–7.
62. Fort DA, Remsing RC, Swatloski RP, Moyna P, Moyna G, Rogers RD. Can ionic liquids dissolve wood? Processing and analysis of lignocellulosic materials with 1–n–butyl–3–methylimidazolium chloride. Green Chem. 2007;9:63–9.
63. Kilpelainen I, Xie HB, King A, Granstrom M, Heikkinen S, Argyropoulos DS. Dissolution of wood in ionic liquids. J Agric Food Chem. 2007;55:9142–8.
64. Zavrel M, Bross D, Funke M, Buchs J, Spiess AC. High – throughput screening for ionic liquids dissolving (ligno–) cellulose. Biores Technol. 2009;100:2580–7.
65. Pu YQ, Jiang N, Ragauskas AJJ. Ionic liquid as a green solvent for lignin. Wood Chem Technol. 2007;27:23–33.
66. Pinkert A, Marsh KN, Pang SS. Alkanolamine ionic liquids and their inability to dissolve crystalline cellulose. Ind Eng Chem Res. 2010;49:11809–13.
67. Lee SH, Doherty TV, Linhardt RJ, Dordick JS. Ionic liquid–mediated selective extraction of lignin from wood leading to enhanced enzymatic cellulose hydrolysis. Biotechnol Bioeng. 2009;102(5):1368–76.
68. Liu Q, Janssen MHA, van Rantwijk F, Sheldon RA. Room – temperature ionic liquids that dissolve carbohydrates in high concentration. Green Chem. 2005;7:39–42.
69. Binder JB, Raines RT. Simple chemical transformation of lignocellulosic biomass into furans for fuels and chemicals. J Am Chem Soc. 2009;131(5):1979–85.
70. Froschauer C, Hummel M, Iakovlev M, Roselli A, Schottenberger H, Sixta H. Separation of hemicellulose and cellulose from wood pulp by means of ionic liquid/cosolvent systems. Biomacromolecules. 2013;14:1741–50.
71. Dadi AP, Varanasi S, Schall CA. Enhancement of cellulose saccharification kinetics using an ionic liquid pretreatment step. Biotechnol Bioeng. 2006;95:904–10.
72. Shang S, Zhu L, Fan J. Physical properties of silk fibroin/cellulose blend films regenerated from the hydrophilic ionic liquid. Carbohydr Polym. 2011;86:462–8.
73. Xiao W, Yin W, Xia S, Ma P. The study of factors affecting the enzymatic hydrolysis of cellulose after ionic liquid pretreatment. Carbohydr Polym. 2012;87:2019–23.
74. Kamiya N, Matsushita Y, Hanaki M, Nakashima K, Narita M, Goto M, Takahashi H. Enzymatic in situ saccharification of cellulose in aqueous – ionic liquid media. Biotechnol Lett. 2008;30(6):1037–40.
75. Yang Z. Ionic liquids and proteins: academic and some practical interactions. In: Maria P, editor. Ionic liquids in biotransformations and organocatalysis. Hoboken: Wiley; 2012. p. 15–72.

76. Lozano P, Bernal B, Bernal JM, Pucheault M, Vaultier M. Stabilizing immobilized cellulase by ionic liquids for saccharification of cellulose solutions in 1-butyl-3-methylimidazolium chloride. Green Chem. 2011;13:1406–10.
77. Zaks A, Klibanov AM. The effect of water on enzyme action in organic media. J Biol Chem. 1988;263:8017–21.
78. Yang F, Li L, Qiang L, Tan W, Liu W, Xian M. Enhancement of enzymatic in situ saccharification of cellulose in aqueous-ionic liquid media by ultrasonic intensification. Carbohydr Polym. 2012;81:311–16.
79. Abboud JL, Kamlet MJ, Taft RW. Regarding a generalized scale of solvent polarities. J Am Chem Soc. 1977;99(25):8325–7.
80. Hallett JP, Welton T. How polar are ionic liquids? ECS Trans. 2009;16:33–8.
81. Lungwitz R, Strehmel V, Spange S. The dipolarity/polarisability of 1-alkyl-3-methylimidazolium ionic liquids as function of anion structure and the alkyl chain length. New J Chem. 2010;34:1135–40.
82. Reichardt C. Polarity of ionic liquids determined empirically by means of solvatochromic pyridinium N-phenolate betaine dyes. Green Chem. 2005;7:339–51.
83. Zhao H, Jones CL, Baker GA, Xia S, Olubajo O, Person VN. Regenerating cellulose from ionic liquids for an accelerated enzymatic hydrolysis. J Biotechnol. 2009;139(1):47–54.
84. Park S, Kazlauskas R. Improved preparation and use of room temperature ionic liquids in lipase-catalyzed enantio- and regioselective acylations. J Org Chem. 2001;66:8395–401.
85. Oehlke A, Hofmann K, Spange S. New aspects on polarity of ionic liquids as measured by solvatochromic probes. New J Chem. 2006;30:533–6.
86. Lau RM, Sorgedrager MJ, Carrea G, van Rantwijk F, Secundo F, Sheldon RA. Dissolution of *Candida antarctica* lipase B in ionic liquids: effects on structure and activity. Green Chem. 2004;6:483–7.
87. Sheldon RA, Lau RM, Sorgedrager MJ, van Rantwijk F, Seddon KR. Biocatalysis in ionic liquids. Green Chem. 2002;4:147–51.
88. Henderson WA. Crystallization kinetics of glyme – LiX and PEO – LiX polymer electrolytes. Macromolecules. 2007;40:4963–71.
89. Irimescu R, Kato K. Investigation of ionic liquids as reaction media for enzymatic enantioselective acylation of amines. J Mol Catal B: Enzym. 2004;30:189–94.
90. Lee SH, Ha SH, Lee SB, Koo YM. Adverse effect of chloride impurities on lipase-catalyzed transesterifications in ionic liquids. Biotechnol Lett. 2006;28:1335–9.
91. Seddon KR, Stark A, Torres MJ. Influence of chloride, water, and organic solvents on the physical properties of ionic liquids. Pure Appl Chem. 2000;72:2275–87.
92. Heintz A, Klasen D, Lehmann JK. Excess molar volumes and viscosities of binary mixtures of methanol and the ionic liquid 4-methyl-N-butylpyridinium tetrafluoroborate at 25, 40, and 50-C. J Sol Chem. 2002;31:467–76.
93. Zhao H, Baker GA, Holmes S. New eutectic ionic liquids for lipase activation and enzymatic preparation of biodiesel. Org Biomol Chem. 2010;9(6):1908–16.
94. Greaves TL, Drummond CJ. Ionic liquids as amphiphile selfassembly media. Chem Soc Rev. 2008;37:1709–26.
95. Qiu Z, Texter J. Ionic liquids in microemulsions. Curr Opin Colloid Interface Sci. 2008;13:252–62.
96. Lozano P, De Diego T, Gmouh S, Vaultier M, Iborra JL. Dynamic structure – function relationships in enzyme stabilization by ionic liquids. Biocatal Biotransform. 2005;23:169–76.
97. Baker GA, Heller WT. Small-angle neutron scattering studies of model protein denaturation in aqueous solutions of the ionic liquid 1-Butyl-3-methylimidazolium chloride. Chem Eng J. 2009;147:6–12.
98. De Diego T, Lozano P, Gmouh S, Vaultier M, Iborra JL. Understanding structure-stability relationships on *Candida antarctica* lipase B in ionic liquids. Biomacromolecules. 2005;6:1457–64.

99. Madeira R, Sorgedrager MJ, Carrea G, van Rantwijk F, Secundo F, Sheldon RA. Dissolution of *Candida antarctica* lipase B in ionic liquids: effects on structure and activity. Green Chem. 2004;6:483–7.
100. De Diego T, Lozano P, Gmouh S, Vaultier M, Iborra JL. Fluorescence and CD Spectroscopic analysis of the alpha – chymotrypsin stabilization by the ionic liquid, 1 – Ethyl – 3 – methylimidazolium Bis[(trifluoromethyl)sulfonyl]amide. Biotechnol Bioeng. 2004;88:916–24.
101. Fujita K, MacFarlane DR, Forsyth M, Moshizawa-Fujita M, Murata K, Nakamura N, Ohno H. Solubility and stability of cytochrome c in hydrated ionic liquids: effect of oxo acid residues and kosmotropicity. Biomacromolecules. 2007;8:2080–6.
102. Lou WY, Zong MH, Smith TJ, Wu H, Wang JF. Impact of ionic liquids on papain: an investigation of structure–function relationships. Green Chem. 2006;8:509–12.
103. Micaêlo NM, Soares CM. Protein structure and dynamics in ionic liquids. Insights from molecular dynamics simulation studies. J Phys Chem B. 2008;112:2566–72.
104. Baker SN, McCleskey TM, Pandey S, Baker GA. Fluorescence studies of protein thermostability in ionic liquids. Chem Commun. 2004;8:940–1.
105. Fan Y, Yan J, Zhang S, Li J, Chen D, Duan P. Fluorescence spectroscopic analysis of the interaction of papain with ionic liquids. Appl Biochem Biotechnol. 2012;168:592–603.
106. Heller WT, O'Neill HM, Zhang Q, Baker GA. Characterization of the influence of the ionic liquid 1-butyl-3-methylimidazolium chloride on the structure and thermal stability of green fluorescent protein. J Phys Chem B. 2010;114:13866–71.
107. Fitzpatrick PA, Steinmetz ACU, Ringe D, Klibanov AM. Enzyme crystal structure in a neat organic solvent. Proc Natl Acad Sci U S A. 1993;90:8653–7.
108. Klibanov AM. Improving enzymes by using them in organic solvents. Nature. 2001;409:241–6.
109. Dupont J. On the solid, liquid and solution structural organization of imidazolium ionic liquids. Braz Chem Soc. 2004;15:341–50.
110. Yang Z. Hofmeister effects: an explanation for the impact of ionic liquids on biocatalysis. J Biotechnol. 2009;144(1):12–22.
111. Constatinescu D, Herrmann C, Weingärtner H. Patterns of protein unfolding and protein aggregation in ionic liquids. Phys Chem Chem Phys. 2010;12:1756–63.
112. Weingärtner H, Cabrele C, Herrmann C. How ionic liquids can help to stabilize native proteins. Phys Chem Chem Phys. 2012;14:415–26.
113. Anderson JL, Ding J, Welton T, Armstrong DW. Characterizing ionic liquids on the basis of multiple solvation interactions. J Am Chem Soc. 2002;24:14247–54.

Part V
Ionic Liquids for Absorption and Biodegradation of Organic Pollutant in Multiphase Systems

Chapter 12
Absorption of Hydrophobic Volatile Organic Compounds in Ionic Liquids and Their Biodegradation in Multiphase Systems

Solène Guihéneuf, Alfredo Santiago Rodriguez Castillo, Ludovic Paquin, Pierre-François Biard, Annabelle Couvert, and Abdeltif Amrane

Abstract The coupling of absorption in a gas-liquid contactor and biodegradation in a two-phase partitioning bioreactor (TPPB) has been shown to be a promising technology for the removal of hydrophobic volatile organic compounds. The choice of the organic phase is crucial, and consequently only two families of compounds comply with the requested criteria, silicone oils and ionic liquids. These latter solvents appear especially promising owing to their absorption capacity towards hydrophobic compounds and their low volatility, as well as the possibility of IL tailoring, allowing a fine-tuning of their physicochemical properties, leading to a wide range of products with various characteristics. Some results on common ionic liquids are highlighted in this chapter: biodegradation rates reported by some authors show that phenol biodegradation in the presence of ILs is up to 40 % higher than those obtained in other multiphase reactors; there is a strong affinity of toluene and DMDS for imidazolium salts, $[C_4Mim][PF_6]$ or $[C_4Mim][NTf_2]$. Performance improvements may be expected from the tailoring of ionic liquid structure, especially towards toxicity reduction. Positive results recorded after cell acclimation to target compounds let expect an important gain from more complex acclimation strategies, including microbial acclimation to both ionic liquids and pollutants.

Keywords Absorption • Activated sludge • Biodegradation • Ionic liquids • Two-phase partitioning bioreactors • Toxicity • Volatile organic compounds • Separation

---

S. Guihéneuf • L. Paquin
Université de Rennes 1, Sciences Chimiques de Rennes, UMR CNRS 6226, Groupe Ingénierie Chimique & Molécules pour le Vivant (ICMV), Bât. 10A, Campus de Beaulieu, Avenue du Général Leclerc, CS 74205, 35042 Rennes Cedex, France

Université européenne de Bretagne, Rennes Cedex, France

A.S.R. Castillo • P.-F. Biard • A. Couvert • A. Amrane (✉)
Ecole Nationale Supérieure de Chimie de Rennes, CNRS, UMR 6226, Avenue du Général Leclerc, CS 50837, 35708 Rennes Cedex 7, France

Université européenne de Bretagne, Rennes Cedex, France
e-mail: abdeltif.amrane@univ-rennes1.fr

Z. Fang et al. (eds.), *Production of Biofuels and Chemicals with Ionic Liquids*, Biofuels and Biorefineries 1, DOI 10.1007/978-94-007-7711-8_12,

12.1 Introduction

When absorption is used to remove pollutants present in the atmosphere, an aqueous phase is generally employed, either water or water containing reagents (acid, basis, oxidant, etc.). More rarely, some organic phases (solvents) are implemented, but their cost implicates their recycling, and then, their regeneration. The use of this type of liquid phase becomes primordial when gaseous pollutants belong to hydrophobic compounds family (i.e. toluene, benzene, xylene, etc.) since their affinity for the liquid phase in which they have to be absorbed must be important. This means that there is an important issue in finding **new absorbents**, displaying high absorption potentialities facing many volatile organic compounds (VOC), and able to be regenerated. For this purpose, absorbent regeneration could be considered after VOC biodegradation; the latter low cost process appears therefore promising. This implies that the solvent will have to fulfil several conditions, especially the absence of biodegradability.

Among the solvents available on the market, a wide number appears biodegradable, even if an acclimation time is often needed. Indeed, various compounds having very low degradation rates or totally refractory towards microorganisms are described as bio-recalcitrant. However, a total absence of biodegradation, even after an acclimation time is required for the proposed process. Therefore and based on biodegradability and biocompatibility criteria, among the available solvents two classes can be selected: silicone oils and **ionic liquids** [1]. These latter appear especially promising owing to their solvent capacity and their low volatility (saturated vapour pressure close to zero), as well as the possibility of IL tailoring to fit the characteristics required for specific applications [2]. Only few reports are available dealing with the use of ILs for pollutants removal; phenol biodegradation has been investigated in the presence of an IL [3]; while the previous works of Quijano et al. [4, 5] are the only reports dealing with the affinity of ILs for the absorption of hydrophobic odorous compounds and their biodegradation in a multiphase bioreactor. These works led to promising results for toluene after cell acclimation to the VOC, while for DMDS more complex strategies, including acclimation to ILs, should be subsequently considered.

The potential of this class of compounds for the absorption of hydrophobic VOCs and the subsequent biodegradation of these compounds in a multiphase bioreactor involving ILs (containing the absorbed VOC) as an organic phase and an aqueous phase containing microorganisms are discussed thereafter. The process considered to implement such solvents in a whole operation is schematically described in Fig. 12.1.

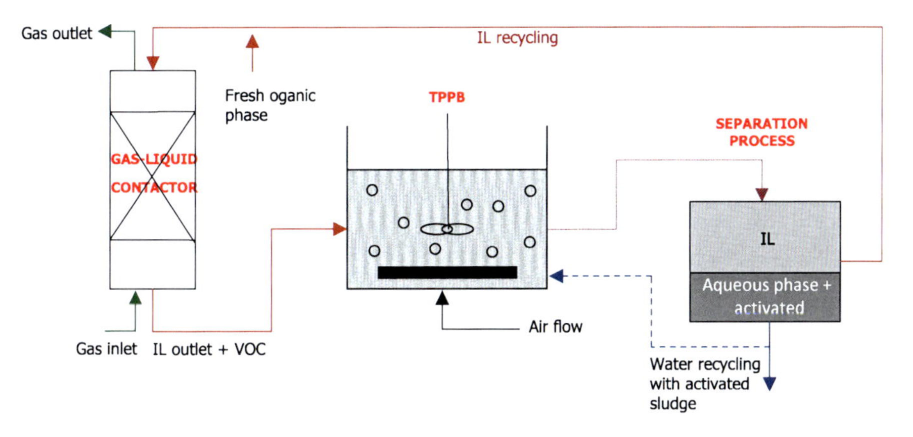

Fig. 12.1 Hybrid absorption-biodegradation process with regeneration of the organic phase

12.2 Screening and Choice of the NAPL – Ionic Liquids (ILs)

Solvent choice is the first issue to be considered since it will determine the whole process. The selected non-aqueous phase liquid (NAPL) should not add pollution, must be non-flammable, and its chemical and thermal properties must fulfil those required with the aim of its recycling [6]. The NAPL must be liquid and not very viscous in a range of temperature between 5 and 40 °C. To make the separation from water after biodegradation step feasible, the considered NAPL should be water-immiscible and should not lead to a stable emulsion.

Several NAPL have been previously used either in a two-phase partitioning bioreactor or as absorbents in gas-liquid contactors (scrubber, airlift, bubble column, etc.). Most of them display very low degradation rates or are refractory towards microorganisms and are described as bio-recalcitrant. However, recalcitrance is not enough for the proposed process, since it means a biodegradation of the considered NAPL after an acclimation time. According to some authors [7], five classes of NAPL are potentially non-biodegradable: HMN (2,2,4,4,6,8,8-Heptamethylnonane) owing to the presence of terminal methyl groups, fluorocarbon FC 40, some polymers like the polyisobutylene which contains many terminal methyl groups, silicone oils, especially polymethylsiloxane, and ionic liquids. However, HMN seems biodegradable by some acclimated microbial communities [8]; fluorocarbons are toxic toward humans or the environment [9], while polymer viscosity may induce a too high energy consumption in the TPPB (stirring).

Among the available solvents, only silicone oils and ionic liquids appear therefore really relevant [1, 10]. Even if silicone oils are interesting NAPL candidates, especially polydimethylsiloxane, for hydrophobic VOCs removal, owing to their biocompatibility, their non-biodegradability [10], and have often been implemented in TPPB [11–13], ionic liquids seem promising.

Ionic liquids have been recognized for about a century, but have only started receiving closer attention in the last two decades. Historically, an ionic liquid is an organic salt with a melting point below 100 °C [14]. They are composed by an association between an organic cation containing one or more hetero-atom(s) (nitrogen, phosphorus or sulfur) and an inert anion or Lewis acid [15], namely the counter-ion, leading to a neutral compound [16]. Since the first chloroaluminate "molten salt", many efforts have been made about ionic liquids to lower their melting points (development of RTILS, "Room-Temperature Ionic Liquids") and to improve their stability towards air and water.

Their low vapor pressure and non-flammability [17, 18] makes them particularly interesting class of solvents for 'green chemistry' or absorption. However, based on recent data, these assumptions have been progressively reconsidered [19–23]. In addition, they are generally thermically stable (decomposition temperatures >150–200 °C), chemically or electrochemically inert.

Their interest is not only due to their remarkable physicochemical properties (lipophilicity, viscosity, density, etc.) but also for their recyclability. However, these properties are usually presented as applicable to all ILs are not so "universal" and the large number of possible combinations of a structural point of view suggests that some ILs are not as harmless [24].

So, ILs can be designed for specific applications [2, 18]. Hence, it is possible to fine-tune IL physicochemical properties by means of modifying the substituent groups or the identity of the cation/anion pair [17, 18].

One of the first reviews on ILs (synthesis, applications, etc.) was published in 1999 and related the general methods of synthesis and the first applications of ILs based chloroaluminates [25]. Below are shown the structures of most common ILs (Fig. 12.2).

As they are readily tunable, ILs could be selected as NAPL for TPPB. The physico-chemical characteristics of the ionic liquid (viscosity, hydrophobicity toxicity, etc.), as well as possible biodegradability, vary according to the considered radical.

12.3 Absorption of Hydrophobic VOCs

12.3.1 Solubility

There is a lack of investigations on VOCs affinity towards ILs. Some results are available dealing with the partition coefficient between an ionic liquid and water [18]:

Polar organic compounds, such as dichloromethane and trichlorobenzene, are soluble in ILs. Huddleston [18] determined the partition coefficient between a given ionic liquid and water, specific to these compounds (log D) (Eq. 12.1):

$$log\ D = log \frac{[Compound]_{IL}}{[Compound]_{Water}} \tag{12.1}$$

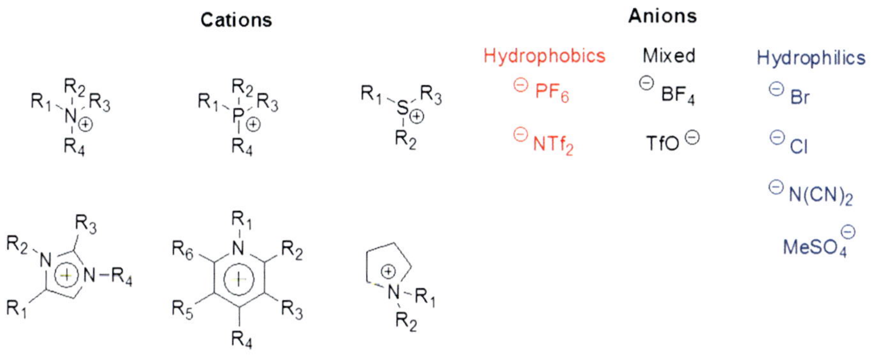

Fig. 12.2 Structures of the most common ILs. Anions hydrophobic and hydrophilic tendencies

For instance, a value of 1.8 and 2.4 were found for toluene – water [18] and $[C_6Mim][PF_6]$ – water [26]. Huddleston [18] showed that ILs can solubilize apolar or lowly polar compounds. The literature dealing with the solubility of organic compounds in ionic liquids remains limited [27–31]. However, a trend seems to emerge showing that polar compounds are more easily soluble than apolar ones. Huddleston [18] also explored the replacement of usual solvents by ionic liquids in the liquid-liquid extraction of VOCs; they found that the affinity of a solute (the VOC) for ILs increases with the augmentation of the log P_{OW} value of the considered VOC.

Regarding the Henry's constants, the scarce available results concern toluene and sulfur VOC, dimethyl sulfide (DMS), dimethyl disulfide (DMDS); they were measured for two ILs, $[C_4Mim][PF_6]$ and $[C_4Mim][NTf_2]$ and are compared in Fig. 12.3 to some other solvents [4, 32]. For instance, the Henry's constants of DMDS and toluene in water are 123 and 615 $Pa.m^3.mol^{-1}$, respectively; they were found to be significantly lower in ILs, with ratio of the partition coefficients in water and the considered IL, $[C_4Mim][PF_6]$ and $[C_4mim][NTf_2]$, of 284 and 448 for toluene and 33 and 37 for DMDS, respectively [4]. If compared to the often used silicone oil (polydimethylsiloxane, PDMS), partition coefficients were similar for DMDS, while toluene showed a higher affinity for $[C_4Mim][NTf_2]$ if compared to $[C_4Mim][PF_6]$ and PDMS [4].

In conclusion, compared to the most often used non-aqueous phase liquid (NAPL), namely silicone oils, especially polydimethylsiloxane, for hydrophobic VOCs removal [10], which have been often implemented in TPPB [11–13], ionic liquids appear especially promising owing to their solvent capacities.

12.3.2 *Diffusion*

Molecular diffusion is the transport phenomena caused by a concentration gradient [33]. It must not be mistaken for convection which is caused by the bulk motion.

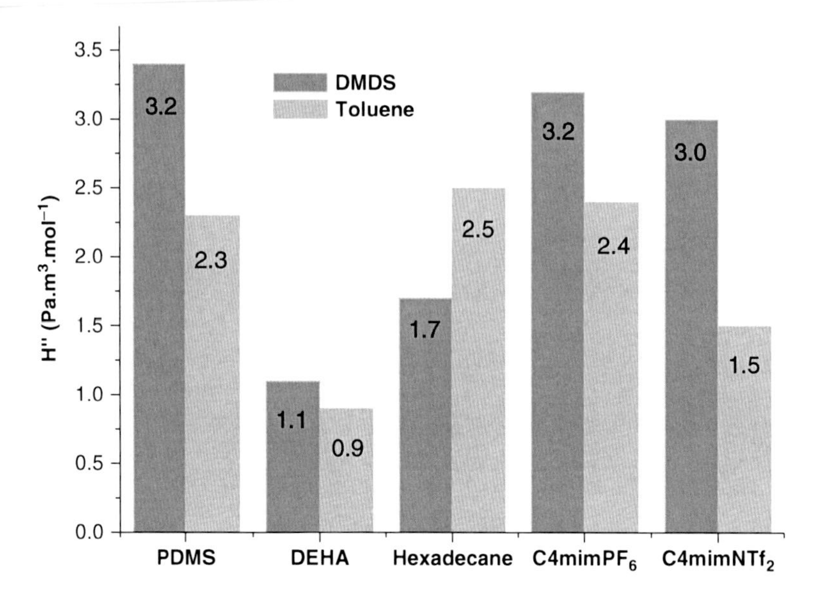

Fig. 12.3 Partition coefficients of DMDS and toluene in various organic phases at 25 °C

The diffusive flux can be described by the Fick's first law at steady state. It postulates that the flux (J in $mol.m^2.s^{-1}$) is proportional to the spatial concentration gradient (from large to low concentration areas):

$$J = -D_{i,j}\frac{\partial C}{\partial x} \tag{12.2}$$

$D_{i,j}$ is the diffusion coefficient (or diffusivity in $m^2.s^{-1}$) of the compound *i* in the solvent *j*.

In gas-liquid mass transfer operations, the removal efficiency depends directly on the mass transfer rate, the absorption rate (which accounts for the solubility) and on the contact time. According to various mass transfer theories (Higbie or Danckwerts theories for example), the mass transfer rate in both phases are proportional to the square root of the diffusion coefficient in each phase [34]. A low mass transfer rate in the liquid phase can significantly hamper the VOC removal and is therefore a key point to control for a liquid ionic selection.

With traditional solvents (water and organic solvents), the diffusion coefficient of non ionic solute increases with the temperature and decreases with the solvent dynamic viscosity (μ_j) and the solute molar volume ($V_{m,i}$):

$$D_{i,j} \propto \frac{T^{\alpha}}{\mu_j^{\beta} V_{m,i}^{\gamma}} \text{ with } \alpha, \beta \text{ and } \gamma \text{ three positive coefficients} \quad (12.3)$$

Several theories are currently applied to calculate diffusion coefficients in traditional solvents (Stokes-Einstein, Wilke-Chang, Arnold, Hayduk-Laudie, Schreibel, etc.) [27, 33]. Except the Arnold theories, $\alpha = \beta = 1$ and $1/3 < \gamma < 0.6$. The solvent molecule size is also sometime taken into account. Thereby, it is admitted that the diffusion coefficient increases with the loading of the solvent in solute. For mass transfer operations, the infinite dilution diffusion coefficient, with an order of magnitude of 10^{-9} $m^2.s^{-1}$, is generally used in the computation [35].

Several experimental techniques have been applied for the determination of the diffusion coefficients of various solutes (CO_2, alkanes, alkenes and hydrofluorocarbons) in ionic liquids. All these techniques are based on two methods: the thermogravimetric method or the manometric method. These methods enable to investigate at the same time the solubility and the diffusion properties of a solute in a given solvent. They are usually conducted in static mode to avoid convection contributions. The manometric method is based on the measurement of the pressure decay in a thermo-controlled cell chamber which contains a layer of the investigated solvent. An accurate amount of gas, often provided by a pressurized feed chamber, is introduced rapidly at the beginning of the measurement. The diffusion coefficient is determined by fitting the pressure decay to a one-dimensional diffusion model for solute uptake into the liquid [28, 36, 37]. An alternative manometric method consists to immobilize the solvent in a membrane which separates the feed and the cell chambers [27, 29, 30]. This technique is often called the two-cells methods or the lag-time technique. The thermogravimetric method is a relatively recent method based on the measurement of the solute uptake in the investigated solvent using a microbalance [31, 38–40]. Buoyancy corrections are necessary to take into account the expansion of the solvent due to the solute absorption during the experiment. The feed gas is usually pure.

For both techniques, vacuum is applied to the cell chamber before the gas sample introduction. Depending on the thickness of the solvent layer, several dimensional diffusion models have been used to determine by fitting the diffusion coefficient (thin-film model, semi-infinite model, etc.).

Diffusion coefficients of small solutes such as VOCs in ionic liquid are usually one or two order of magnitude lower than in traditional organic solvents (in the range 10^{-11}–10^{-10} $m^2.s^{-1}$) (Table 12.1) mainly because of the high viscosity of the RTILs. Diffusivity drops more significantly in RTILs than in traditional solvents when the temperature decreases (lower molecular agitation and larger viscosity). Several studies demonstrate that this evolution follows the Arrhenius law, with activation energies typically larger than traditional solvents in the range 10–25 kJ. mol^{-1} [27, 36, 38].

Table 12.1 Viscosity and molar volumes of the various solvent investigated by Scovazzo et al. and Camper et al. Values of the diffusivities at 303 K for several solutes [27–30]

RTIL (solvent j)	μ_j (cP)	$V_{m,j}$ (cm^3/mol)	$D_{i,RTIL}$ at 303 K (10^{10} m^2.s^{-1}) CO$_2$	Ethene	Propene	1-Butene	Butadiene	Methane	Propane
$[C_2Mim][NTf_2]$	26	258	6.6	5.1	3.3	2.7	3.7		
$[C_2Mim][NTf_2]^a$	27.8		8.1	7.4	4.3				5.1
$[C_2Mim][TfO]$	45	188	5.2	4.5	2.6	2.2	3.0		
$[C_2Mim][BETI]$	77	294	4.5	2.5	1.7	1.2	1.7		
$[C_4Mim][PF_6]$	176	209	2.7	2.0	1.1	0.8	1.2		
$[C_4SO_2Mim][TfO]$	554	210	1.1	0.7	0.5	0.3	0.4		
$[P_{2444}][DEP]$	207	323	3.5	2.1	1.2	0.78	1.46		
$[P_{(14)666}][DCA]$	213	578	3.0	2.2	1.36	1.08	1.67		
$[P_{(14)666}][NTf_2]$	243	804	6.2	2.9	2.6	1.98	2.9		
$[P_{(14)666}][Cl]$	1,316	590	3.0	2.1	1.61	1.15	1.73		
$[P_{(14)444}][DBS]$	3,011	731	1.7	1.06	0.65	0.41	0.65		
$[N_{(4)111}][NTf_2]$	71	289.6	4.87	2.29	1.73	1.37	1.95	NQ[b]	1.16
$[N_{(4)113}][NTf_2]$	85	315.4	4.83	2.67	2.02	1.33	2.13	NQ	1.09
$[N_{(6)111}][NTf_2]$	100	324.5	4.38	3.19	1.39	1.39	2.14	NQ	1.02
$[N_{(6)113}][NTf_2]$	126	353.1	3.72	NQ	1.59	1.33	2.12	1.22	1.02
$[N_{(6)222}][NTf_2]$	167	365.8	4.68	1.90	1.56	1.02	1.46	NQ	0.88
$[N_{(10)111}][NTf_2]$	173	393.2	4.60	3.07	2.37	1.67	2.63	2.64	1.32
$[N_{(10)113}][NTf_2]$	183	426.4	3.78	3.44	1.73	1.59	2.06		
$[N_{(1)444}][NTf_2]$	386	383.5	3.41		0.90	0.62	1.13	3.41	
$[N_{(1)888}][NTf_2]$	532	600.6	3.43	2.70	1.58	1.37	2.33	NQ	1.01

[a]Values of Camper et al. [28]
[b]NQ means Not Quantifiable

Scovazzo and coworkers investigated by the lag-time technique the solubility and diffusion of CO_2 and several VOCs (ethylene, propylene, 1-butene, butadiene, methane, butane) at 303 K in five imidazolium, five phosphonium and nine ammonium based RTILs, covering a large range of viscosities (from 10 to 3,000 cP) [27, 29, 30]. They found that the value of α depends on the kind of cation (0.66 for the imidazolium, 0.35 for the phosphonium and 0.59 for the ammonium based RTILs) whereas the Stokes-Einstein and the Arnold theories predict respectively 1 and 0.5. γ was equal to 1.04 for imidazolium, 1.26 for phosphonium and 1.27 for ammonium based RTILs. Therefore, diffusivity in RTILs is less dependent on viscosity and more dependent on solute size than predicted by the conventional Stokes-Einstein model. As mentioned by Morgan, the deviation between RTILs and traditional organic solvents may result from the physical situation of small solutes diffusing in an universe of large IL molecules solvents [27].

At identical viscosities, diffusivities are larger in phosphonium, then in ammonium, then in imidazolium based RTILs. This may be explained by the increased molar volume of the phosphonium based ILs ($\approx$600 cm^3.mol^{-1} compared to approximately 400 and 200 $cm^3.mol^{-1}$ for respectively ammonium and imidazolium based ILs) which can allow for faster diffusion rates. Skrzypczak and Neta suggest that this trend is due to the increased number and length of the aliphatic chains present on the phosphonium cation [41]. Because the chains are flexible and can move more rapidly than the whole cation, they enable a more rapid diffusion of solutes from one void to another in the phosphonium-based ILs [41]. Therefore, the amount of free volume in an anionic liquid could be a better indicator of diffusivity than viscosity.

1,3-butadiene diffuses faster for an identical molar volume than the other alkene due to a possible weak complexation of the conjugated double bonds with the positively charged cation which facilitates the transport [30]. All these conclusions suggest that finding a universal correlation to determine diffusivities for all classes of RTILs and solute is utopist, even if similar trends are observed for different kind of RTILs. Moreover, it is also important to consider the effect of other impurities and large solute concentrations on diffusion. It has been shown that water and other co-solvents increase very significantly the viscosity and consequently the diffusivity. Therefore, the diffusivities should be measured for any couple RTIL-solute to assess the mass transfer rate.

Camper et al. measured the diffusivities of several VOCs (ethane, ethane, propane, propene) and CO_2 on $[C_2Mim][NTf_2]$ for various temperatures using the manometric method and a semi-infinite model [28]. They confirmed the same trends and the order of magnitude found by Morgan et al. with the same solute-RTIL couple, even if the values of Morgan were 20 % lower (Table 12.1).

Shiflett and Yokozeki investigated the solubility and the diffusion of hydrofluorocarbons in several RTILs (mainly imidazolium based RTILs) by the thermogravimetric method in isothermal mode for pressure up to 20 bar [31, 39, 40]. The effective measured diffusion coefficients increase with the pressure and the temperature applied in the chamber. Indeed, a larger pressure applied increases the solute uptake in the solution. Therefore, the determined diffusion coefficients are not determined at infinite dilution. The values found for the various diffusion coefficients are all included between $2 \cdot 10^{-11}$ and $8 \cdot 10^{-11}$ $m^2.s^{-1}$ at 298.15 K.

Except for the few VOCs presented before, the most investigated solute remains CO_2. Even if this compound is out of the scope of this review, this study is worthwhile since it enables a comparison between the different measurement techniques. Hou and Baltus investigated CO_2 solubility and diffusivity in five imidazolium based RTILs by the manometric method and using a transient thin-liquid film model [36]. Their results are consistent with the results of Camper et al. and Morgan et al. deduced with manometric methods (Table 12.2) [27, 28]. However, the values found using the thermogravimetric method were considerably smaller (five times) [38]. Hou and Baltus suggest that the thermogravimetric method present a larger uncertainty due to several buoyancy corrections necessary and they questioned the fact that the measurements

Table 12.2 Values of CO_2 diffusion coefficients in three RTILs reported in the literature

RTIL	$D_{i,RTIL}$ (10^{10} $m^2.s^{-1}$)	Refs.	Method
Water	22.5 (303 K)	[34]	
$[C_2Mim][NTf_2]$	6.6 (303 K)	[27]	Lag-time technique (manometric method)
	8.1 (303 K)	[28]	Pressure decay technique (semi-infinite model)
$[C_4Mim][BF_4]$	1.7 ± 0.6 (323 K, 1–10 bar)	[38]	Thermogravimetric method
	4.8 (323 K)	[36]	Pressure decay technique (transient thin liquid film model)
	1.8 (323 K, 20 bar)	[42]	Expansion measurement with a cathethometer
$[C_4Mim][PF_6]$	1.2 ± 0.3 (323 K, 1–10 bar)	[38]	Thermogravimetric method
	0.8 (323 K, 20 bar)	[42]	Expansion measurement with a cathethometer

were performed at large pressure which might influence the density and the viscosity of the RTIL [36]. Whatever, the difference between the two methods remains controversial and poorly commented in the literature. Some systematic and/or random errors would be responsible for the discrepancy among research groups. Further investigations to try to understand the difference between both methods would be interesting.

12.3.3 Interactions

For conventional solvents, molecules solubilization is generally expressed in terms of dielectric constant of solvent. However, these criteria could not be applied to ionic liquids. Several studies have been conducted to establish a specific ranking based on interactions between ILs and solutes [43]. This model considered interactions involving hydrogen bond and polarizability. According to these authors, ionic liquids are polar solvents but they have completely different behaviors from conventional solvents, so it is impossible to compare these two kinds of solvents only on their polarity.

Most of the time, the π interactions between aromatic cations and solutes are prevailing. Imidazolium ring is a good electron acceptor and due to the strong electron delocalization, the nitrogen does not form H bonds easily. On the contrary, a pyridinium cation is a good electron donor [44].

Some ionic liquids have been functionalized (TSILs: Task Specific Ionic Liquids) for a specific purpose. For example, ureas or thioureas groups were introduced in few ionic liquids in order to capture CO_2, H_2S or heavy metals [45]. RTILs (Room Temperature Ionic Liquids) have also been developed to perform extraction of metals [46] or elements such as Uranium [47]. Bifunctionalized ionic liquids can be used to optimize the extraction of Europium [48]. The $[C_4Mim]$ $[NTf_2]$ was proved to be effective in extraction of inorganic acids (HNO_3, HCl, $HReO_4$, $HClO_4$) from aqueous phases [49].

Several authors have studied the distribution of organic compounds, such as aniline, benzene derivatives and organic acids or organic anions (phenolates) between an ionic liquid phase (usually [C_xMim] [PF_6, BF_4 or NTf_2]) and an aqueous phase. As expected, the most lipophilic compounds are more soluble in the ionic liquid phase [50–61]. In addition, it has been shown that aromatic compounds were more soluble in ionic liquids than their aliphatic counterparts, probably due to π-stacking interactions [43, 62–64].

Some studies have reported that the physical and chemical properties of ionic liquids are due to intra-molecular hydrogen interactions and Van Der Waals interactions. These studies explain the particular geometries adopted by ILs [65, 66]. Hydrogen bonds define the cation position versus the anion but also the distance between the two poles of an ionic liquid [67]. The structural organization of an IL can be explained by a combination of both types of interaction in the liquid phase, Coulomb and Van der Waals interactions [68, 69]. We observe that these interactions have a direct influence on the melting point, viscosity and ionic liquids enthalpy of vaporization. These interactions between the anion and the cation can also explain the variable hydrophobicity observed. It is essentially depending on the nature of the anions. The self-organization of ionic liquids may also influence their potential extraction and gas absorption [65, 66, 70, 71].

The [C_xMim] [PF_6] were applied to the extraction of anionic dyes [72], to the identification or the extraction of organic pollutants in soil or in water [54, 73, 74] or to extract and separate bioactive molecules from plants [75–77]. They are known to solubilize some natural polymers (cellulose, BSA, etc.), sugars or amino acids [46, 78–82]. They constitute the overwhelming majority of the ionic liquids used for the extraction of organic molecules.

N-methylimidazoliums functionalized by carboxylic chains (CH_2CO_2H) and associated with fluorinated anions such as BF_4^- or PF_6^- were synthesized in order to trap organic compounds (chlorophenyl, amines) [83]. Some ionic liquids (mainly alkylimidazoliums) are described for the absorption of volatile organic compounds such as benzene, toluene, phenols, anilines or of sulphur heterocycles [84, 85]. Some studies report the solubility of several hydrocarbons (benzene, toluene, xylene, heptane, hexadecane, methanol, acetonitrile or chloroform) in ionic liquids such as [C_xMim] [PF_6] [86]. This work has shown that aromatic hydrocarbons, methanol and acetonitrile are soluble or partially soluble in [C_xMim] [PF_6], while aliphatic hydrocarbons are immiscible. It has been shown that for a fixed cation, the anion could affect the interaction between ILs and VOCs. Hard anions (NO_3^-,$MeSO_3^-$, etc.) are worse hydrogen bonds acceptor than softer anions such as $B(CN)_4^-$ and offer low affinities with VOCs [87, 88].

Some studies have demonstrated that the ability of the anion to accept hydrogen bonds was related to the distribution of organic compounds between the ionic liquid phase and an aqueous phase [89]. A study was published by

Milota, and describes a process for absorption of gaseous VOCs (MeOH, formaldehyde, phenol, acrolein, acetaldehyde and propionaldehyde) in an ionic liquid absorption column (Tetradecyl (trihexyl) phosphonium dicyanamide) [90, 91].

Predictive models have been developed by Chen, to improve understanding of interactions between ionic liquids and organic compounds [92]. Oliferenko, conducted a study based on 48 ionic liquids and 23 industrial gases (alkanes, alkenes, fluoroalkanes…). The most soluble compounds described in this study are butadiene and butene (high polarizable molecules) [93]. The absorption value of gas was measured for several ILs. The CO_2 absorption is the most widely described [94], as also discussed above (see Sect. 12.3.2). Thus, Jalili, found that H_2S was better absorbed than CO_2 in IL [95]. Henry constants of few gases absorbed by [C_4Mim] [PF_6] were published by Safamirzaei, and the solubility of gases (CO_2, ethylene, ethane, CH_4, Ar, O_2, CO, H_2, N_2) has been reported [96, 97]. It seems that it is the nature of the anion which is crucial for the absorption of gas by ILs [98]. Thus, supported imidazoliums are well known to have a good absorption capacity for CO_2 and this characteristic is exacerbated when the cation is functionalized with an amino acid or an amine [99].

It also appears that the solubility of CO_2 increase with the molecular weight of the ILs [94]. These authors have studied the selectivity of absorption of different gas versus CO_2 (H_2, N_2, O_2, CH_4, H_2S, etc.). By reducing the size and molecular mass of ionic liquids, it is possible to trap the most volatile gases. Increasing the pressure also improves the absorption of gases in ILs [96]. Many analyzes can be conducted to study the interaction between VOCs and ionic liquids but the most common is the FTIR spectroscopy to visualize the characteristic peaks of the functions present on VOCs (alcohols, aldehydes, etc.) and potential shifts induced by the ionic liquid/VOC interactions [100].

12.4 Biodegradation in Multiphase Bioreactors

Microorganisms are able to assimilate a wide range of the available solvents, particularly alkanes, ketones or hydroxylated solvents (carboxylic acid, aldehydes, etc.). Solvents containing long alkyl chains or alcohol, ester or carboxylic groups are also biodegradable, and are precursors of beta oxidation. Other solvents like phthalates or plasticiser compounds (for example adipates) can also be degraded by various microorganisms such as *Rhodococcus* or *Sphingomonas* [101, 102]. However, various authors showed that biodegradability decreases with the presence of long alkyl chains or hydroxyl, ester and acid groups on the molecule [7, 11]. Previously, Alexander [103] reported that high molecular weight compounds, with lot of ramifications, are biologically recalcitrant. Besides, the type, the number, and the position of the substitutes on simple organic molecules influence their biodegradability. Various compounds having very low degradation rates or totally refractory towards microorganisms are described as bio-recalcitrant. However, a total absence of biodegradation, even after an acclimation time, is required for the proposed process.

Therefore, among the available solvents only silicone oils and ionic liquids [1] comply with the non-biodegradability criterion; while these latter appear especially promising owing to their solvent capacity and their low volatility [17, 18], as well as the possibility of IL tailoring to fit the characteristics required for specific applications [2]. In addition to fine-tuning their physicochemical properties, other ILs properties such as microbial toxicity are also related to the IL structure, showing that suitable ILs for microbial application can likely be designed. ILs have been therefore selected for implementation in the proposed process.

ILs applications in biotechnology have mainly focused on enzymatic catalysis [104]; a versatile battery of reactions being successfully performed in the presence of ILs, including transesterification, perhydrolysis, enantioselective reduction of ketones, and ammoniolysis [16, 105]. On the other hand, ILs toxicity has been reported as a key drawback for whole-cell applications [106, 107]. Studies on ILs toxicity (acute toxicity tests) are usually based on bioluminescence using microorganisms such as *Vibrio fischeri* or *Photobacterium phosphoreum* [108–110]; however, these microorganisms are not representative of the microbial cells commonly used in bioprocesses. Recently, Azimova et al. [111] observed that the IL toxic concentration for a *Pseudomonas* strain was up to 700 times higher than those for *V. fischeri*. Regarding microorganisms commonly used in biotechnology, there are contradictory reports in the literature. Successful whole-cell processes in the presence of ILs have been reported (e.g. synthesis of ketones and alcohols, lactic acid and antibiotic production) [52, 112–114], but also toxic effects of ILs towards yeasts and bacteria can be found [115, 116].

In addition, ILs biodegradability is a fundamental aspect that must be addressed before applying ILs in a whole-cell process. The non-biodegradability of the solvent is a required characteristic during a biotechnological process; being this particularly important when the solvent must be continuously reused [9]. Reports on ILs biodegradability are controversial in the literature. Most of the authors, working on imidazolium-based ILs, reported that ILs are not biodegradable [107, 109, 117]; while some authors observed ILs biodegradation by bacteria and fungi [118, 119]. These apparently contradictory reports clearly indicate that further evidences are necessary to a better understanding of ILs toxicity and biodegradability, which constitute the base for whole-cell biotechnological applications.

Based on these considerations, the regeneration of the ILs can be envisaged by biodegradation. For this purpose, activated sludge can be used to assimilate the VOC absorbed in the IL, enabling subsequent IL recycling to allow its use for a new cycle of VOC absorption. Multiphase bioreactors are frequently encountered in environmental biotechnology; several configurations involving gas/solid/liquid or gas/liquid/liquid phases have been reported. The use of such multiphase systems for the biodegradation of numerous pollutants (e.g. ethylacetate, phenol, toluene, benzene, xylene, and volatile organic contaminants) is well documented in the literature [120–122].

12.4.1 Two-Phase Partitioning Bioreactor (TPPB)

Multiphase bioreactors for environmental applications are known as two-phase partitioning bioreactors (TPPB – Fig. 12.4). TPPB are based on the addition of a non-aqueous liquid phase (NAPL) offering a high affinity for the target pollutants to be removed [123] in order to improve pollutant mass transfer from the gaseous to the liquid phase, and hence to improve the subsequent biodegradation kinetics. In most cases, pure microorganism strains, microbial consortium or activated sludge are implemented in this kind of reactors, with some possible drawbacks, namely possible NAPL toxicity towards microorganisms or on the contrary NAPL assimilation by the microorganisms.

As a consequence, among the only classes of solvents fulfilling the required characteristics, silicone oils and ionic liquids, ILs appear especially promising. It is worth noting that possible ILs toxicity has been reported as a key drawback for whole-cell applications [106, 107], even if there is not a general agreement regarding toxicity. The non-biodegradability of the NAPL is a required characteristic, owing to its continuous reuse [9]. However, reports on ILs biodegradability are controversial in the literature. These apparently contradictory reports clearly indicate that ILs structure selection is a key step in process development since it influences all preponderant properties.

All TPPB studies are based on the addition of a NAPL, either a liquid solvent or a solid polymer, with a high affinity for the target pollutants to be removed [123] in order to overcome two main limitations encountered during pollutant destruction: (i) the high toxicity of some pollutants resulting in cell inhibition and (ii) a limited substrate delivery to the microbial community in the case of pollutants with low affinity for water [1, 9]. NAPL addition improves the transfer of the target compounds from the gaseous phase to the liquid phase, and hence enhances their subsequent biodegradation. In most cases, pure microorganism strains, microbial consortium or activated sludge are implemented in this kind of reactors, with some possible drawbacks, namely possible NAPL toxicity towards microorganisms or on the contrary NAPL assimilation by the microorganisms.

VOC biodegradation performances depend on the presence and the assimilation potential of various microbial agents such as bacteria, micro-algae, fungi or yeasts contained in the aqueous phase [124]. there is a selective partitioning of the pollutant between water and NAPL in a TPPB; hydrophobic (or toxic) compounds are delivered to the aqueous phase at sub-inhibitory levels for microorganisms in case of toxic compounds or at the solubilisation limit in case of hydrophobic compounds.

For Déziel et al. [7] three mechanisms can be involved in the consumption of hydrophobic or toxic compounds:

1. VOC consumption in the aqueous phase. The degradation rate depends on the mass transfer rate from the organic to the aqueous phase, reported as lower than the VOC biodegradation rate [125, 126].

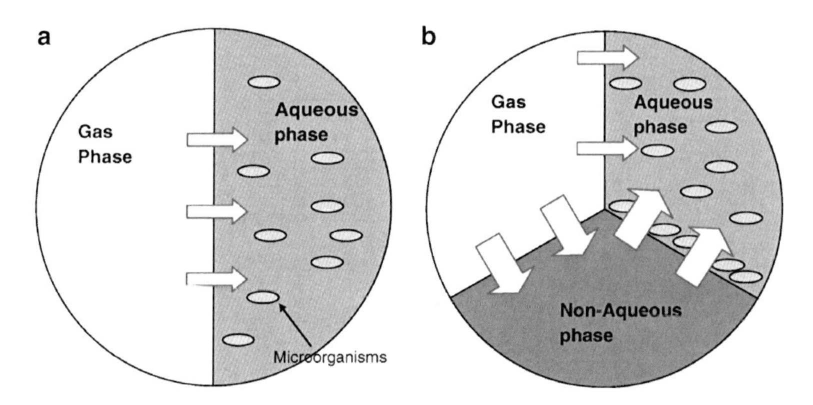

Fig. 12.4 Schematic description of mass transfer in conventional bioreactors (**a**) and in two-phase partitioning bioreactors TPPB (**b**)

2. VOC removal takes mainly place at the liquid-liquid (organic-aqueous) interface, since VOC can be directly assimilated at the interface after biofilm development between both liquid phases [127], or microorganisms accumulation at the interface [11, 125].
3. If the microorganisms are surfactant or emulsifier-producers, the formation of small droplets or micelles can lead to a reduction of the surface tension of the aqueous phase and to an increase of the interfacial area until microemulsion formation [128]. This phenomenon improves substrate availability for the microorganisms mainly located at the organic-aqueous interface.

   There is a general agreement that the major part of hydrophobic VOC uptake is achieved at the interface of both liquid phases [7, 129].

TPPB involving ILs as NAPL can take advantage of the high absorption capacity of ILs for a wide number of pollutants [46]. As already specified, there is a lack of data dealing with the use of ILs for pollutants removal; Baumann et al. [3] showed efficient phenol biodegradation rates in the presence of ILs in multiphase reactors [9]. Regarding hydrophobic VOCs, microorganisms acclimation led to interesting results for toluene, while further work is needed to optimise DMDS degradation in the presence of ILs [5].

ILs characteristics (non-inflammability, low vapor pressure, etc.) make them attractive solvents for chemical processes [130] and several kind of bioreactors, such as multiphase bioreactors, involve the use of ILs, especially for enzymatic transformations [2]. Biphasic devices involving IL/water systems are available; but they are mainly used for organic compounds extraction (or the extraction of compounds formed *in situ* during chemical transformations) or macromolecules (proteins) from the aqueous phase thanks to ILs [114, 131–133]. For instance, ILs can be used for fuel desulfurization [134], biofuels production [135], the extraction of food waste [73] or can be also included as part of solar cells [136].

To separate valuable molecules or to develop vectors for molecules solubilization, membrane processes have been described involving membranes modified with ILs [137].

In view of the implementation of ILs in TPPB and their subsequent recycling owing to their cost, their non-biodegradability and their biocompatibility are important parameters. Biodegradability is discussed thereafter, while toxicity towards microorganisms is discussed in the following sub-section (cf. 4.2).

Recently, some works have been conducting dealing with ILs biodegradation by a bacterial consortium. Docherty et al. [138] have examined the various steps and the degradation products of some imidazolium- and pyridinium-based ILs. It was shown a capacity of the microorganisms to degrade such ILs substituted by an octyl or a hexyl group. Contrarily, imidazolium- and pyridinium-based ILs with a butyl substitution seem to be not assimilated by activated sludge. Docherty et al. [138] also observed that the removal rate of these compounds increases with the length of the carbon chain, and hence they also highlight the ease and the rapidity of the assimilation of the pyridinium-based ILs with a long carbon chain. Contrarily, no biodegradation by activated sludge of the imidazolium-based ILs was observed in the case of [C_4Mim][Br], [C_4Mim][PF_6] and [C_4Mim][BF_4] anions. However, the addition of an ester group in the alkyl chain results in the biodegradation of the liquid ionic [138].

The biodegradability of the basic constituents of ILs was examined by Stolte et al. [139]. For example, they showed that 1,2-dimethylimidazole is completely assimilated by a bacterial consortium after 31 days, while the 1-methylimidazole and the 1-butylimidazole are not degraded. These authors also reported that [C_4Mim][Cl] is not degraded after 31 days of contact time with active sludge, while 11 % degradation has been observed for [C_6Mim][Cl].

In conclusion, the alkyl chain added to the cation has a non-negligible impact on the degradation, while the anion impact appears not clear; and finally a total absence of biodegradability was only shown for [C_4Mim][PF_6] [109].

12.4.2 Toxicity of ILs Towards Microorganisms or Biocompatibility

These criteria have to be considered as soon as possible in the ILs structure development. First studies concerning ILs toxicity were reported in the 1950s and since the 1990s many publications have dealt with this topic as ILs have become popular in green chemistry [110, 140, 141]. A good knowing of IL toxicity is necessary to develop an industrial application [110].

Pham et al. published a quite exhaustive report summarizing many of the results described in literature [110]. In 2010, another review treated of ILs ecotoxicity (in water and soils) and toxicity on humans [142]. It is crucial to consider that toxicity is dependent of the biological target's nature. Indeed, an IL can be harmless towards a specific microorganism and very toxic towards another one.

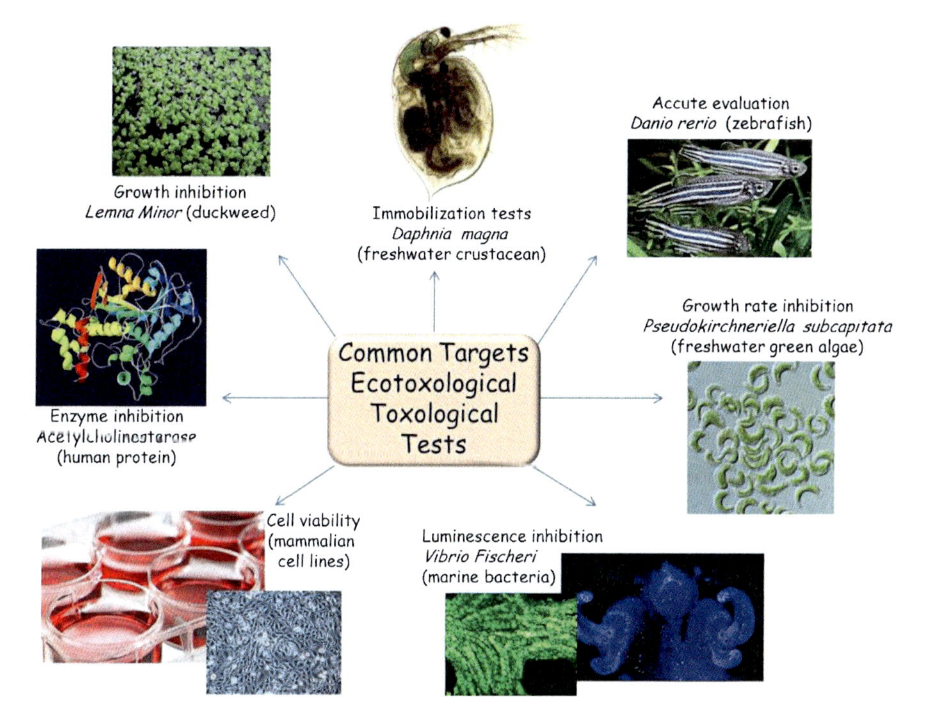

Fig. 12.5 Biological targets used for ILs toxicity evaluation

To assess regarding ILs toxicity towards microorganisms, microbial activity in the presence or not of ILs can be compared. For instance, Zhang et al. [143] showed that the activity of *Aureobasidium pullulans* was not affected by the presence of $[C_4Mim][PF_6]$, contrarily to some other solvents like hexane or dibutylphtalate. Matsumoto et al. [52] examined the toxicity of 1-R-3-methylimidazolium hexafluorophosphate ILs with R = butyl, hexyl or octyl, on a lactic acid bacteria, *Lactobacillus delbruekii*; they observed that lactic acid production decreases with the augmentation of the alkyl chain. However, the study realized on various lactic acid bacteria showed that the IL tolerance is related to the acid production (microbial activity in the presence of IL < microbial activity in the presence of water).

Figure 12.5 presents the biological targets used for ILs' toxicity evaluation [110].

12.4.2.1 Toxicity Towards Microorganisms

A lot of these evaluations have been practiced on *Danio rerio* (zebrafish) or *Caenorhabditis elegans* with concentrations between 1 and 100 $mg.L^{-1}$. Before proceeding, ILs must be dried (water quantity is currently determined by a Karl-Fischer titration in ppm) and highly pure (determination by HPLC, NMR,. . .) [144].

Here are enunciated some microorganisms used to evaluate ILs toxicity: *Danio rerio* (zebrafish), *Scenedesmus vacuolatus* (green algae), *Lemna minor* (marine plant), *Caenorhabditis elegans* (earthworm), *Physa acuta* (aquatic snail), *Oocystis submarine* (Baltic algae), *Cyclotella meneghiniana* (Baltic algae), *Daphnia magna* (marine plant), *Lactobacillus* (bacteria), *L. rhamnosus* (bacteria), *Vibrio fischeri* (bacteria), *E. Coli* (bacteria), *Pichia pastoris* (bacteria), *Bacillus cereus* (bacteria) [51, 52, 140, 144, 145].

A high quantity of tests has been practiced on *Vibrio fischeri*, a marine bacteria [146–148]. In general, ILs are more toxic than acetonitrile, an organic solvent known for its toxicity [149]. They also have observed that pyridiniums are among the most toxic ILs and morpholiniums seems to be the less toxic [150]. Ammoniums are toxic towards Zebrafish [145] and other structures as choline-based ILs, morpholiniums, tropiniums, quinuclidiniums or alkylpyrrolidiniums were developed to improve biocompatibility (in association with bromide anion) [149]. Alkyl chain length and the number of carbon atoms (>4 carbons) play an important role in toxicity. Longer chains exhibit higher toxicities [107, 108, 140, 144, 151]. The chain's position on the ring does not influence the toxicity (e.g. methyl's position on dimethylimidazoliums) [146].

Other evaluations on aquatic microorganism *Dreissena polymorpha*, a mollusc [152, 153], and *Daphnia magna*, a crustacean [154], showed similar tendencies (long alkyl chains increase lipophilicity and intensify toxicity).

Some functional groups like esters increase toxicity [141] whereas other functional groups like nitriles, hydroxyles or ethers reduce it [119, 149, 155]. Some 1-methyl-3-alkoxyalkylimidazoliums have been evaluated on bacterial targets have been found harmless, except if the cation wears long alkyl chains (>6–7 carbons) [106]. Moreover, aromatic groups like phenols show higher toxicity than a butyl chain [149].

Alvarez-Guerra et al. established a model comparing 30 anions and 64 cations on *Vibrio fischeri* [147].

Perfluoroanions as PF_6^- in water at 50 °C can produce HF, *via* hydrolysis, and this acid can be very harmfull towards microorganisms [18, 119, 156–158]. The hydrolysis of PF_6^-'s hydrolysis was studied by Swatloski et al. and these authors proposed anions in principle non toxics but they confer an hydrophilic profile [159]. It was shown that $RfBF_3^-$ anions were very stable towards hydrolysis and produce less quantity of HF [160]. Ignat'ev et al. reported the synthesis of $Rf_3PF_3^-$ anions (Rf is perfluoroalkyl chain containing 2 or 3 carbon atoms) less sensitive towards hydrolysis [156] and these ILs exhibit lower viscosities but higher toxicities [161].

All these studies demonstrate that the lipophilicity of the anion is a preponderant criteria [115]. It can be expected that the association with an appropriate cation could modify the global toxicity of the IL and balance the intrinsic anion toxicity [150].

Metabolites produced by degradation of ILs could be toxic too and a metabolisation route by cytochrome P_{450} of C_4Mim cation has been proposed in the literature [44].

To conclude, all these studies show that lipophilicity of ILs is the determinant criteria related to toxicity and a hypothesis is that lipophilic cations can interact with lipidic membranes of microorganisms [150]. Considering this fact, measuring the Log P_{OW} parameter could be a relevant indicator to estimate on ILs toxicity [51, 52, 162]. As we have evocated earlier, toxicity also depends on the biological target [119], so some results in appearance contradictory were reported in literature. As an example, NTf_2^- is not toxic towards some bacteria whereas it is toxic towards some microorganisms (*Scenedesmus vacuolatus* and *Vibrio fischeri*) [150].

The environmental parameters must be considered because it was established that they can influence toxicity values. Indeed, it was published that NaCl content in water has a protective effect in high concentrations. Microorganism's size and the cell membranes' composition also play an important role since they are implicated in interactions between the IL and the biological target [142].

In aquatic environment, interactions with DOM (Dissolved Organic Mattercould) influence toxicity of ILs. Evaluations on *Lemna minor* demonstrated that association with high concentrations of DOM increase ILs toxicity [163].

12.4.2.2 Cytotoxicity

Ranke and co-workers are pioneers in cytotoxic evaluation of ILs. Their first study of cytotoxicity of ILs was reported in 2002 and concerned 1-butyl-3-methylimidazolium chloride [164]. J774A.1 macrophage cells were used and the authors determined the LC-50 (lethal concentration leaving 50 % cells alive): 0.50 mg. mL^{-1} after 48 h of incubation. They also showed that an augmentation of the incubation time increases the cytotoxicity of BmimCl. These results highlighted the importance of wider evaluations of ILs cytotoxicities.

Jastorff et al. proposed a SAR study based on the alkylimidazolium scaffold. They showed that long alkyl chains increase the global toxicity of imidazoliums (evaluation on IPC-81 leukemia cell lines) [44]. The year after, the same team proposed a detailed biological study of methyl- and ethylimidazoliums on IPC-81 (leukemia cells) and C_6 (glioma cells) rat cell lines [151]. Again, they observed that long alkyl chains increase toxicity against cell lines involved. In this case, the anion (here PF_6^-, BF_4^- and Br^-) seems to have a few influence whereas another study published by Stepnowski et al. showed that the anion bulk, especially NTf_2^-, has a preponderant role in toxicity towards HeLa cell lines [165]. These results involved that the cytotoxicity of ILs depended on several parameters: IL structure, concentrations and cell lines To rationalize the cytotoxic behavior of alkylimidazoliums, Ranke's team studied the cellular distribution of some ILs and correlated cytotoxicity and lipophilicity [150, 162, 166]. Indeed, most hydrophobic ILs are more toxic and exhibit higher cellular sorption ($C_8MimBF_4 > C_6MimBF_4 > C_4MimBF_4$). $C_{10}MimBF_4$ was too toxic and further evaluations were not purchased. After HPLC analysis, they determined cellular distribution (~80 % in cytosol, ~12 % in membrane and ~8 % in nucleus). It is commonly assumed that cytotoxicity of ILs is partially due to interactions between ILs and cell lipidic membranes [167].

On another hand, this same team demonstrated that lipophilic properties of the anion side chains and his chemical stability are widely involved in cytotoxicity but the anions tested do not exhibit intrinsic cytotoxicity towards IPC-81 cell lines [168].

Another group investigated cytotoxicity of ILs. Their studies based on varied ILs (pyrrolidiniums, piperidiniums, imidazoliums...) showed the same tendencies (longer alkyl chains involve higher cytotoxicities) and pyridiniums seems to be more toxic than other ILs [169]. They also showed that functionalized ILs (ethers,...) exhibit significantly lower cytotoxicities [170].

Several exhaustive reports were published in 2010 and 2011 and summarize all known studies concerning cytotoxicity of ILs towards mammalian cell lines (HeLa, CaCo-2, IPC-81, HT-29, C6, MCF-7, NCI60, V79). They confirmed the interdependence between lipophilicity, structure, concentration and cytotoxicity [110, 142, 171]. About 230 ILs with various structures were evaluated on IPC-81 (mammalian cell lines derived from a model of acute myelogenic leukemia) and a QSTR profile was established, confirming the tendencies previously mentioned [172]. Some results are summarized in the next table (see Table 12.3) [162, 172].

However, cytotoxic potential of ILs is a crucial parameter to develop industrial processes, it is not a latent obstacle. For example, some pharmaceuticals (e.g. lidocaine) have been grafted on ILs to increase solubility or biodisponibility [173, 174]. In the aim to set up industrial applications, moreover in pharmaceutical industry, establishing cytotoxic profile of ILs is necessary and will represent an important field of search for the future.

12.4.3 ILs Biodegradability

Some methods are available to assess the biodegradability of an IL, which are not detailed in this chapter. However, Coleman et al. [141] proposed an exhaustive summary; for more than 60 % possible biodegradation, a compound can be classified as biodegradable.

However and beforehand to a biodegradability study, possible biosorption of ILs should be assessed. Experiments using 4 g/L thermically inactivated sludge (autoclaved) showed a constant IL concentration (10 g/L initial amount) throughout the experiments which lasted 30 days, indicating an almost negligible biosorption of the considered imidazolium IL, $[C_4Mim][PF_6]$ and $[C_4Mim][NTf_2]$. Biodegradability tests were then performed on these ILs showing constant IL values throughout the short- and long-term (5 days and 1 month) experiments performed.

These results are in agreement with those of other authors [107, 109]. Indeed, with an imidazolium-based cation, irrespective of the considered anion, such as Br^-, Cl^-, NTf_2^- and BF_4^-, these authors did not observe any biodegradation using activated sludge. Similarly, Stolte et al. [139] reported that activated sludge is not able, even after 31 days, to assimilate $[C_4Mim]$ cations.

Table 12.3 $Log_{10}(EC_{50})$ (EC_{50} (μM): concentration for 50 % effect of the evaluated substance) concerning IPC-81

Cation	Anion	Log_{10} EC_{50}	Cation	Anion	Log_{10} EC_{50}
C_4Mim	Cl	3.55	$C_{10}Mim$	PF_6	1.5
C_4Mim	Br	3.43	$C_{10}Mim$	Cl	1.34
C_4Mim	BF_4	3.12	$C_{10}Mim$	BF_4	0.77
C_4Mim	PF_6	3.1	C_4Iq	Br	2.32
PhMim	BF_4	>3	C_4Iq	BF_4	2.16
PhMim	Cl	>3	C_6Iq	BF_4	1.07
PhMim	PF_6	>3	C_8Iq	BF_4	0.14
C_6Mim	BF_4	2.98	C_8Iq	Br	−0.03
C_6Mim	PF_6	2.91	C_4Mpyrr	Cl	>4.3
C_6Mim	Cl	2.85	C_4Mpyrr	Br	3.77
C_8Mim	Cl	2.01	C_4Mpyrr	BF_4	2.9
C_8Mim	PF_6	1.96	C_8Mpyrr	Cl	2.59
C_8Mim	BF_4	1.59	C_8Mpyrr	BF_4	1.82
C_4Pyr	Br	3.9	C_4Pyr	BF_4	3.16

Boethling [175] presented some factors used in the design of biodegradable compounds: the presence of potential sites of enzymatic hydrolysis (for example, ester and amides groups in the molecule) or the introduction of oxygen in the form of hydroxyl, aldehyde or carboxylic acid groups or even the presence of unsubstituted linear alkyl chains (especially in molecules with more than four carbon atoms) and phenyl rings [175]. Most of authors who designed ILs follow these factors, showing that some might be or not completely applied to ILs [108, 109, 117, 138, 139]. Therefore, primary biodegradation of ILs have been studied with different anions, cations and side chains configurations.

An absence of biological degradation was found that for imidazolium-based ILs with short alkyl chains (<6 carbon atoms) and short functionalized side chains, and that the introduction of functional groups with a higher chemical reactivity did not improve their biodegradability. For example, butylmethylimidazolium salts were found to be poorly biodegradable, while an impact on the biodegradability was observed related to the counter-ion considered, leading to the following order for their biodegradability [117]:

$$\text{Octyl-OSO}_3 > N(CN)_2^- = Cl^- = Br^- = PF_6^- = NTf_2^-$$

Gathergood et al. [117] also studied the biodegradability of 3-methyl-1-(propyloxycarbonyl) imidazolium and it varies according to the following order:

$$\text{Octyl-OSO}_3 > N(CN)_2^- > NTf_2^- > Br^- > PF_6^- = BF_4^-$$

However, octyl chains in imidazolium cations have shown to be highly biodegradable and the introduction of –OH and –COOH into the octyl chains improves primary degradation of ILs [108, 138, 139].

In addition, ILs containing an ester or an amide group in the alkyl side chain (>4 carbon atoms) were found to be biodegradable and that biodegradability increases slightly if the alkyl chain length increases for the lowest alkyl esters and then remain nearly invariable [117].

Nevertheless, for pyridinium with long side chains (6–8 carbon atoms), Docherty et al. [138] have observed complete mineralization of ILs, but those with short side chains (<4 carbon atoms) are not mineralized. Therefore, pyridinium-based ILs might be considered as readily biodegradable. A link between the length of the alkyl chain and the metabolization rate of pyridinium compounds is also discussed.

Yu et al. [176] showed that the presence of benzene cycles also increases the biodegradability of ILs, but no information dealing with their toxicity is presented. It should be specified that the less toxic IL is, *a priori,* more easily biodegradable since it does not attack the microorganisms involved in its degradation [177].

Other structural modifications would allow reducing the biodegradability of ILs: the presence of halogens (particularly chlorine and fluorine), alkyl ramified side chains (trisubstituted nitrogen or quaternary carbons), functional groups such as nitro, amino or arylamino, polycyclical structures (indole, etc.), heterocyclic structures, aliphatic ethers. Concerning the anions, it has been showed that the alkylsulfates increase the biodegradability of ILs [178].

Identification of the biodegradation products of some ILs, such as imidazoliums and pyridiniums, have been the purpose of several studies [118, 119, 139, 140, 179–182]. The related metabolites generally result from a first oxidation reaction on a lateral chain of the cation, followed by water or CO_2 extrusion.

The objective of most of the available biodegradability studies is to develop 'green' ILs and hence biodegradable [183]. However and owing to their high cost, for possible implementation in a TPPB the subsequent recycling of the IL should be considered in view of reducing cost process, and thus the objective is an absence of biodegradability (and toxicity) facing the considered microorganisms.

12.4.4 VOC Biodegradation in TPPBs

Biological processes play an important role in the treatment of VOC. According to the compound or the family of compounds to be removed, the biomass does not contain the same microbial species. At lab-scale, bacteria are mainly used; commercial or strains isolated for their potential can be implemented. Some recent applications also involve fungi, which are more tolerant than bacteria to low water activities and acidic pH and their important enzyme complex [184], especially aromatic compounds [185]. However, owing to the low resistance of pure strains facing actual effluents variability, industrial applications involve mainly activated sludge. The use of multiphase systems for VOCs degradation has been the subject to numerous studies. These systems have been tested to treat various VOCs, including toluene [186], benzene [187], hexane [11] etc., using mainly bacteria belonging to various species, like *Pseudomonas* or *Mycobacterium* [186, 188, 189], microbial consortium [190] or activated sludge [191, 192]. Various organic phases have been implemented for this purpose, including hexadecane [187], dodecane [193] or silicone oil [13, 194, 195].

There are scarce results regarding the use of ILs as NAPLs for VOCs absorption. To our knowledge, only the imidazolium salts, $[C_4Mim][PF_6]$ and $[C_4Mim][NTf_2]$, have been implemented in a TPPB for toluene and DMDS biodegradation [4, 5]; activated sludge was considered, which was beforehand acclimated or not to the target VOC.

In the absence of activated sludge acclimation, there was a clear toxic effect of both ionic liquids, especially in the presence of $[C_4Mim][NTf_2]$, since higher biodegradation rates were recorded in the absence of IL; the trend was especially pronounced in the case of toluene. Cell acclimation was therefore needed, which clearly improved biodegradation rates for both VOCs in the control and in the presence of IL. The improvement was especially significant for toluene, in the presence of both ionic liquids, and the most striking result was observed after acclimation, since biodegradation rates were nearly similar for the control deprived of IL and in the presence of 5 % $[C_4Mim][PF_6]$, 0.49 and 0.48 $g.m^{-3}.h^{-1}$ respectively.

Neither biosorption nor biodegradability has been observed for these two ILs. However, $[C_4Mim][NTf_2]$ has a toxic effect since even after cell acclimation, biodegradation rates remained lower than those observed in the absence of this IL. Promising results have been recorded for toluene in the presence of $[C_4Mim][PF_6]$ after cell acclimation, while both ILs appeared toxic regarding microorganisms involved in DMDS assimilation. From this, more complex strategies, including acclimation to IL, should be subsequently considered.

12.5 Conclusion

Hydrophobic ILs show interesting specific properties which make them attractive for the development of chemical and biochemical processes. These properties can be designed by selecting appropriates anion and cation. Regarding their synthesis, the general rule is a short and simple process, even if various possible functionalizations can lead to complex synthetic scheme. Among the wide number of structure, alkylimidazoliums are the most studied ILs.

In the case of an implementation in a bioreactor, the RTIL parameters to control are viscosity, safety towards the microorganisms contained in the reactor and an absence of ecotoxicity, an absence of biodegradability and a high affinity for the targeted VOC. They should display a high hydrophobicity to allow an easy separation from the aqueous phase and hence an efficient recycling (low losses during successive recycling cycles for a low water solubility, below 2 %). Furthermore, a synthesis at a moderate cost and in high amounts is also required for the selected ILs.

Accordingly and among the tested ILs, some promising results have been obtained using $[C_4Mim][PF_6]$ and $[C_4Mim][NTf_2]$, showing that they can be an alternative to the most often implemented organic phase, silicone oil. Indeed, the partition coefficients for some model VOCs, toluene and DMDS, appear similar to those observed with silicone oil. Inhibitory tests for glucose consumption have shown that after 1 day lag phase, biodegradation rates are comparable to those observed in the absence of IL. In addition, neither biosorption nor biodegradation by activated sludge have been observed for $[C_4Mim][PF_6]$ and $[C_4Mim][NTf_2]$.

In addition, significant rates of toluene biodegradation have been found in the presence of $[C_4Mim][PF_6]$, similar to those observed in the absence of IL; however only after activated sludge acclimation. Contrarily, $[C_4Mim][NTf_2]$ shows a toxic effect even after an acclimation phase, since low degradation rates have been observed. At the opposite, a toxic effect of IL has been observed regarding DMDS, even after an acclimation time.

This toxic effect can limit the use of ILs in multiphasic bioreactors, but the promising results recorded for toluene suggest that more investigations are needed, especially regarding the acclimation strategy, and in particular to the considered IL in addition to VOC acclimation.

In conclusion, these compounds can be an alternative to silicone oils, but further works are needed to confirm their relevance for implementation in multiphase bioreactors.

Acknowledgement The authors want to thanks the French National Research Agency (ANR - Blank program) for the financial support of this work.

References

1. Darracq G, Couvert A, Couriol C, Amrane A, Le Cloirec P. Removal of hydrophobic VOC in an integrated process coupling absorption and biodegradation – selection of an organic liquid phase. Water Air Soil Pollut. 2012;223:4969–97.
2. Quijano G, Couvert A, Amrane A. Ionic liquids: applications and future trends in bioreactor technology. Bioresour Technol. 2010;101:8923–30.
3. Baumann MD, Daugulis AJ, Jessop PG. Phosphonium ionic liquids for degradation of phenol in a two-phase partitioning bioreactor. Appl Microbiol Biotechnol. 2005;67:131–7.
4. Quijano G, Couvert A, Amrane A, Darracq G, Couriol C, Le Cloirec P, Paquin L, Carrié D. Potential of ionic liquids for VOC absorption and biodegradation in multiphase systems. Chem Eng Sci. 2011;66:2707–12.
5. Quijano G, Couvert A, Amrane A, Darracq G, Couriol C, Le Cloirec P, Paquin L, Carrié D. Toxicity and biodegradability of ionic liquids: new perspectives towards whole-cell biotechnological applications. Chem Eng J. 2011;174:27–32.
6. Bruce LJ, Daugulis AJ. Solvent selection strategies for extractive biocatalysis. Biotechnol Prog. 1991;61:116–24.
7. Déziel E, Commeau Y, Villemur R. Two-liquid-phase bioreactors for enhanced degradation of hydrophobic/toxic compounds. Biodegradation. 1999;10:219–33.
8. Rontani JF, Giusti G. Study of the biodegradation of poly-branched alkanes by a marine bacterial community. Mar Chem. 1986;20:197–205.
9. Quijano G, Hernandez M, Thalasso F, Muñoz R, Villaverde S. Two-phase partitioning bioreactor in environment biotechnology. Appl Microbiol Biotechnol. 2009;84:829–46.

10. Darracq G, Couvert A, Couriol C, Amrane A, Thomas D, Dumont E, Andres Y, Le Cloirec P. Silicone oil: an effective absorbent for hydrophobic volatile organic compounds (VOC) removal. J Chem Technol Biotechnol. 2010;85:309–13.
11. Muñoz R, Arriaga S, Hernandez S, Guieysse B, Revah S. Enhanced hexane biodegradation in a two phase partitioning bioreactor: overcoming pollutant transport limitations. Process Biochem. 2006;41:1614–19.
12. Aldric JM, Thonart P. Performance of a water/silicone oil two-phase partitioning bioreactor using *Rhodococcus erythropolis* T902.1 to remove volatile organic compounds from gaseous effluents. J Chem Technol Biotechnol. 2008;83:1401–8.
13. Mahanty B, Parkshirajan K, Dasu VV. Biodegradation of pyrene by *Mycobacterium frederiksbergense* in a two-phase partitioning bioreactor system. Bioresour Technol. 2008;99:2694–8.
14. Wilkes JS. A short history of ionic liquids-from molten salts to neoteric solvents. Green Chem. 2002;4(2):73–80.
15. Yau HM, Chan SJ, George SRD, Hook JM, Croft AK, Harper JB. Ionic liquids: just Molten salts after all? Molecules. 2009;14(7):2521–34.
16. Yang Z, Pan W. Ionic liquids: green solvents for non aqueous biocatalysis. Enzyme Microbial Technol. 2005;37:19–28.
17. Earle MJ, Seddon KR. Ionic liquids, green solvents for the future. Pure Appl Chem. 2000;72:1391–8.
18. Huddleston JG, Visser AE, Reichert WM, Willauer HD, Broker GA, Rogers RD. Characterization and comparison of hydrophilic and hydrophobic room temperature ionic liquids incorporating the imidazolium cation. Green Chem. 2001;3:156–64.
19. Earle MJ, Katdare SP, Seddon KR. Paradigm confirmed: the first use of ionic liquids to dramatically influence the outcome of chemical reactions. Org Lett. 2004;6(5):707–10.
20. Smiglak M, Reichert WM, Holbrey JD, Wilkes JS, Sun L, Thrasher JS, Kirichenko K, Singh S, Katritzky AR, Rogers RD. Combustible ionic liquids by design: is laboratory safety another ionic liquid myth? Chem Commun. 2006;24:2554–6.
21. Earle MJ, Esperança JMSS, Gilea MA, Lopes JNC, Rebelo LPN, Magee JW, Seddon KR, Widegren JA. The distillation and volatility of ionic liquids. Nature. 2006;439(7078):831–4.
22. Liaw H-J, Huang S-K, Chen H-Y, Liu S-N. Reason for ionic liquids to be combustible. Procedia Eng. 2012;45:502–6.
23. Diallo A-O, Morgan AB, Len C, Marlair G. An innovative experimental approach aiming to understand and quantify the actual fire hazards of ionic liquids. Energy Environ Sci. 2013;6 (3):699–710.
24. Diallo AO, Len C, Morgan AB, Marlair G. Revisiting physico-chemical hazards of ionic liquids. Sep Purif Technol. 2012;97:228–34.
25. Welton T. Room-temperature ionic liquids. Solvents for synthesis and catalysis. Chem Rev. 1999;99:2071–84.
26. Carda-Broch S, Berthod A, Armstrong DW. Solvent properties of the 1-butyl-3-methylimidazolium hexafluorophosphate ionic liquid. Anal Bioanal Chem. 2003;375(2):191–9.
27. Morgan D, Ferguson L, Scovazzo P. Diffusivities of gases in room-temperature ionic liquids: data and correlations obtained using a lag-time technique. Ind Eng Chem Res. 2005;44 (13):4815–23.
28. Camper D, Becker C, Koval C, Noble R. Diffusion and solubility measurements in room temperature ionic liquids. Ind Eng Chem Res. 2005;45(1):445–50.
29. Ferguson L, Scovazzo P. Solubility, diffusivity, and permeability of gases in phosphonium-based room temperature ionic liquids: data and correlations. Ind Eng Chem Res. 2007;46 (4):1369–74.
30. Condemarin R, Scovazzo P. Gas permeabilities, solubilities, diffusivities, and diffusivity correlations for ammonium-based room temperature ionic liquids with comparison to imidazolium and phosphonium RTIL data. Chem Eng J. 2009;147(1):51–7.

31. Shiflett MB, Yokozeki A. Solubility and diffusivity of hydrofluorocarbons in room-temperature ionic liquids. AICHE J. 2006;52(3):1205–19.
32. Vuong MD, Couvert A, Couriol C, Amrane A, Le Cloirec P, Renner C. Determination of the Henry's constant and the mass transfer velocity of VOCs in solvents. Chem Eng J. 2009;150:430.
33. Basmadjian D. Mass transfer and separation processes: principles and applications. New York: CRC Press; 2007.
34. Roustan M. Transferts gaz-liquide dans les procédés de traitement des eaux et des effluents gazeux. Paris: Lavoisier; 2003.
35. Danckwerts PV. Gas absorption with instantaneous reaction. Chem Eng Sci. 1968;23 (9):1045–51.
36. Hou Y, Baltus R. Experimental measurement of the solubility and diffusivity of CO_2 in room-temperature ionic liquids using a transient thin-liquid-film method. Ind Eng Chem Res. 2007;46(24):8166–75.
37. Shi W, Sorescu DC, Luebke DR, Keller MJ, Wickramanayake S. Molecular simulations and experimental studies of solubility and diffusivity for pure and mixed gases of H_2, CO_2, and Ar absorbed in the ionic liquid 1-n-hexyl-3-methylimidazolium bis(trifluoromethylsulfonyl) amide ([hmim][Tf2N]). J Phys Chem B. 2010;114(19):6531–41.
38. Shiflett MB, Yokozeki A. Solubilities and diffusivities of carbon dioxide in ionic liquids: [bmim][PF6] and [bmim][BF4]. Ind Eng Chem Res. 2005;44(12):4453–64.
39. Shiflett MB, Harmer MA, Junk CP, Yokozeki A. Solubility and diffusivity of difluoromethane in room-temperature ionic liquids. J Chem Eng Data. 2006;51(2):483–95.
40. Shiflett MB, Harmer MA, Junk CP, Yokozeki A. Solubility and diffusivity of 1,1,1,2-tetrafluoroethane in room-temperature ionic liquids. Fluid Phase Equilibr. 2006;242 (2):220–32.
41. Skrzypczak A, Neta P. Diffusion-controlled electron-transfer reactions in ionic liquids. J Phys Chem A. 2003;107(39):7800–3.
42. Gong Y, Wang H, Chen Y, Hu X, Ibrahim A-R, Tanyi A-R, Hong Y, Su Y, Li J. A high-pressure quartz spring method for measuring solubility and diffusivity of CO_2 in ionic liquids. Ind Eng Chem Res. 2013;52(10):3926–32.
43. Crowhurst L, Mawdsley PR, Perez-Arlandis JM, Salter PA, Welton T. Solvent-solute interactions in ionic liquids. Phys Chem Chem Phys. 2003;5(13):2790–4.
44. Jastorff B, Störmann R, Ranke J, Mölter K, Stock F, Oberheitmann B, Hoffmann W, Hoffmann J, Nüchter M, Ondruschka B. How hazardous are ionic liquids? Structure-activity relationships and biological testing as important elements for sustainability evaluation. Green Chem. 2003;5(2):136–42.
45. Visser AE, Swatloski RP, Reichert WM, Mayton R, Sheff S, Wierzbicki A, Davis Jr JH, Rogers RD. Task-specific ionic liquids incorporating novel cations for the coordination and extraction of Hg^{2+} and Cd^{2+}: Synthesis, characterization, and extraction studies. Environ Sci Technol. 2002;36(11):2523–9.
46. Zhao H, Xia S, Ma P. Use of ionic liquids as 'green' solvents for extractions. J Chem Technol Biotechnol. 2005;80(10):1089–96.
47. Visser AE, Rogers RD. Room-temperature ionic liquids: new solvents for f-element separations and associated solution chemistry. J Solid State Chem. 2003;171(1):109–13.
48. Sun X, Ji Y, Hu F, He B, Chen J, Li D. The inner synergistic effect of bifunctional ionic liquid extractant for solvent extraction. Talanta. 2010;81(4):1877–83.
49. Gaillard C, Mazan V, Georg S, Klimchuk O, Sypula M, Billard I, Schurhammer R, Wipff G. Acid extraction to a hydrophobic ionic liquid: the role of added tributylphosphate investigated by experiments and simulations. Phys Chem Chem Phys. 2012;14(15):5187–99.
50. Huddleston JG, Rogers RD. Room temperature ionic liquids as novel media for "clean" liquid-liquid extraction. Chem Commun. 1998;16:1765–6.
51. Matsumoto M, Mochiduki K, Fukunishi K, Kondo K. Extraction of organic acids using imidazolium-based ionic liquids and their toxicity to *Lactobacillus rhamnosus*. Sep Purif Technol. 2004;40:97–101.

52. Matsumoto M, Mochiduki K, Kondo K. Toxicity of ionic liquids and organic solvents to lactic acid-producing bacteria. J Biosci Bioeng. 2004;98:344–7.
53. Fan J, Fan Y, Pei Y, Wu K, Wang J, Fan M. Solvent extraction of selected endocrine-disrupting phenols using ionic liquids. Sep Purif Technol. 2008;61(3):324–31.
54. Koylecki T, Sawinski W, Sokoowski A, Ludwig W, Polowczyk I. Extraction of organic impurities using 1-butyl-3-methylimidazolium hexafluorophosphate [BMIM][PF6]. Pol J Chem Technol. 2008;10(1):79–83.
55. Egorov VM, Smirnova SV, Pletnev IV. Highly efficient extraction of phenols and aromatic amines into novel ionic liquids incorporating quaternary ammonium cation. Sep Purif Technol. 2008;63(3):710–15.
56. Nakamura K-I, Kudo Y, Takeda Y, Katsuta S. Partition of substituted benzenes between hydrophobic ionic liquids and water: evaluation of interactions between substituents and ionic liquids. J Chem Eng Data. 2011;56(5):2160–7.
57. Katsuta S, Nakamura K, Kudo Y, Takeda Y. Mechanisms and rules of anion partition into ionic liquids: phenolate ions in ionic liquid/water biphasic systems. J Phys Chem B. 2012;116 (2):852–9.
58. Katsuta S, Nakamura K, Kudo Y, Takeda Y, Kato H. Partition behavior of chlorophenols and nitrophenols between hydrophobic ionic liquids and water. J Chem Eng Data. 2011;56 (11):4083–9.
59. Rosatella AA, Branco LC, Afonso CAM. Studies on dissolution of carbohydrates in ionic liquids and extraction from aqueous phase. Green Chem. 2009;11(9):1406–13.
60. Tomé LIN, Catambas VR, Teles ARR, Freire MG, Marrucho IM, Coutinho JAP. Tryptophan extraction using hydrophobic ionic liquids. Sep Purif Technol. 2010;72:167–73.
61. Ha SH, Mai NL, Koo Y-M. Butanol recovery from aqueous solution into ionic liquids by liquid-liquid extraction. Process Biochem. 2010;45(12):1899–903.
62. Hanke CG, Johansson A, Harper JB, Lynden-Bell RM. Why are aromatic compounds more soluble than aliphatic compounds in dimethylimidazolium ionic liquids? A simulation study. Chem Phys Lett. 2003;374(1):85–90.
63. Arce A, Earle MJ, Katdare SP, Rodriguez H, Seddon KR. Application of mutually immiscible ionic liquids to the separation of aromatic and aliphatic hydrocarbons by liquid extraction: a preliminary approach. Phys Chem Chem Phys. 2008;10(18):2538–42.
64. Calvar N, Dominguez I, Gomez E, Dominguez Ã. Separation of binary mixtures aromatic+ aliphatic using ionic liquids: influence of the structure of the ionic liquid, aromatic and aliphatic. Chem Eng J. 2011;175:213–21.
65. Dong K, Song Y, Liu X, Cheng W, Yao X, Zhang S. Understanding structures and hydrogen bonds of ionic liquids at the electronic level. J Phys Chem B. 2012;116(3):1007–17.
66. Dong K, Zhang S. Hydrogen bonds: a structural insight into ionic liquids. Chem-Eur J. 2012;18(10):2748–61.
67. Housaindokht MR, Hosseini HE, Googheri MSS, Monhemi H, Najafabadi RI, Ashraf N, Gholizadeh M. Hydrogen bonding investigation in 1-ethyl-3-methylimidazolium based ionic liquids from density functional theory and atoms-in-molecules methods. J Mol Liq. 2013;177:94–101.
68. Gomes M, Lopes J, Padua A. Thermodynamics and micro heterogeneity of ionic liquids. Top Curr Chem. 2010;290:161–83.
69. Roland CM, Bair S, Casalini R. Thermodynamic scaling of the viscosity of van der Waals, H-bonded, and ionic liquids. J Chem Phys. 2006;125(12):124508/124501–124508/124508.
70. Greaves TL, Drummond CJ. Ionic liquids as amphiphile self-assembly media. Chem Soc Rev. 2008;37(8):1709–26.
71. Kempter V, Kirchner B. The role of hydrogen atoms in interactions involving imidazolium-based ionic liquids. J Mol Struct. 2010;972(1–3):22–34.
72. Pei YC, Wang JJ, Xuan XP, Fan J, Fan M. Factors affecting ionic liquids based removal of anionic dyes from water. Environ Sci Technol. 2007;41(14):5090–5.

73. Wang Y-X, Cao X-J. Extracting keratin from chicken feathers by using a hydrophobic ionic liquid. Process Biochem. 2012;47(5):896–9.
74. Khodadoust AP, Chandrasekaran S, Dionysiou DD. Preliminary assessment of imidazolium-based room-temperature ionic liquids for extraction of organic contaminants from soils. Environ Sci Technol. 2006;40(7):2339–45.
75. Bonny S, Paquin L, Carrié D, Boustie J, Tomasi S. Ionic liquids based microwave-assisted extraction of lichen compounds with quantitative spectrophotodensitometry analysis. Anal Chim Acta. 2011;707(1):69–75.
76. Louros CLS, Claudio AFM, Neves CMSS, Freire MG, Marrucho IM, Pauly J, Coutinho JAP. Extraction of biomolecules using phosphonium-based ionic liquids+ K_3PO_4 aqueous biphasic systems. Int J Mol Sci. 2010;11(4):1777–91.
77. Tang B, Bi W, Tian M, Row KH. Application of ionic liquid for extraction and separation of bioactive compounds from plants. J Chromatogr B. 2012;904:1–21.
78. Sashina ES, Novoselov NP, Kuz'mina OG, Troshenkova SV. Ionic liquids as new solvents of natural polymers. Fibre Chem. 2008;40(3):270–7.
79. Wang J, Pei Y, Zhao Y, Hu Z. Recovery of amino acids by imidazolium based ionic liquids from aqueous media. Green Chem. 2005;7(4):196–202.
80. Zhao H, Baker GA, Song Z, Olubajo O, Crittle T, Peters D. Designing enzyme-compatible ionic liquids that can dissolve carbohydrates. Green Chem. 2008;10(6):696–705.
81. Huaxi L, Zhuo L, Jingmei Y, Changping L, Yansheng C, Qingshan L, Xiuling Z. Liquid-liquid extraction process of amino acids by a new amide-based functionalized ionic liquid. Green Chem. 2012;14(6):1721–7.
82. Absalan G, Akhond M, Sheikhian L. Partitioning of acidic, basic and neutral amino acids into imidazolium-based ionic liquids. Amino Acids. 2010;39(1):167–74.
83. Cai Y, Zhang Y, Peng Y, Lu F, Huang X, Song G. Carboxyl-functional ionic liquids as scavengers: case studies on benzyl chloride, amines, and methanesulfonyl chloride. J Comb Chem. 2006;8(5):636–8.
84. Anugwom I, Mäki-Arvela P, Salmi T, Mikkola J-P. Ionic liquid assisted extraction of nitrogen and sulphur-containing air pollutants from model oil and regeneration of the spent ionic liquid. J Environ Prot. 2011;2(6):796–802.
85. Ma J, Hong X. Application of ionic liquids in organic pollutants control. J Environ Manage. 2012;99:104–9.
86. Matsumoto M, Inomoto Y, Kondo K. Selective separation of aromatic hydrocarbons through supported liquid membranes based on ionic liquids. J Membr Sci. 2005;246(1):77–81.
87. Blahut A, Dohnal V, Vrbka P. Interactions of volatile organic compounds with the ionic liquid 1-ethyl-3-methylimidazolium tetracyanoborate. J Chem Thermodyn. 2012;47:100–8.
88. Blahut A, Dohnal V. Interactions of volatile organic compounds with the ionic liquid 1-Butyl-1-methylpyrrolidinium dicyanamide. J Chem Eng Data. 2011;56(12):4909–18.
89. Ventura SPM, Neves CMSS, Freire MG, Marrucho IM, Oliveira J, Coutinho JAP. Evaluation of anion influence on the formation and extraction capacity of ionic-liquid-based aqueous biphasic systems. J Phys Chem B. 2009;113(27):9304–10.
90. Milota M, Mosher P, Li K. VOC and HAP removal from dryer exhaust gas by absorption into ionic liquids. For Prod. 2007;57(5):73–7.
91. Milota M, Mosher P, Li K. RTIL absorption of organic emissions from press and dry exhaust. For Prod. 2008;58(4):97–101.
92. Chen C-C, Simoni LD, Brennecke JF, Stadtherr MA. Correlation and prediction of phase behavior of organic compounds in ionic liquids using the nonrandom two-liquid segment activity coefficient model. Ind Eng Chem Res. 2008;47(18):7081–93.
93. Oliferenko AA, Oliferenko PV, Seddon KR, Torrecilla JS. Prediction of gas solubilities in ionic liquids. Phys Chem Chem Phys. 2011;13(38):17262–72.
94. Ramdin M, de Loos TW, Vlugt TJH. State-of-the-art of CO_2 capture with ionic liquids. Ind Eng Chem Res. 2012;51(24):8149–77.

95. Jalili AH, Mehdizadeh A, Shokouhi M, Ahmadi AN, Hosseini-Jenab M, Fateminassab F. Solubility and diffusion of CO_2 and H_2S in the ionic liquid 1-ethyl-3-methylimidazolium ethylsulfate. J Chem Thermodyn. 2010;42(10):1298–303.
96. Anthony JL, Maginn EJ, Brennecke JF. Solubilities and thermodynamic properties of gases in the ionic liquid 1-N-butyl-3-methylimidazolium hexafluorophosphate. J Phys Chem B. 2002;106(29):7315–20.
97. Sharma A, Julcour C, Kelkar AA, Deshpande RM, Delmas H. Mass transfer and solubility of CO and H_2 in ionic liquid. Case of [Bmim][PF6] with gas-inducing stirrer reactor. Ind Eng Chem Res. 2009;48(8):4075–82.
98. Safamirzaei M, Modarress H. Application of neural network molecular modeling for correlating and predicting Henry's law constants of gases in [bmim][PF6] at low pressures. Fluid Phase Equilibr. 2012;332:165–72.
99. Kolding H, Fehrmann R, Riisager A. CO_2 capture technologies: current status and new directions using supported ionic liquid phase (SILP) absorbers. Sci China Chem. 2012;55 (8):1648–56.
100. Gao T, Andino JM, Alvarez-Idaboy JR. Computational and experimental study of the interactions between ionic liquids and volatile organic compounds. Phys Chem Chem Phys. 2010;12(33):9830–8.
101. Nalli S, Cooper DG, Nicell JA. Metabolites from the biodegradation of di-ester plasticizers by *Rhodococcus rhodochrous*. Sci Total Environ. 2006;366:286–94.
102. Liang DW, Zhang T, Fang HHP, He J. Phtalates biodegradation in the environment. Appl Microbiol Biotechnol. 2008;80:183–98.
103. Alexander M. Non-biodegradable and other recalcitrant molecules. Biotechnol Bioeng. 1973;15:611–47.
104. Park S, Kazlauskas RJ. Biocatalysis in ionic liquids – advantages beyond green technology. Curr Opin Biotechnol. 2003;14:432–7.
105. Roosen C, Muller P, Greiner L. Ionic liquids in biotechnology: applications and perspectives for biotransformations. Appl Microbiol Biotechnol. 2008;81:607–14.
106. Pernak J, Sobaszkiewicz K, Mirska I. Anti-microbial activities of ionic liquids. Green Chem. 2003;5(1):52–6.
107. Romero A, Santos A, Tojo J, Rodriguez A. Toxicity and biodegradability of imidazolium ionic liquids. J Hazard Mater. 2008;151(1):268–73.
108. Docherty KM, Kulpa Jr CF. Toxicity and antimicrobial activity of imidazolium and pyridinium ionic liquids. Green Chem. 2005;7(4):185–9.
109. Garcia MT, Gathergood N, Scammells PJ. Biodegradable ionic liquids. Part II. Effect of the anion and toxicology. Green Chem. 2005;7:9–14.
110. Pham TPT, Cho CW, Yun YS. Environmental fate and toxicity of ionic liquids: a review. Water Res. 2010;44:352–72.
111. Azimova MA, Morton SA, Frymier PD. Comparison of three bacterial toxicity assays for imidazolium-derived ionic liquids. J Environ Eng. 2009;135:1388–92.
112. Brautigam S, Dennewald D, Schurmann M, Lutje-Spelberg J, Pitner WR, Weuster-Botz D. Whole-cell biocatalysis: evaluation of new hydrophobic ionic liquids for efficient asymmetric reduction of prochiral ketones. Enzyme Microbial Technol. 2009;45:316.
113. Pfruender H, Jones R, Weuster-Botz D. Water immiscible ionic liquids as solvents for whole cell biocatalysis. J Biotechnol. 2006;124:190.
114. Cull SG, Holbrey JD, Vargas-Mora V, Seddon KR, Lye GJ. Room-temperature ionic liquids as replacements for organic solvents in multiphase bioprocess operations. Biotechnol Bioeng. 2000;69(2):227–33.
115. Ganske F, Bornscheuer UT. Growth of *Escherichia coli*, *Pichia pastoris* and *Bacillus cereus* in the presence of the ionic liquids [BMIM][BF4] and [BMIM][PF6] and organic solvents. Biotechnol Lett. 2006;28(7):465–9.
116. Sendovski M, Nir N, Fishman A. Bioproduction of 2-phenylethanol in a biphasic ionic liquid aqueous system. J Agric Food Chem. 2010;58:2265.

117. Gathergood N, Garcia MT, Scammells PJ. Biodegradable ionic liquids: Part I. Concept, preliminary targets and evaluation. Green Chem. 2004;6(3):166–75.
118. Esquivel-Viveros A, Ponce-Vargas F, Esponda-Aguilar P, Prado-Barragan LA, Gutiierrez-Rojas M, Lye GJ, Huerta-Ochoa S. Biodegradacion de [bmim][PF6] utilizando Fusarium sp. Rev Mex Ing Quim. 2009;8(2):163–8.
119. Pham TPT, Cho C-W, Jeon C-O, Chung Y-J, Lee M-W, Yun Y-S. Identification of metabolites involved in the biodegradation of the ionic liquid 1-butyl-3-methylpyridinium bromide by activated sludge microorganisms. Environ Sci Technol. 2009;43(2):516–21.
120. Jianping W, Yu C, Xiaoqiang J, Dongyan C. Simultaneous removal of ethyl acetate and ethanol in air streams using a gas–liquid–solid three-phase flow airlift loop bioreactor. Chem Eng J. 2005;106:175.
121. Muñoz R, Villaverde S, Guieysse B, Revah S. Two-phase partitioning bioreactor for treatment of volatile organic compounds. Biotechnol Adv. 2007;25:410–22.
122. Rehmann L, Prpich JP, Daugulis AJ. Remediation of PAH contaminated soils: application of a solid–liquid two-phase partitioning bioreactor. Chemosphere. 2008;73:804.
123. Rols JL, Condoret JS, Fonade C, Goma G. Mechanisms of enhanced oxygen transfer in fermentation using emulsified oxygen-vectors. Biotechnol Bioeng. 1990;35:427–35.
124. Rehmann L, Daugulis AJ. Biodegradation of biphenyl in a solid-liquid two-phase partitioning bioreactor. Biochem Eng J. 2007;36:195–201.
125. Efroymson RA, Alexander M. Biodegradation by an Arthrobacter species of hydrocarbons partitioned into an organic solvent. Appl Environ Microbiol. 1991;57:1441–7.
126. Bouchez M, Blanchet D, Vandecasteele JP. An interfacial uptake mechanism for the degradation of pyrene by *Rhodococcus* strain. Microbiology. 1997;143:1087–93.
127. Guieysse B, Cirne MDDTG, Mattiasson B. Microbial degradation of phenanthrene and pyrene in a two-liquid-phase-partitioning bioreactor. Appl Microbiol Biotechnol. 2001;56:796–802.
128. Desai JD, Banat IM. Microbial production of surfactants and their commercial potential. Microbiol Mol Biol Rev. 1997;61:47–64.
129. Ascon-Cabrera MA, Lebeault JM. Selection of xenobiotic-degrading microorganisms in a biphasic aqueous-organic system. Appl Environ Microbiol. 1993;59:1717–24.
130. Brennecke JF, Maginn EJ. Ionic liquids: innovative fluids for chemical processing. AICHE J. 2001;47(11):2384–9.
131. Freire MG, Neves CMSS, Marrucho IM, Lopes JNC, Rebelo LPN, Coutinho JAP. High-performance extraction of alkaloids using aqueous two-phase systems with ionic liquids. Green Chem. 2010;12(10):1715–18.
132. Gubicza L, Belafi-Bako K, Feher E, Frater T. Waste-free process for continuous flow enzymatic esterification using a double pervaporation system. Green Chem. 2008;10 (12):1284–7.
133. Oppermann S, Stein F, Kragl U. Ionic liquids for two-phase systems and their application for purification, extraction and biocatalysis. Appl Microbiol Biotechnol. 2011;89(3):493–9.
134. Ge J, Zhou Y, Yang Y, Xue M. Catalytic oxidative desulfurization of gasoline using ionic liquid emulsion system. Ind Eng Chem Res. 2011;50(24):13686–92.
135. De Diego T, Manjon A, Lozano P, Vaultier M, Iborra JL. An efficient activity ionic liquid-enzyme system for biodiesel production. Green Chem. 2011;13(2):444–51.
136. Gamstedt H, Hagfeldt A, Kloo L. Photoelectrochemical studies of ionic liquid-containing solar cells sensitized with different polypyridyl-ruthenium complexes. Polyhedron. 2009;28 (4):757–62.
137. Ng YS, Jayakumar NS, Hashim MA. Behavior of hydrophobic ionic liquids as liquid membranes on phenol removal: experimental study and optimization. Desalination. 2011;278(1):250–8.
138. Docherty KM, Dixon JK, Kulpar JCF. Biodegradation of imidazolium and pyridinium ionic liquids by an activated sludge microbial community. Biodegradation. 2007;18:481–93.

139. Stolte S, Abdulkarim S, Arning J, Blomeyer-Nienstedt A-K, Bottin-Weber U, Matzke M, Ranke J, Jastorff B, Thöming J. Primary biodegradation of ionic liquid cations, identification of degradation products of 1-methyl-3-octylimidazolium chloride and electrochemical wastewater treatment of poorly biodegradable compounds. Green Chem. 2008;10(2):214–24.
140. Zhao D, Liao Y, Zhang Z. Toxicity of ionic liquids. Clean Soil Air Water. 2007;35(1):42–8.
141. Coleman D, Gathergood N. Biodegradation studies of ionic liquids. Chem Soc Rev. 2010;39 (2):600–37.
142. Frade RFM, Afonso CAM. Impact of ionic liquids in environment and humans: an overview. Hum Exp Toxicol. 2010;29(12):1038–54.
143. Zhang F, Ni Y, Sun Z, Zheng P, Lin W, Zhu P, Ju N. Asymmetric reduction of ethyl 4-chloro-3-oxobutanoate to ethyl (S)-4-chloro-3-hydroxybutanoate catalyzed by *Aureobasidium pullulans* in an aqueous/ionic liquid biphase system. Chin J Catal. 2008;29(6):582.
144. Swatloski RP, Holbrey JD, Memon SB, Caldwell GA, Caldwell KA, Rogers RD. Using *Caenorhabditis elegans* to probe toxicity of 1-alkyl-3-methylimidazolium chloride based ionic liquids. Chem Commun. 2004;6:668–9.
145. Pretti C, Chiappe C, Pieraccini D, Gregori M, Abramo F, Monni G, Intorre L. Acute toxicity of ionic liquids to the zebrafish (*Danio rerio*). Green Chem. 2006;8(3):238–40.
146. Papaiconomou N, Estager J, Traore Y, Bauduin P, Bas C, Legeai S, Viboud S, Draye M. Synthesis, physicochemical properties, and toxicity data of new hydrophobic ionic liquids containing dimethylpyridinium and trimethylpyridinium cations. J Chem Eng Data. 2010;55 (5):1971–9.
147. Alvarez-Guerra M, Irabien A. Design of ionic liquids: an ecotoxicity (*Vibrio fischeri*) discrimination approach. Green Chem. 2011;13(6):1507–16.
148. Ventura SPM, Marques CS, Rosatella AA, Afonso CAM, Gonçalves F, Coutinho JAP. Toxicity assessment of various ionic liquid families towards *Vibrio fischeri* marine bacteria. Ecotoxicol Environ Saf. 2012;76(2):162–8.
149. Viboud S, Papaiconomou N, Cortesi A, Chatel G, Draye M, Fontvieille D. Correlating the structure and composition of ionic liquids with their toxicity on *Vibrio fischeri*: a systematic study. J Hazard Mater. 2012;215–216:40–8.
150. Stolte S, Arning J, Bottin-Weber U, Müller A, Pitner W-R, Welz-Biermann U, Jastorff B, Ranke J. Effects of different head groups and functionalised side chains on the cytotoxicity of ionic liquids. Green Chem. 2007;9(7):760–7.
151. Ranke J, Müller K, Stock F, Bottin-Weber U, Poczobutt J, Hoffmann J, Ondruschka B, Filser J, Jastorff B. Biological effects of imidazolium ionic liquids with varying chain lengths in acute *Vibrio fischeri* and WST-1 cell viability assays. Ecotoxicol Environ Saf. 2004;58 (3):396–404.
152. Kulacki KJ, Lamberti GA. Toxicity of imidazolium ionic liquids to freshwater algae. Green Chem. 2008;10(1):104–10.
153. Costello DM, Brown LM, Lamberti GA. Acute toxic effects of ionic liquids on zebra mussel (*Dreissena polymorpha*) survival and feeding. Green Chem. 2009;11(4):548–53.
154. Bernot RJ, Brueseke MA, Evans-White MA, Lamberti GA. Acute and chronic toxicity of imidazolium-based ionic liquids on *Daphnia magna*. Environ Toxicol Chem. 2005;24 (1):87–92.
155. Samori C, Sciutto G, Pezzolesi L, Galletti P, Guerrini F, Mazzeo R, Pistocchi R, Prati S, Tagliavini E. Effects of imidazolium ionic liquids on growth, photosynthetic efficiency, and cellular components of the diatoms *Skeletonema marinoi* and *Phaeodactylum tricornutum*. Chem Res Toxicol. 2011;24(3):392–401.
156. Ignat'ev NV, Welz-Biermann U, Kucheryna A, Bissky G, Willner H. New ionic liquids with tris (perfluoroalkyl) trifluorophosphate (FAP) anions. J Fluor Chem. 2005;126(8):1150–9.
157. Jacquemin J, Husson P, Padua AAH, Majer V. Density and viscosity of several pure and water-saturated ionic liquids. Green Chem. 2006;8(2):172–80.
158. Chiappe C, Pieraccini D. Ionic liquids: solvent properties and organic reactivity. J Phys Org Chem. 2005;18(4):275–97.

159. Swatloski RP, Holbrey JD, Rogers RD. Ionic liquids are not always green: hydrolysis of 1-butyl-3-methylimidazolium hexafluorophosphate. Green Chem. 2003;5:361–3.
160. Zhou ZB, Matsumoto H, Tatsumi K. Low-melting, low-viscous, hydrophobic ionic liquids: 1-alkyl (alkyl ether)-3-methylimidazolium perfluoroalkyltrifluoroborate. Chem-Eur J. 2004;10(24):6581–91.
161. Steudte S, Stepnowski P, Cho C-W, Thöming J, Stolte S. (Eco) toxicity of fluoro-organic and cyano-based ionic liquid anions. Chem Commun. 2012;48(75):9382–4.
162. Ranke J, Müller A, Bottin-Weber U, Stock F, Stolte S, Arning J, Störmann R, Jastorff B. Lipophilicity parameters for ionic liquid cations and their correlation to in vitro cytotoxicity. Ecotoxicol Environ Saf. 2007;67(3):430–8.
163. Larson JH, Frost PC, Lamberti GA. Variable toxicity of ionic liquid-forming chemicals to Lemna minor and the influence of dissolved organic matter. Environ Toxicol Chem. 2008;27 (3):676–81.
164. Hassoun EA, Abraham M, Kini V, Al-Ghafri M, Abushaban A. Cytotoxicity of the ionic liquids, 1-N-butyl-3-methylimidazolium chloride. Res Commun Pharm Toxicol. 2002;7 (1&2):23–31.
165. Stepnowski P, Skladanowski AC, Ludwiczak A, Laczynska E. Evaluating the cytotoxicity of ionic liquids using human cell line HeLa. Hum Exp Toxicol. 2004;23(11):513–17.
166. Ranke DJ, Cox M, Müller A, Schmidt C, Beyersmann D. Sorption, cellular distribution, and cytotoxicity of imidazolium ionic liquids in mammalian cells-influence of lipophilicity. Toxicol Environ Chem. 2006;88(2):273–85.
167. Jeong S, Ha SH, Han S-H, Lim M-C, Kim SM, Kim Y-R, Koo Y-M, So J-S, Jeo T-J. Elucidation of molecular interactions between lipid membranes and ionic liquids using model cell membranes. Soft Matter. 2012;8(20):5501–6.
168. Stolte S, Arning J, Bottin-Weber U, Matzke M, Stock F, Thiele K, Uerdingen M, Welz-Biermann U, Jastorff B, Ranke J. Anion effects on the cytotoxicity of ionic liquids. Green Chem. 2006;8(7):621–9.
169. Salminen J, Papaiconomou N, Kumar RA, Lee J-M, Kerr J, Newman J, Prausnitz JM. Physicochemical properties and toxicities of hydrophobic piperidinium and pyrrolidinium ionic liquids. Fluid Phase Equilibr. 2007;261(1):421–6.
170. Kumar RA, Papaiconomou N, Lee JM, Salminen J, Clark DS, Prausnitz JM. In vitro cytotoxicities of ionic liquids: effect of cation rings, functional groups, and anions. Environ Toxicol. 2009;24(4):388–95.
171. Arning J, Matzke M. Toxicity of ionic liquids towards mammalian cell lines. Curr Org Chem. 2011;15(12):1905–17.
172. Fatemi MH, Izadiyan P. Cytotoxicity estimation of ionic liquids based on their effective structural features. Chemosphere. 2011;84(5):553–63.
173. Dobler D, Schmidts T, Klingenhöfer I, Runkel F. Ionic liquids as ingredients in topical drug delivery systems. Int J Pharm. 2013;441:620–7.
174. Hough WL, Smiglak M, Rodríguez H, Swatloski RP, Spear SK, Daly DT, Pernak J, Grisel JE, Carliss RD, Soutullo MD. The third evolution of ionic liquids: active pharmaceutical ingredients. New J Chem. 2007;31(8):1429–36.
175. Boethling RS. Designing biodegradable chemicals. In: Designing safer chemicals, ACS symposium series, vol. 640. Washington, DC: American Chemical Society; 1996. p. 156–71.
176. Yu G, Zhao D, Wen L, Yang S, Chen X. Viscosity of ionic liquids: database, observation, and quantitative structure-property relationship analysis. AICHE J. 2012;58(9):2885–99.
177. Wells AS, Coombe VT. On the freshwater ecotoxicity and biodegradation properties of some common ionic liquids. Org Process Res Dev. 2006;10(4):794–8.
178. Harjani JR, Singer RD, Garcia MT, Scammells PJ. The design and synthesis of biodegradable pyridinium ionic liquids. Green Chem. 2008;10(4):436–8.
179. Zhang C, Wang H, Malhotra SV, Dodge CJ, Francis AJ. Biodegradation of pyridinium-based ionic liquids by an axenic culture of soil Corynebacteria. Green Chem. 2010;12(5):851–8.

180. Ford L, Harjani JR, Atefi F, Garcia MT, Singer RD, Scammells PJ. Further studies on the biodegradation of ionic liquids. Green Chem. 2010;12(10):1783–9.
181. Docherty KM, Joyce MV, Kulacki KJ, Kulpa CF. Microbial biodegradation and metabolite toxicity of three pyridinium-based cation ionic liquids. Green Chem. 2010;12(4):701–12.
182. Neumann J, Grundmann O, Thöming J, Schulte M, Stolte S. Anaerobic biodegradability of ionic liquid cations under denitrifying conditions. Green Chem. 2010;12(4):620–7.
183. Morrissey S, Pegot B, Coleman D, Garcia MT, Ferguson D, Quilty B, Gathergood N. Biodegradable, non-bactericidal oxygen-functionalised imidazolium esters: a step towards 'greener' ionic liquids. Green Chem. 2009;11(4):475–83.
184. Qi B, Moe WM, Kinney KA. Biodegradation of volatile organic compounds by five fungal species. Appl Microbiol Biotechnol. 2002;58:689.
185. Woertz JR, Kinney KA, McIntosh NDP, Szaniszlo PJ. Removal of toluene in a vapor-phase bioreactor containing a strain of the dimorphic black yeast *Exophiala lecanii-corni*. Biotechnol Bioeng. 2001;75:558.
186. Daugulis AJ, Boudreau NG. Removal and destruction of high concentration of gaseous toluene in a two-phase partitioning bioreactor by *Alcaligenes xylosoxidans*. Biotechnol Lett. 2003;25:1421–4.
187. Nielsen DR, Daugulis AJ, Mclellan PJ. Transient performance of a two-phase partitioning bioscrubber treating a benzene-contamination gas stream. Environ Sci Technol. 2005;39:8971–7.
188. Césario MT, Brandsma JB, Boon MA, Tramper J, Beeftink HH. Ethene removal from gas by recycling a water-immiscible solvent through a packed absorber and a bioreactor. J Biotechnol. 1998;62:105–18.
189. MacLeod CT, Daugulis AJ. Interfacial effects in a two-phase partitioning bioreactor: degradation of polycyclic aromatic hydrocarbons (PAHs) by a hydrophobic *Mycobacterium*. Process Biochem. 2005;40:1799–805.
190. Marcoux J, Déziel E, Villemur R, Lépine F, Bisaillon JG, Beaudet R. Optimization of high molecular weight polycyclic aromatic hydrocarbons' degradation in a two-liquid-phase bioreactor. J Appl Microbiol. 2000;88:655–62.
191. Tomei MC, Annesini RS, Daugulis AJ. Biodegradation of 4-nitrophenol in a two-phase sequencing batch reactor: concept demonstration, kinetics and modelling. Appl Microbiol Biotechnol. 2008;80:1105–12.
192. Darracq G, Couvert A, Couriol C, Amrane A, Le Cloirec P. Absorption and biodegradation of hydrophobic VOCs: determination of Henry's constants and biodegradation levels. Water Sci Technol. 2009;59:1315–22.
193. Janikowski TB, Velicogna D, Punt M, Daugulis A. Use of a two-phase partitioning bioreactor for degrading polycyclic aromatic hydrocarbons by *Sphingomonas sp*. J Appl Microbiol Biotechnol. 2002;59:368–76.
194. Aldric JM, Gillet S, Delvigne F, Blecker C, Lebeau F, Wathelet JP, Manigat G, Thonart P. Effect of surfactants and biomass on the gas/liquid mass transfer in an aqueous-silicone oil two-phase partitioning bioreactor using *Rhodococcus erythropolis* T902.1 to remove VOCs from gaseous effluents. J Chem Technol Biotechnol. 2009;84:1274–83.
195. Aldric JM, Lecomte JP, Thonat P. Study on mass transfer of isopropylbenzene and oxygen in a two-phase partitioning bioreactor in the presence of silicone oil. Appl Biochem Biotechnol. 2009;153:67–79.

Editors' Biography

Prof. Dr. Zhen FANG is the leader and founder of biomass group, Xishuangbanna Tropical Botanical Garden, Chinese Academy of Sciences. He is also an adjunct full Professor of Life Sciences, University of Science and Technology of China. He is the inventor of "fast hydrolysis" process. He is specializing in thermal/biochemical conversion of biomass, nanocatalyst synthesis and its applications, pretreatment of biomass for biorefineries. He obtained his PhDs from China Agricultural University (Biological and Agricultural Engineering, 1991, Beijing) and McGill University (Materials Engineering, 2003, Montreal).

Richard L Smith, Jr. is Professor of Chemical Engineering, Graduate School of Environmental Studies, Research Center of Supercritical Fluid Technology, Tohoku University, Japan. Professor Smith has a strong background in physical properties and separations and obtained his Ph.D. in chemical engineering from the Georgia Institute of Technology (USA). His research focuses on developing green chemical processes, especially those that use water and carbon dioxide as the solvents in their supercritical state. He has expertise in physical property measurements and in separation techniques with ionic liquids and has more than 200 scientific papers, patents and reports in the field of chemical engineering. Professor Smith is the Asia Regional

Z. Fang et al. (eds.), *Production of Biofuels and Chemicals with Ionic Liquids*,
Biofuels and Biorefineries 1, DOI 10.1007/978-94-007-7711-8,

Editor for the *Journal of Supercritical Fluids* and has served on editorial boards of major international journals associated with properties and energy.

Xinhua Qi is Professor of Environmental Science, Nankai University, China. Professor Qi obtained his Ph.D. from the department of environmental science, Nankai University, China. Professor Qi has a strong background in environmental treatment techniques in water and in chemical transformations in ionic liquids. His research focuses on the catalytic conversion of biomass into chemicals and biofuels with ionic liquids. Professor Qi had published more than 50 scientific papers, books and reports with a number of papers being in top-ranked international journals.